BARRY'S
ADVANCED
CONSTRUCTION
OF BUILDINGS

BARRY'S ADVANCED CONSTRUCTION OF BUILDINGS

Stephen Emmitt

and

Christopher A. Gorse

Blackwell
Publishing

© 2006 by Stephen Emmitt and Christopher A. Gorse

Editorial offices:
Blackwell Publishing Ltd, 9600 Garsington Road, Oxford OX4 2DQ, UK
    Tel: +44 (0)1865 776868
Blackwell Publishing Inc., 350 Main Street, Malden, MA 02148-5020, USA
    Tel: +1 781 388 8250
Blackwell Publishing Asia Pty Ltd, 550 Swanston Street, Carlton, Victoria 3053, Australia
    Tel: +61 (0)3 8359 1011

First published 2006 by Blackwell Publishing Ltd

ISBN-10: 1-4051-1054-6
ISBN-13: 978-1-4051-1054-9

Library of Congress Cataloging-in-Publication Data

Emmitt, Stephen.
    Barry's advanced construction of buildings / Stephen Emitt and Christopher A. Gorse.– Ist ed.
        p.   cm.
    Revision of volumes 3, 4 & 5 of Construction of buildings by Robin Barry.
    Includes bibliographical references and index.
    ISBN-13: 978-1-4051-1054-9 (alk. paper)
    ISBN-10: 1-4051-1054-6 (alk. paper)
    1. Building.   I. Gorse, Christopher A.   II. Barry, R. (Robin) Construction of buildings.
    III. Title.
TH146.E47 2006
690 – dc22   2005017736

Contents

Preface

In the Preface to *Barry's Introduction to Construction of Buildings* we set out our aims and objectives to revise the five-volume series and present it in a two-volume format. In this 'advanced' volume our target audience is mainly second year (Level 2) students of architecture and construction. Emphasis in the book shifts from mainly loadbearing construction to framed construction and associated technologies, although the principles and many of the construction techniques described in the introductory volume are still relevant and complementary to the techniques described here. This volume incorporates revised material from *Barry 3, 4* and part of 5, supported with new illustrations and photographs to help aid understanding of the technologies described. In keeping with the introductory volume, we have also added new material and new chapters to reflect ongoing changes in how we construct buildings.

One of the biggest changes since the first edition of the *Construction of Buildings* was published in 1958 is the growth in pre-fabrication and off-site assembly. With increased sophistication of information communication technologies (ICT), computer facilitated manufacturing and lean production techniques, emphasis has shifted towards the controlled environment of the factory. This is most noticeable in the mass production of high quality building components but is also reflected in volumetric buildings, manufactured at a central location, transported to site, craned into position and 'plugged-in' to the service connections. Chapter 8 describes the key functional/performance requirements for the main elements common to off-site production. We have also included a new chapter devoted to alternative methods of construction, Chapter 12, in which some of the less common approaches are described. Some of these techniques are the antithesis of industrial production, with emphasis on use of local (sustainable) materials and labour. The aim here is to show a small range of materials and techniques that may be used as a discrete system or used in conjunction with more familiar techniques and materials. Hopefully, by looking at both ends of the spectrum, we have maintained a degree of balance of the book.

Readers familiar with earlier editions of the series will recognise a subtle shift towards the inclusion of some 'sustainable' construction techniques. Many of these (e.g. energy conservation) are a reflection of, and response to, new building regulation, codes and associated legislation; others (e.g. rainwater recycling) are related to a more general concern for our environment. We felt it important to start to integrate environmentally responsible ideals within the main text of the *Barry* series, rather than separate them out as something unusual or detached from everyday practice.

We have tried to keep the *Barry* series topical, informative and above all else useful for students studying construction technology. In doing this we have tried to maintain a balanced and objective style to both volumes. The choice of materials and construction technologies is related to many inter-related factors, such as the characteristics of the site, the performance requirements of the client, prevailing legislation, market conditions and time. Thus the choice of the most 'appropriate' materials and technologies for a given situation is a relatively complex decision-making process that is related to a specific context. This two-volume series provides essential information in helping readers to make an informed decision for a specific context, be it a student assignment or a building project. We hope that the spirit of the *Barry* series continues to inform and stimulate those interested in and engaged in the construction of buildings.

Stephen Emmitt
Christopher Gorse

Acknowledgements

We are indebted to many people who have, in a variety of ways, helped us to realise the revisions to this work. Julia Burden was instrumental in giving us the chance to take on this project and once again Blackwell Publishing have retained their collective faith in our ability to deliver, for which we are very grateful. Over the years our students have continued to be the inspiration for writing books and they deserve our heartfelt credit for helping to keep our feet firmly on the ground by asking the 'why' and the 'how' questions.

Our colleagues and friends in industry and academia also deserve acknowledgement, in particular we would like to thank:

Shaun Long at Cameron Taylor Group, Peter Dracup at Taylor Woodrow, Mike Armstrong at Mace Ltd, Gordon Throup at Big Sky Construction Services Ltd, Roo Modasia at HBG Ltd, Karen Makin at Roger Bullivant Ltd, Joanne Bridges at Yorkon, David Highfield, David Johnston, David Roberts, Robert Ellis at Leeds Metropolitan University, Peter Galley at Carillion, Nigel Smith at Miller Construction, Simon Taylor at Bovis Lend Lease and Stewart Jones at NHBC and Age Concern.

We would also like to extend our gratitude to Roger Bullivant Ltd, Yorkon Ltd and Styro Stone Ltd. for information and photographs supplied.

1 Introduction

In *Barry's Introduction to Construction of Buildings* we provided an introductory chapter that set out some of the basic requirements and conditions relevant to all building projects, regardless of size and complexity. In this volume the emphasis shifts from domestic to larger-scale buildings, primarily residential, commercial and industrial buildings constructed with loadbearing frames. Many of the principles and techniques set out in the introductory volume are, however, still appropriate. In addition to the more familiar, dare we say conventional, approaches to construction we have included in this volume details of alternative approaches. Primarily grounded in a desire to build in a more environmentally friendly way, these techniques appear to be attracting more interest and application. In this introductory chapter we start to address some additional, yet related, issues, again with the aim of providing some context to the chapters that follow.

1.1 The function and performance of buildings

In *Barry's Introduction to Construction of Buildings* we set out some of the fundamental functional and performance requirements of buildings. We continue the theme here with some additional requirements applicable to the construction of buildings, regardless of type, size or complexity.

Structure and fabric

Structure and building fabric have a very special relationship. It is the combined performance of the structure and building fabric, together with the integration of services, which determines the overall performance of the building during its life. In loadbearing construction the materials forming the structural support also provide the fabric and hence the external and internal finishes. In framed structures the fabric is independent of the structure, with the fabric applied to the loadbearing structural frame.

Loading

Buildings need to accommodate the loads and forces acting on them if they are to resist collapse. One of the most important considerations is how forces are transferred within the structure. Buildings are subject to three types of loading:

❑ Dead loads. Dead loads remain relatively constant throughout the life of a building, unless it is remodelled at a future date. These loads comprise the combined weight of the materials used to construct the building. Loads are transferred to the ground via the foundations. Because the weight of individual components is known, the dead load can be easily calculated.

❏ Live loads. Unlike dead loads, the live loads acting on a building will vary. Life loads comprise the weight of people using the building, the weight of furniture and equipment etc. Seasonal changes will result in (temporary) live loading from rainfall and snow. Structural design calculations assume an average maximum live load based on the use of the building (plus a safety factor). If the building use changes, then it will be necessary to check the anticipated live loading against that used at the design stage.

❏ Wind loads. All buildings are subject to wind loading. Maximum wind loads (gusts) are determined by considering the maximum recorded wind speed in a particular location and adding a safety factor. Wind loading is an important consideration for both permanent and temporary structures. It is also an important consideration when designing and installing temporary weather protection to protect building workers and work in progress from the elements.

When the total loading has been calculated for the proposed building it is then possible to design the building structure, fabric and foundations.

Structural frames

Timber, steel and reinforced concrete are the main materials used for structural frames. The benefits of one material over another need to be considered against a wide variety of design and performance parameters, such as the following:

❏ Extent of clear span required
❏ Height of the building
❏ Extent of anticipated loading
❏ Fire resistance and protection
❏ Embodied energy and associated environmental impact
❏ Ease of fixing the fabric to the frame (constructability)
❏ Availability of materials and labour skills
❏ Extent of prefabrication desired
❏ Site access (restrictions)
❏ Erection programme and sequence
❏ Maintenance and ease of adaptability
❏ Ease of disassembly and re-use of materials
❏ Life cycle costs

In some cases it is common to use one material only for the structural frame, for example timber. In other situations it may be beneficial to use a composite frame construction, for example concrete and steel.

Dimensional stability

Stability of the building as a whole will be determined by the independent movement of different materials and components within the structure over time – a complex interaction determined by the dimensional variation of individual components when subjected to changes in moisture content, changes in temperature and not forgetting changes in loading:

❑ Moisture movement. Dimensional variation will occur in porous materials as they take up or, conversely, lose moisture through evaporation. Seasonal variations in temperature will occur in temperate climates and affect many building materials. Indoor temperature variations should also be considered.

❑ Thermal movement. All building materials exhibit some amount of thermal movement because of seasonal changes in temperature and (often rapid) diurnal fluctuations. Dimensional variation is usually linear. The extent of movement will be determined by the temperature range the material is subjected to, its coefficient of expansion, its size and its colour. These factors are influenced by the material's degree of exposure, and care is required to allow for adequate expansion and contraction through the use of control joints.

❑ Loading. Dimensional variation will occur in materials that are subjected to load. Deformation under load may be permanent; however, some materials will return to their natural state when the load is removed. Thus live and wind loads need to be considered too.

Understanding the different physical properties of materials will help in detailing the junctions between materials and with the design, positioning and size of control joints. Movement in materials can be substantial and involve large forces. If materials are restrained in such a way that they cannot move then these forces may exceed the strength of the material and result in some form of failure. Control joints, sometimes described as 'movement joints' or 'expansion joints', are an effective way of accommodating movement and associated stresses.

Designers and builders must understand the nature of the materials and products they are specifying and building with. This includes the materials' scientific properties; structural properties; characteristics when subjected to fire; interaction with other materials; anticipated durability for a given situation; life cycle cost; service life; maintenance requirements; recycling potential; environmental characteristics such as embodied energy; health and safety characteristics; and last but not least their aesthetic properties if they are to be seen when the building is complete. With such a long list of considerations it is essential that designers and builders work closely with manufacturers and consult independent technical reports. A thorough understanding of materials is fundamental to ensuring feasible constructability and disassembly strategies. Consideration should be given to the service life of materials and manufactured products, since any assembly is only as durable as the shortest service life of its component parts.

Flexibility and the open building concept

The vast majority of buildings will need to be adjusted or adapted in some way to accommodate the changing needs of the building users and owners. In domestic construction this may entail the addition of a small extension to better accommodate a growing family, conversion of unused roof space into living accommodation or the addition of a conservatory. Change of building owner often means that the kitchen or bathroom (which may be functional and in a good state of repair) will be upgraded or replaced to suit the taste and needs of the new building owners. Thus what was perfectly functional to one building user is not to another, necessitating the need for alterations.

In commercial buildings a change of tenant can result in major building work, as, for example, internal partition walls are moved to suit different spatial demands. Change of retailer will also result in a complete refitting of most shop interiors. These are just a few examples of the amount of alterations and adaptations made to buildings, which, if not planned and managed in a strategic manner, will result in a considerable amount of waste. Emphasis should be on re-using and recycling materials as they are disassembled and, if possible, the flexibility of internal space.

Although these are primarily design considerations, the manner in which materials and components are connected can have a major influence on the ease, or otherwise, of future alterations.

Flexibility

Designing and detailing a building to be flexible in use presents a number of challenges, some of which may be known and foreseen at the briefing stage but many of which cannot be predicted. Thought should be given to the manner in which internal, non-loadbearing walls are constructed and their ease of disassembly and re-use. Similarly, the position of services and the manner in which they are fixed to the building fabric need careful thought at the design and detailing stage. For example, a flexible house design would have a structural shell with non-loadbearing internal walls (moveable partitions, folding walls etc.), zoned under floor space heating (allowing for flexible use of space) and carefully positioned wet and electrical service runs (in a designated service zone or service wall).

Open building *— learn def'n — see overleaf for systems*

The open building concept aims to provide buildings that are relatively easy to adapt to changing needs with minimum waste of materials and little inconvenience to building users. The main concept is based on taking the entire life cycle of a building and the different service lives of the building's individual components into account. Since an assembly of components is dependent upon the service life of its shortest-living element, it may be useful to view the building as a system of time-dependent levels. Terminology varies a little, but the use of a three-level system, primary, secondary and tertiary, is becoming common. Described in more detail the levels are:

❑ The primary system. Service life of approximately 50–100 years. This comprises the main building elements such as the loadbearing structure, the external fabric, building services structure etc. The primary system is a long-term investment and difficult to change without considerable cost and disruption. Sometimes described as the building 'shell'.

❑ The secondary system. Service life of approximately 15–50 years. This comprises elements such as internal walls, floor and ceiling finishes, building services installations, doors and mechanical vertical circulation systems such as lifts and escalators. The secondary system is a medium-term investment and should be capable of adaptation through disassembly and reassembly.

❑ The tertiary system. Service life of approximately 5–15 years. This comprises elements such as fittings and furniture and equipment associated with the building use, e.g. office equipment. The tertiary system is a short-term investment and elements should be capable of being changed without any major building work.

The shorter the service life of components, the greater the need for replacement, hence the need for easy and safe access. Applying this strategy to a development of apartments, the structure and external fabric would be the primary system. The secondary system would include kitchens, bathrooms and services. The tertiary system would cover items such as the furniture and household appliances. If a discrete, modular system is used then it is relatively easy to replace the kitchen or bathroom without major disruption and recycle the materials. This 'plug in' approach is certainly not a new concept but has started to become a more realistic option as the sector has started to adopt off-site production.

Maintenance and repair

It is currently estimated that over 60% of the building stock in England is more than 40 years old. Approximately six million houses are classified as unfit to live in because of problems with damp and inadequate thermal insulation. Combined with the desire of building owners to upgrade their properties, this means that a large proportion of building work is concerned with existing buildings. Many of the principles described in this current *Barry* series will, of course, be relevant to work to existing buildings. For readers concerned with restoration and repair work some of the earlier editions of *Barry* may be useful in helping to describe some of the main techniques used at the time.

Security

Security of buildings and their contents (goods and people) has become a primary concern for the vast majority of building sponsors and owners. In residential developments the primary concern is with theft of property, with emphasis on the integrity of doors and windows. In commercial developments the concern is for the safety of the people using the building and for the security of the building's contents. Doctors' surgeries and hospitals have experienced an increase in attacks on staff and patients, leading to the installation of active security measures in an attempt to deter crime. Theft from retail stores and warehouses continues to be a major concern for businesses. Where buildings are located away from housing areas it may be possible to enclose the site with a secure fence and controlled entrance gates, but for the buildings located in urban and semi-urban locations isolating the building from its neighbours is rarely a realistic option. Vandalism and the fear of terrorist attacks are additional security concerns, leading to changes in the way buildings are designed and constructed. Measures may be passive, active or a combination of both.

Passive measures

A passive approach to security is based on the concept of inherent security measures, where careful consideration at the design and detailing phase can make a major difference to the security of the building and its contents. Building layout and the positioning of, for example, doors and windows to benefit from natural surveillance needs to be combined with the specification of materials and components that match the necessary functional requirements. The main structural materials and the method of construction will have a significant impact on the resistance of the structure to forced entry. For example, consideration should be given to the ease with which external cladding may be removed and/or broken through, and depending on the estimated risk an alternative form of construction may be more appropriate.

Ram raiding, the act of driving a vehicle through the external fabric of the building to create an unauthorised means of access and egress for the purposes of theft, has become a significant problem for the owners of commercial and industrial premises. Concrete and steel bollards, set in robust foundations and spaced at close centres around the perimeter of the building, are one means of providing some security against ram raiding, especially where it is inappropriate to construct a secure perimeter fence.

Active measures

Active security measures, such as alarms and monitoring devices, may be deployed in lieu of passive measures or in addition to inherent security features. For new buildings active measures should be considered at the design stage to ensure a good match between passive and active security. Integration of cables and mounting and installation of equipment should also be considered early in the detailed design stage. Likewise, when applying active security measures to existing buildings care should be taken to analyse and utilise any inherent features. Some of the active measures include:

- ❏ Intruder alarm systems
- ❏ Entrance control systems in foyers/entrance lobbies
- ❏ Coded door access
- ❏ CCTV monitoring
- ❏ Security personnel patrols

Health, safety and wellbeing

Worldwide the construction sector has a poor health and safety record. Various approaches have been taken to try to improve the health, safety and wellbeing of everyone involved in construction. These include more stringent legislation, better education and training of workers, and better management practices. Similarly, a better understanding of the sequence of construction (a combination of constructability principles and detailed method statements) has helped to identify risk hazards and minimise or even eliminate them. This also applies to future demolition of the building, with a detailed disassembly strategy serving a similar purpose. There are four main, inter-related stages to consider. They are:

- ❏ Prior to construction. The manner in which a building is designed and detailed, i.e. the materials selected and their intended relationship to one another, will have a significant bearing on the safety of operations during construction. Extensive guidance is available on the Safety in Design homepage (www.safetyindesign.org).
- ❏ During construction. Ease of constructability will have a bearing on safety during production. Off-site manufacturing offers the potential of a safer environment, primarily because the factory setting is more stable and easier to control than the constantly changing construction site. However, the way in which work is organised and the attitude of workers toward safety will also have a significant bearing on accident prevention.
- ❏ During use. Routine maintenance and repair is carried out throughout the life of a building. Even relatively simple tasks such as changing a light bulb can become a potential hazard if the light fitting is difficult to access. Elements of the building with short service lives (and/or with high maintenance requirements) must be accessed safely.
- ❏ Demolition and disassembly. Attention must be given to the workers who at some time in the future will be charged with disassembling the building. Method statements and guidance on a suitable and safe disassembly strategy are required.

1.2 Environmental concerns

There is an extensive literature concerning the environmental impact of building materials, construction activities and the use (and misuse) of buildings during their lifetime. We know that we must do more to respect our planet and build in a way that has a positive impact on our environment. Unfortunately the reality is that we could do a lot more in this regard. The choices made by parties to the design and construction process will colour the environmental impact of a building during its construction, use and eventual recycling, and so care should be taken to consider the whole life of the building. From a construction perspective consideration should be given to the method of construction, maintenance and repair, future adaptability of the structure and the recycling of materials as and when the building is demolished or substantially remodelled. This is particularly important at the detailing and specification stage when materials and components are selected. Adopting the open building concept may be one way forward, but there are many ways in which we can improve the relationship between our artificial environment and our natural one. For example, detailing buildings so as to reduce unnecessary waste during production not only helps to reduce land fill, it also saves time and money.

Climate change

There continues to be considerable speculation as to the future impact of climate change. In the UK the general consensus is that the average temperature will continue to rise, as will the amount of rainfall and the average wind speed. The message from the weather forecasters is wetter, warmer and windier. This has given rise to a number of concerns about the suitability of the existing building stock and also to the technologies being employed for the erection of new buildings. How, for example, do these predicted changes impact on the way in which we detail the external fabric of buildings? Are existing Codes, Standards and building practices adequate? The general consensus is that we should adopt a cautious approach, although we would urge against over-detailing, which is wasteful.

Some concern has been expressed about new buildings, especially homes that are built from lightweight materials, such as timber-framed, steel-framed, modular and other lightweight construction systems. The fear is that with an expected increase in temperatures the internal temperature of lightweight construction may become too high during the summer, thus necessitating air-conditioning (increased energy demands) and/or better shading and natural ventilation. Buildings constructed of heavy walls, with small windows and sun shading devices (e.g. shutters, verandas) are less susceptible to temperature fluctuation. However, there are plenty of places around the world that have a warmer climate than the UK and where lightweight construction is used successfully. The answer to the problem is not so much about the type of construction used, rather the manner in which the building is designed to respond better to its immediate environment (e.g. verandas and shading devices). Passive design that uses little or no energy for heating and cooling should be given full consideration before design commences.

Energy efficiency and environmental performance
The environmental performance of buildings has long been a cause for concern but it is an area in which it is difficult for the building owner to get reliable information. Designers and builders must make a greater effort to provide buildings with:

- ❑ Lower running costs
- ❑ Enhanced air quality and natural daylight
- ❑ Use of low-allergy materials
- ❑ Use of environmentally friendly materials
- ❑ Water efficiency (and recycling) measures
- ❑ Ease of adaptation and alteration

If these factors are addressed at the conceptual and detailed design phases then the initial cost of the construction is likely to be similar to a project that is less energy efficient and less environmentally friendly. Add to this the considerable cost savings over the life of the building and it is difficult to understand why buildings are still being constructed with such scant regard for the whole life performance of the constructed works.

1.3 New methods and products

An exciting feature of construction is the amount of innovation and change constantly taking place in the development of new materials, methods and products: many of which are used in conjunction with the more established technologies. Some of the more obvious areas of innovative solutions are associated with changing regulations (e.g. air tightness requirements), changing technologies (e.g. new cladding systems), the trend towards greater use of off-site production (e.g. volumetric system build), advances in building services (e.g. provision of broadband) and a move to the use (and re-use) of recycled materials (e.g. thermal insulation manufactured from recycled material). Many of the changes are, however, quite subtle as manufacturers make gradual technical 'improvements' to their product portfolio. The gradual innovations are often brought about by the use of new production plant and/or are triggered by competition from other manufacturers, with manufacturers seeking to maintain and improve market share through technical innovation. In the vast majority of cases this results in building products with improved performance standards.

Combined with changing fashions in architectural design and manufacturers' constant push toward the development of new materials and products, we are faced with a very wide range of systems, components and products from which to choose. All contributors to the design and erection of buildings, from clients and architects to contractors and specialist sub-contractors, will have their own attitude to new products. Some are keen to use new products and/or new techniques, while others are a little more cautious and tend to stick to what they know. Whatever one's approach, it is important to keep up to date with the latest product developments and investigate those products and methods that may well prove to be beneficial. Maintaining relationships with product manufacturers is one way of achieving this, indeed we would urge readers to visit manufacturers and talk to them about their products. This should be balanced against independent research reports relating to specific or generic product types.

Whatever approach is taken to the use of innovative materials, components and structural systems it is important to remember that compliance is required with the Building Regulations and appropriate Codes and Standards. This applies equally to buildings constructed on site and to those produced in whole or in part in factories.

The Building Regulations and supporting guidance (*Approved Documents* in England and Wales, and in Northern Ireland; *Guidance Documents* in Scotland) are structured in such a way as to encourage the adoption of innovative approaches to the design and construction of buildings. This is done through setting performance standards, which must be achieved or bettered by the proposed construction. Acceptance of innovative proposals is in the hands of the building control body handling the application, thus applicants must submit sufficient information on the innovative proposal to allow an accurate assessment of its performance. This is done by supplying data on testing, certification, technical approvals, CE marking and compliance with the Construction Products Directive (CPD), Eurocodes and Standards, calculations, detailed drawings and written specifications where appropriate.

1.4 Product selection and specification

Both the quality and the long-term durability of a building depend upon the selection of suitable building products and the manner in which they are assembled. This applies to buildings constructed on the site and to off-site production. The majority of people contributing to the design and construction of a building are, in some way or another, involved in the specification of building products, i.e. making a choice as to the most appropriate material or component for a particular situation. Architects and engineers will specify products by brand name (a prescriptive specification) or through the establishment of performance criteria (a performance specification), which is discussed in more detail later. Contractors and sub-contractors will be involved in the purchase and installation of the named product or products that match the specified performance requirements, i.e. they will also be involved in assessing options and making a decision. Similarly, designers working for manufacturers producing off-site units will also be involved in material and product selection; here the emphasis will be on secure lines of supply.

The final choice of product and the manner in which it is built into the building will have an effect on the overall quality and performance of the building. Traditionally, the factors affecting choice of building products have been the characteristics of the product (its properties, or 'fitness for purpose'), its initial cost and its availability. However, a number of other factors are beginning to influence choice, some of which are dependent on legislation, others of which are also dependent upon product safety (during construction, use and replacement/recycling) and environmental concerns as to the individual and collective impact of the materials used in the building's construction. Selection criteria for a particular project will cover the following areas, the importance of one over another is dependent on the location of the product and the type of project:

- ❑ Aesthetics
- ❑ Availability
- ❑ Compatibility (with other products)
- ❑ Compliance with legislation
- ❑ Cost (whole life costs) ‿ maintenance, useful life,
- ❑ Durability
- ❑ Ease of installation (buildability)
- ❑ Environmental impact
- ❑ Health and safety
- ❑ Replacement and recyclability
- ❑ Risk (associated with the product and manufacturer)

For very small projects it is common for contractors to select materials and products from the stock held by their local builders' merchant: choice being largely dependent upon what the merchant stocks (availability) and initial cost. For larger projects there is a need to confirm specification decisions in a written document, the specification.

The written specification

Specifications are written documents that describe the requirements to which the service or product has to conform, i.e. its defined quality. It is the written specification, not the drawings, which defines and hence determines the quality of the finished work. The term specification tends to be used in the singular, which is a little misleading. In practice, the work to be carried out will be described in specifications written by the different specialists involved in the construction project. The structural engineer will write the specification for the structural elements, such as foundations and steelwork, whereas the architect will be concerned with materials and finishes. Similarly, there will be an electrical and mechanical specification for the services provision etc. This collection of multi-authored information is known as 'the specification'.

People from different backgrounds will use the written specification for a number of quite different tasks. It will be used during the pre-contract phase to help prepare costings and tenders. During the contract, operatives and the site managers, to check that the work is proceeding in accordance with the defined quality, will read the specification. Post-contract, the document will form a record of materials used and set standards, which is useful for alteration and repair work and as a source of evidence in disputes.

Specifying quality — check NBS specs

Drawings (together with schedules) indicate the quantity of materials to be used and show their finished relationship to each other. It is the written specification that describes the quality of the workmanship, the materials to be used, and the manner in which they are to be assembled. Trying to define quality is a real challenge when it comes to construction, partly because of the complex nature of building activity and partly because of the number of actors who have a stake in achieving quality. The term quality tends to be used in a subjective manner and, of course, is negotiable between the project stakeholders. In terms of the written specification, quality can be defined through the quality of materials and the quality of workmanship. Designers can define the quality of materials they require through their choice of proprietary products or through the use of performance parameters and appropriate reference to standards and codes. Designers do not tell the builder how to construct the building; this is the contractor's responsibility, hence the need for method statements. The specification will set out the appropriate levels of workmanship, again by reference to codes and standards, but it is the people doing the work, and to a certain extent the quality of supervision, that determines the quality of the finished building.

Specification types

There are two ways in which construction work may be specified, either through a performance specification or through a prescriptive specification:

❑ Performance specification is where the designer describes the material and workmanship attributes required (which must be met or bettered), leaving the decision about specific products and standards of workmanship to the contractor. Performance specifications tend to be favoured by contractors because they give more latitude in their choice of products.

❏ Prescriptive specification tends to be favoured by designers. The designer produces the design requirements and specifies in detail the materials to be used by listing proprietary products, methods and standards of workmanship. Changes may be proposed by the contractor but must be approved by the designer before a substitution may be made.

It has been argued that performance specifications encourage innovation, although it is hard to find much evidence to support such a view. The performance approach allows, in theory at least, a degree of choice and hence competition. The advantages of one approach over another is largely a matter of circumstance and personal preference. However, it is common for performance and prescriptive specifications to be used on the same project for different elements of the building.

National Building Specification

Standard formats provide a useful template for specifiers and help to ensure a degree of consistency, as well as saving time. In the UK the National Building Specification (NBS) is widely used because it helps to save time in this way and is familiar to other parties to the design and assembly process. Available as computer software, it helps to make the writing of specifications relatively straightforward because prompts are given to assist the writer's memory. The NBS is an extensive document containing a library of clauses and references to appropriate standards. The effectiveness of the finished document, as with all templates, depends upon the ability of the person writing the specification, i.e. filling in the gaps. Despite the name, the NBS is not a national specification in the sense that it must be used; many design offices use their own particular hybrid specifications that suit them and their type of work.

Co-ordinated Project Information

Co-ordinated Project Information (CPI) is a system that categorises drawings and written information (specifications). CPI is used in British Standards and in the measurement of building works, the Standard Method of Measurement (SMM7). This relates directly to the classification system used in the National Building Specification.

One of the conventions of co-ordinated project information is the 'Common Arrangement of Work Sections' (CAWS), which superseded the traditional subdivision of work by trade sections. CAWS lists around 300 different classes of work according to the operatives who will do the work; indeed the system was designed to assist the dissemination of information to sub-contractors. This allows bills of quantities to be arranged according to CAWS. The system also makes it easy to refer items coded on drawings, in schedules and in bills of quantities back to the written specification. Thus the written specification is the central document in the information chain.

2 Scaffolding and Associated Work

With the exception of some modular build projects, scaffolding is required to provide a safe and convenient working area above ground level. Scaffolding and associated temporary supports are also required for work on existing buildings, including repair and maintenance projects. A number of different types of scaffolding systems are available to suit different circumstances. Special structures have been designed to support the facade of buildings in facade retention schemes. These are described in this chapter and there is a brief look at refurbishment and demolition.

2.1 Scaffolding

There is a limit to the safe working height at which a worker can access the building work from ground level. Therefore some form of temporary support is required to provide a safe and convenient working surface. This is known as scaffold or scaffolding. Scaffolding is used on new build projects and for work to existing structures, including maintenance and repair work. The temporary structure needs to be structurally safe yet also capable of rapid erection, disassembly and reuse.

Functional requirements

The primary functional requirements for scaffolding are to:

❑ Provide a safe horizontal working platform and
❑ Provide safe horizontal and vertical access to buildings

The scaffold may be owned and maintained by a contractor, although it is more common for the scaffolding to be hired from a scaffolding sub-contractor. Temporary structures must be designed to suit their purpose. Certified structural engineers should be used to design the temporary works and scaffolding, and the structure must be checked on a regular basis to ensure it remains safe throughout its use. Scaffolding and temporary works should always be checked before use following extreme weather conditions, e.g. strong winds.

Inspections and maintenance

A competent person must inspect the scaffolding and associated temporary supports prior to use. The inspection must be recorded in the site log. The

Table 2.1   Scaffold board thickness and span

Thickness of scaffold board (mm)	Max. span between bearers (ledgers) (mm)	Min. overhang from bearer (mm)	Max. overhang (mm)
32	1000	50	128
38	1500	50	152
50	2600	50	200
63	3200	50	252

whole of the scaffold must be checked, including the tying in. Sections that are welded, bolted or fabricated off site must also be checked.

Scaffold components

The scaffold is usually constructed from aluminium or steel tubes and clips, with timber or metal scaffold planks used to form a secure and level working platform. Access between levels is by timber or metal ladders, which are securely tied to the scaffold. Scaffolds must comply with BS 5973: 1993 *Access and working scaffolds and special scaffold structures in steel*. The configuration of the tubes, clips and ties is discussed and illustrated below under the different types of scaffold system. Other common components are scaffold boards and edge protection.

Scaffold boards
A standard scaffold board is 225 mm wide by 38 mm thick with a maximum span of 1.5 m. The board is made from sawn softwood. Lightweight metal scaffold boards are used in some systems. Greater spans can be achieved by using thicker boards (Table 2.1); the distances between transoms on which the scaffolding boards span must not exceed the maximum span allowed for each board. Each board must be closely butted together so that there is no chance of the board slipping off the supporting tubes. Each board must overhang the ledger by 50 mm, but the overhang must not exceed four times the thickness of each scaffolding board.

Scaffolding boards are butted together to make a working platform; the minimum working platform depth is three boards. When materials are loaded on to the platform the clear passage for workers should be at least 430 mm. If the materials are to be manoeuvred on the scaffold a distance of 600 mm clear pedestrian passage must be maintained at all times. When laying bricks, the scaffold platform should be at least five boards wide (1150 mm). Hop-up brackets may be used to increase the working height of the lift and increase the working width of the scaffolding platform (Figure 2.1). When using hop-up brackets care must be taken not to overload the scaffold. The cantilevered bracket induces bending moments in the standards.

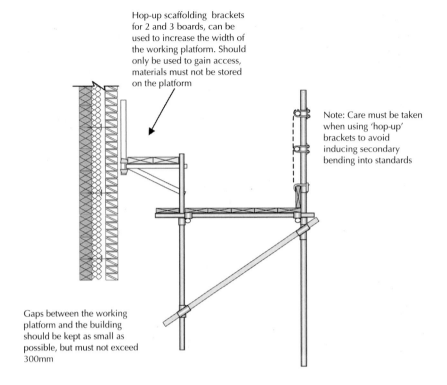

Hop-up scaffolding brackets for 2 and 3 boards, can be used to increase the width of the working platform. Should only be used to gain access, materials must not be stored on the platform

Note: Care must be taken when using 'hop-up' brackets to avoid inducing secondary bending into standards

Gaps between the working platform and the building should be kept as small as possible, but must not exceed 300mm

Figure 2.1   Hop-up brackets.

Toe boards

Toe boards must be used at the end of the scaffolding to ensure that materials and tools do not fall off the scaffold. The toe boards must be a minimum height of 150 mm. The boards also prevent the possibility of people slipping off the edge of the platform. Toe boards may be removed to allow access for materials and workers, but must be replaced immediately afterwards.

Scaffold types

Putlog scaffolds

Putlog scaffolds are erected as the external wall is constructed. The scaffolding uses the external wall as part of the support system (Figure 2.2 and Photograph 2.1). Standards and ledgers are tied to the putlogs. Each putlog has one flat end that rests on the bed or perpendicular joints in the brick or blockwork. The blade end of the putlog is usually placed horizontally and inserted fully into the brickwork joint, ensuring a full bearing is achieved. Where putlog scaffolds are used on refurbishment work, joints may be raked out to insert the blade end. In such works the blade may also be placed vertically. Where the putlog scaffold is used in new works, the putlog is placed on the wall at

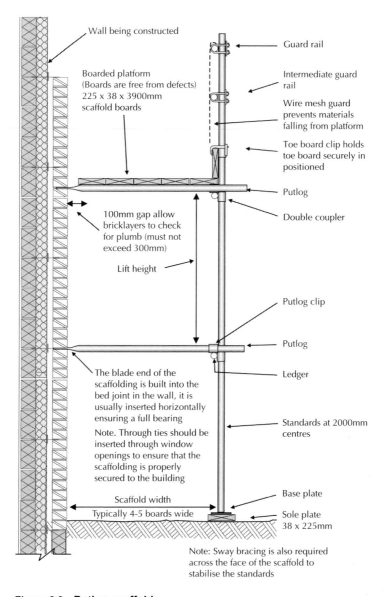

Wall being constructed

Boarded platform
(Boards are free from defects)
225 x 38 x 3900mm
scaffold boards

100mm gap allow
bricklayers to check
for plumb (must not
exceed 300mm)

Lift height

The blade end of the
scaffolding is built into the
bed joint in the wall, it is
usually inserted horizontally
ensuring a full bearing

Note. Through ties should be
inserted through window
openings to ensure that the
scaffolding is properly
secured to the building

Scaffold width
Typically 4-5 boards wide

Guard rail

Intermediate guard
rail

Wire mesh guard
prevents materials
falling from platform

Toe board clip holds
toe board securely in
positioned

Putlog

Double coupler

Putlog clip

Putlog

Ledger

Standards at 2000mm
centres

Base plate

Sole plate
38 x 225mm

Note: Sway bracing is also required
across the face of the scaffold to
stabilise the standards

Figure 2.2   Putlog scaffold.

the required lift height and the wall is constructed around the blade end of
the putlog. While the system uses less scaffolding and is less expensive than
independent scaffolding, it is essential that the erection of the scaffold is coordinated with the sequencing of brickwork. The scaffold lifts must progress at
the same speed as the masonry work. Health and safety requirements call for
competent and certified scaffolding erectors to construct and alter scaffolding,

Ledger

Sway bracing

Putlogs built
into wall

Standards

Scaffold base
plate

Scaffold board
sole plate

Photograph 2.1   Putlog scaffolding.

thus this system is not used as much as it used to be. At one time bricklaying gangs would have a labourer who could also erect the scaffold as the brickwork progressed. With good scheduling and coordination of brickwork and scaffold lifts, the system can still prove economical. Zigzag (sway) bracing is applied diagonally to the face of the scaffold, tying the ledgers and standards together. Plan bracing and ledger bracing should be used where specified.

Independent scaffolding

These scaffolds are erected 'independently' of the building structure, unlike putlog scaffolds, and are tied to the structure through window openings

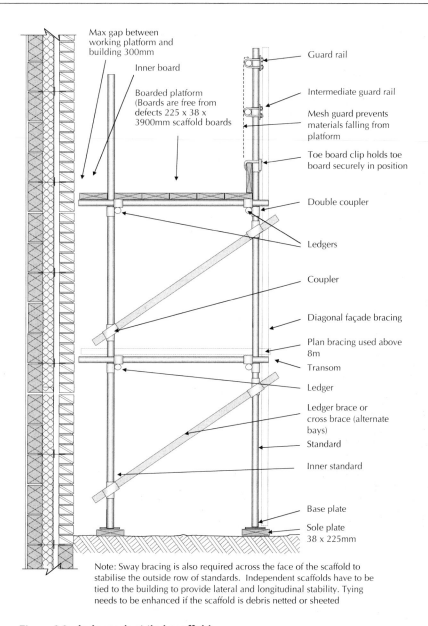

Max gap between working platform and building 300mm

Inner board

Boarded platform (Boards are free from defects 225 x 38 x 3900mm scaffold boards

Guard rail

Intermediate guard rail

Mesh guard prevents materials falling from platform

Toe board clip holds toe board securely in position

Double coupler

Ledgers

Coupler

Diagonal façade bracing

Plan bracing used above 8m

Transom

Ledger

Ledger brace or cross brace (alternate bays)

Standard

Inner standard

Base plate

Sole plate 38 x 225mm

Note: Sway bracing is also required across the face of the scaffold to stabilise the outside row of standards. Independent scaffolds have to be tied to the building to provide lateral and longitudinal stability. Tying needs to be enhanced if the scaffold is debris netted or sheeted

Figure 2.3   Independent tied scaffold.

(Figure 2.3 and Photograph 2.2). Ties are required to ensure horizontal stability is maintained. Independent scaffolds are constructed from two parallel rows of standards tied by transoms, which bridge the width of the scaffold, and ledgers, which run along the length of the scaffold. A space is usually maintained between the scaffold and the building to allow the masonry to progress

Standard

Double coupler

Ledger brace

Transom

Ledger

Inner standard

Photograph 2.2   Independent scaffold.

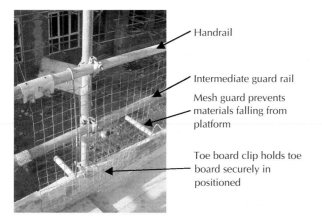

Handrail

Intermediate guard rail

Mesh guard prevents materials falling from platform

Toe board clip holds toe board securely in positioned

A Guard rail and toe board

B Double coupler

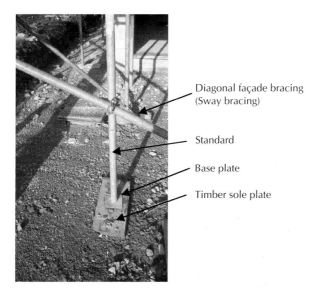

Diagonal façade bracing (Sway bracing)

Standard

Base plate

Timber sole plate

C Standard base plate

Transom or putlog

Ledger

D Scaffold clip

Photograph 2.3   Scaffolding components.

unhindered by the scaffold. The gap allows the brickwork to be checked for plumb and also helps to reduce damage to the brickwork caused by mortar snots splashing off the scaffold and on to the wall. On long stretches of scaffold, continuous diagonal tubes can be used to run from the top to the bottom of the scaffold structure. These act as façade bracing. Bracing is used to resist horizontal loads and to stiffen the structure. The bracing prevents distortion to the rectangular grids. Sway (zigzag) bracing may be applied diagonally to the face of the scaffold, tying the ledgers and standards together. Lateral bracing is also applied across the ledgers.

Proprietary scaffold systems

Proprietary systems are another type of independent scaffold. Proprietary systems rely on the same principles as independent scaffold, but use standard lengths for ledgers, transoms and standards, all of which are capable of being clipped together and dismantled easily and quickly. The standards often come with spigot ends, which allow the next standard to be located over the locating spigot very quickly. The jointing systems vary depending on the manufacturer. Proprietary systems rely on ledgers and transoms having a locating lug or bracket fixed to each end; these ends can be quickly dropped into the clips, sockets or cups, which are fixed at regular intervals on the standards. Components, such as ledgers and transoms, are designed so that they can be interchangeable.

For small scaffolds, bracing may not be required across the width of the scaffold since the frame is very rigid. Bracing must, however, run across the bays in accordance with the manufacturer's instructions. Where loads are increased or hop-ups are used, additional bracing is required. Each proprietary system varies with manufacturer and system, and manufacturer's instructions must be carefully followed to ensure that the scaffold is erected safely.

To aid the flow of work, 'hop-up' scaffolding units can be used to increase the height that the workforce can access at each lift (Figures 2.4, 2.5 and Photograph 2.4). These can be used between the standards or can be used between the internal standard and wall, providing a platform that is closer to the area of work and not reducing the width of the standard platform. Where 'hop-ups' are used externally, additional bracing is usually required.

Proprietary scaffolds have the benefit of rapid erection and disassembly; however, their use is limited to relatively standard operations due to the size of the components. Where loads are known to be considerable and the scaffolding arrangement is complicated due to specific project layout/geometry, then traditional scaffolding designed by a structural engineer may provide a more flexible and appropriate solution.

Lateral stability – tying into the building

Independent, putlog and proprietary scaffold systems must be tied into stable parts of the building structure to ensure that the scaffolding remains stable. Ties must not be linked to fittings and fixings such as balustrades or service pipes. Ties can be locked behind the walls of window openings, between floors, or braced against the cill and head of window openings (Figures 2.6, 2.7, 2.8 and Photograph 2.5).

The scaffold design must always be checked to ensure that it is capable of carrying the load of the scaffold, loads imposed by materials, the workforce and wind loads. Additional ties may be required if the loads exerted on the

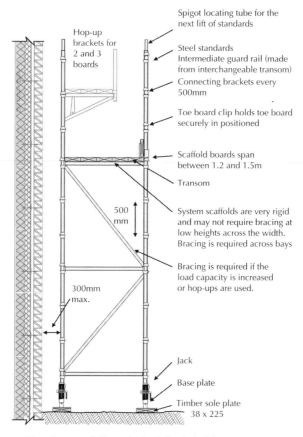

Spigot locating tube for the next lift of standards

Steel standards
Intermediate guard rail (made from interchangeable transom)

Connecting brackets every 500mm

Toe board clip holds toe board securely in positioned

Scaffold boards span between 1.2 and 1.5m

Transom

System scaffolds are very rigid and may not require bracing at low heights across the width. Bracing is required across bays

Bracing is required if the load capacity is increased or hop-ups are used.

Hop-up brackets for 2 and 3 boards

500 mm

300mm max.

Jack

Base plate

Timber sole plate 38 x 225

Note: System scaffolds need to be tied to the building façade to ensure stability

Figure 2.4   Proprietary scaffolding system – based on the sgh cup-locking system (www.sgh.co.uk).

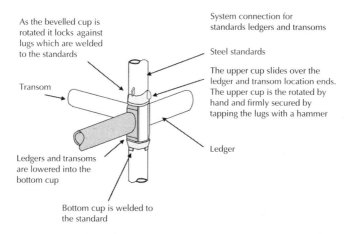

As the bevelled cup is rotated it locks against lugs which are welded to the standards

System connection for standards ledgers and transoms

Steel standards

The upper cup slides over the ledger and transom location ends. The upper cup is the rotated by hand and firmly secured by tapping the lugs with a hammer

Transom

Ledgers and transoms are lowered into the bottom cup

Ledger

Bottom cup is welded to the standard

Figure 2.5   System method of connecting standards, ledgers and transoms – based on cup-locking system (www.sgh.co.uk).

A

B

C Fixing points accommodating
handrails, ledgers and bracing

D Fixing clip

Photograph 2.4   System scaffolding.

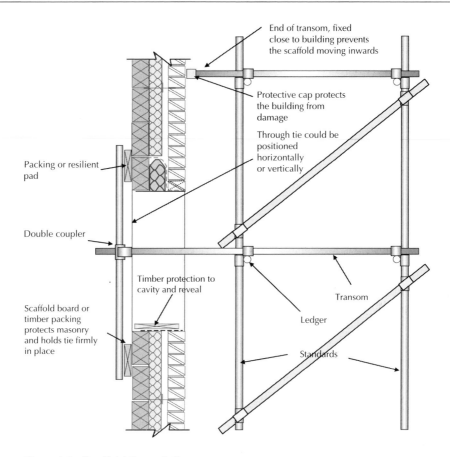

End of transom, fixed
close to building prevents
the scaffold moving inwards

Protective cap protects
the building from
damage

Through tie could be
positioned
horizontally
or vertically

Packing or resilient
pad

Double coupler

Timber protection to
cavity and reveal

Transom

Scaffold board or
timber packing
protects masonry
and holds tie firmly
in place

Ledger

Standards

Figure 2.6   Scaffold through tie.

scaffolding are increased. Where protective sheeting is used (discussed below)
extra ties will be needed to transfer the wind loads. When tying scaffolding to
a building, the points at which the scaffold is tied in must be secure and stable.
Loads are transferred at the tie points and the structure must be capable of
supporting the load.

Ties can, and often do, obstruct work. For example, it may not be possible
to fit windows where a through tie is fitted. Before any ties are removed the
scaffold design must be checked to ensure that it can remain stable while the
tie is temporarily removed or repositioned.

Weather protection – scaffold sheetings

Protective sheeting is tied to the scaffold structure to provide some protection
to the workers and the building works from wind and rain. The sheeting may
also help to protect the public from dust and debris falling from the scaffolding

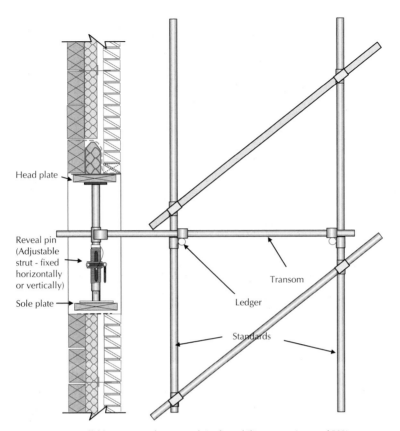

Head plate

Reveal pin (Adjustable strut - fixed horizontally or vertically)

Sole plate

Transom

Ledger

Standards

Note: Scaffolds must not rely on reveal ties for stability – a maximum of 50% reveal ties is permitted

Figure 2.7 Scaffold reveal tie.

as work proceeds. Protective sheeting is usually made from polythene sheet, although other materials such as timber panels, corrugated steel sheeting and natural fabrics may be used. It is common practice to use the external face of the weather protection for advertising. The protective sheeting will be subject to wind load, and the wind loading on the scaffold will be increased considerably. Thus scaffolding must be designed to ensure that the wind loads can be accommodated and safely transferred to the ground.

When scaffolding is sheeted the number of ties used is normally doubled. As a rule of thumb (CIRIA, 1995) the sum of B (the number of bays between each tie) × L (the number of lifts between ties) must be less than 8 for unsheeted scaffolds and less than 3 for sheeted scaffolds. This can be expressed as:

$$\text{For unsheeted scaffolds} \quad B \times L =\leq 8$$
$$\text{For sheeted scaffolds} \quad B \times L =\leq 3$$

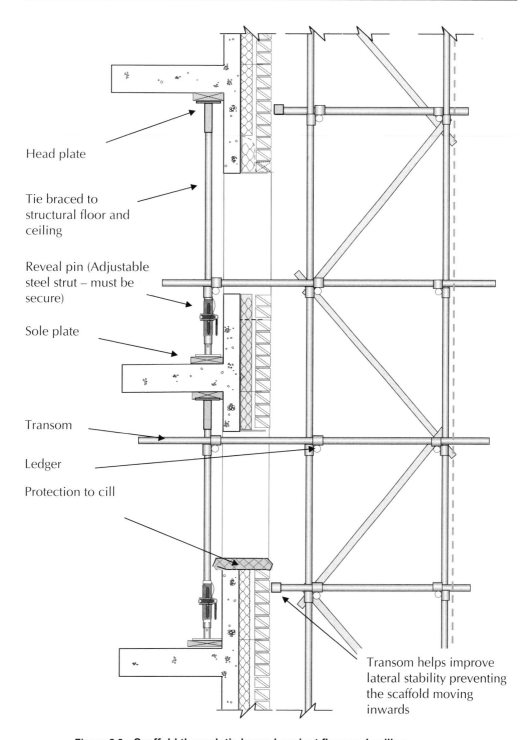

Head plate

Tie braced to
structural floor and
ceiling

Reveal pin (Adjustable
steel strut – must be
secure)

Sole plate

Transom

Ledger

Protection to cill

Transom helps improve
lateral stability preventing
the scaffold moving
inwards

Figure 2.8   Scaffold through tie braced against floor and ceiling.

Photograph 2.5   Through tie.

It is becoming common practice to construct a temporary, sheeted framework over construction works that are sensitive to the weather. These temporary structures help to create a relatively stable 'internal' environment in which the work can proceed without interruption from rain and snow. These temporary works are usually based on scaffold systems and are designed to suit specific circumstances.

Independent scaffold systems

Independent scaffold towers

Independent scaffold towers are usually delivered to site as prefabricated units that simply clip together (Photograph 2.6). Most are fabricated from aluminium, due to its light weight and easy handling properties, but steel is also

Scaffolding partially complete

Staging

Outrigger

Photograph 2.6 Independent scaffold towers with outriggers.

used. When working above one or two lifts, outriggers are usually required to increase the stability of the tower. When using prefabricated towers it is essential that the manufacturer's instructions are rigorously followed. Scaffold towers have a notorious reputation for collapsing and/or falling over when overloaded or used incorrectly.

Birdcage scaffolding

Birdcage scaffolds provide a stage with a large surface area, usually providing access to the whole ceiling area. The birdcage scaffold, with its large working platform, is suited to high level work on ornate or complex ceiling structures, intricate decorative finishes to ceilings and restoration works. Birdcages stand on the floor without being tied to walls and are therefore considered to be a type of independent scaffold system.

The scaffold comprises rows of standards placed at regular intervals in a grid formation connected by ledgers. The ledgers and braces run in both directions connecting and linking all of the standards to form a cage-like scaffold. The decking of the platform at the top of the cage is usually boarded out with two layers of scaffold planks. Each layer of planks runs at 90° to each other. The boards are laid and trapped, ensuring that there are no loose ends or gaps that would present a hazard. Polythene sheeting or dust mats are placed below the scaffold, preventing dust or small objects falling on to those working below the cage. Birdcage scaffolds have only one working deck, which is fully boarded, and are only suitable for light work.

Edge protection and scaffolding for roofs

Accidents occur due to people falling from the eaves of a roof, slipping down pitched roofs and over the eaves, and from people falling over the gable end of roofs. When working on roofs for prolonged periods of time (anything over a few hours) full edge protection should be provided. The edge protection should be designed and constructed to prevent people and materials falling from the edges of the roof. Edge protection normally consists of guardrails and toe boards extended from an existing scaffold or installed and fixed into the perimeter walls or the edge of the roof. If scaffolding is below the eaves, additional edge boards and intermediate rails may be required to prevent people falling over or through the scaffold. Where the roof is particularly steep the edge protection needs to be strong and should be capable of withstanding the force of a worker and materials falling on to the scaffold. If the roof is particularly long, intermediate platforms across the slope of the roof may need to be provided. Typical methods of providing support to the edge of a roof are shown in Figures 2.9, 2.10 and 2.11.

Trussed-out and suspended scaffolds

It is sometimes unsafe, uneconomical or simply not practical to construct a scaffold from the ground. In such situations trussed-out or suspended systems are used.

Trussed-out scaffolding systems
A trussed-out scaffold is one method of scaffolding the higher levels of a building without transferring the load to the ground (Figure 2.12). Another method of high-level scaffolding is to cantilever the loads out of the building using a secured beam (Figure 2.13). Truss and cantilever scaffolds are highly dependent on the strength of the structure. A full structural survey of the building and the design of the scaffold must be undertaken by a competent structural engineer. With such complicated scaffolding systems all steel tubes, fittings and beams must be specially checked before and during use. All anchorage points must be securely fixed to suitable structural components of the building.

Suspended scaffolds
Cradles can be permanently rigged to the roof or may be temporarily installed to provide access at heights when it is not possible or it is simply uneconomical

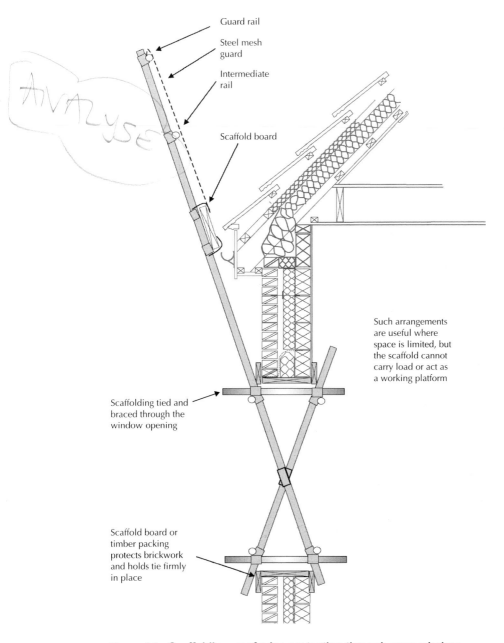

Figure 2.9   Scaffolding: roof edge protection through open window.

to erect a scaffold from the ground. Cradles and other suspended scaffold systems are becoming much more common and can be designed to suit heavy and light duty work. The area directly below the suspended cradle should always be cordoned off, preventing people walking under the cradle being struck by

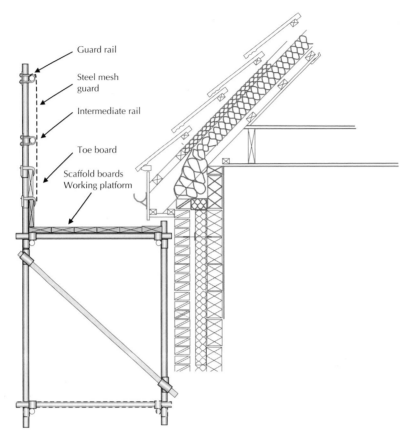

Guard rail

Steel mesh guard

Intermediate rail

Toe board

Scaffold boards
Working platform

Note: Where the projection of the roof is such that workers may fall between the scaffolding rails two or more toe boards may be necessary

Figure 2.10   Scaffolding roofs: working platform below eaves.

falling objects. All personnel using cradles must be trained and certified to use the equipment.

Alternative methods of access from heights

Other methods of gaining access from a height include:

❑ Slung scaffolds
❑ Bosun's chairs
❑ Abseiling
❑ Mobile elevating work platforms (MEWP), e.g. scissor lifts
❑ Mast climbing work platforms (MCWP)
❑ Man-riding skips

- ❏ Hoists (passenger and goods hoists)
- ❏ Ladders
- ❏ Ladder towers
- ❏ Stair towers
- ❏ Vehicle mounted platforms (cherry pickers)

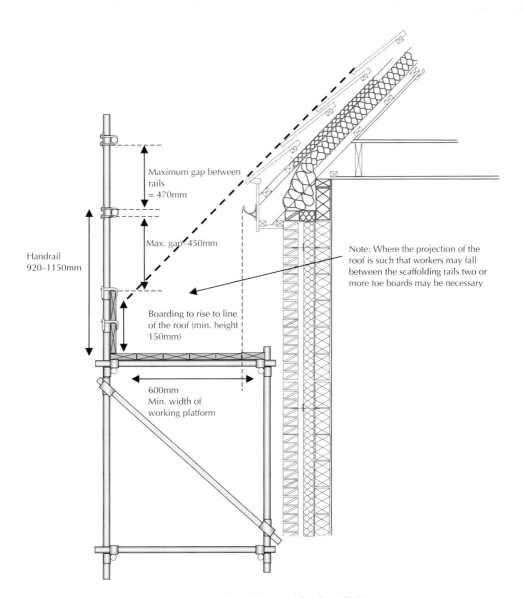

Maximum gap between rails = 470mm

Max. gap 450mm

Handrail 920–1150mm

Boarding to rise to line of the roof (min. height 150mm)

Note: Where the projection of the roof is such that workers may fall between the scaffolding rails two or more toe boards may be necessary

600mm Min. width of working platform

Figure 2.11   Scaffold: dimensions for top lift of scaffold.

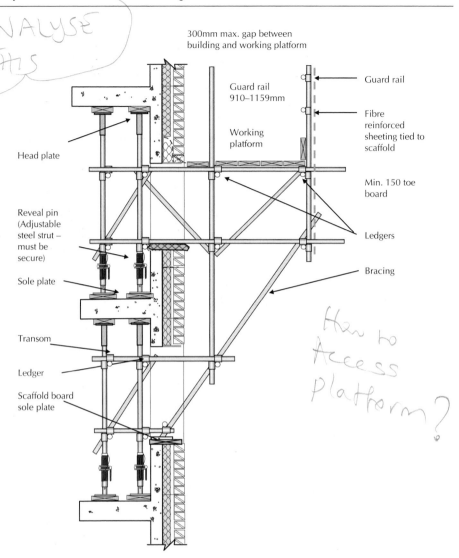

ANALYSE THIS

300mm max. gap between building and working platform

Guard rail

Guard rail 910–1159mm

Fibre reinforced sheeting tied to scaffold

Working platform

Head plate

Min. 150 toe board

Reveal pin (Adjustable steel strut – must be secure)

Ledgers

Sole plate

Bracing

Transom

Ledger

Scaffold board sole plate

How to Access Platform?

Note: Adequacy of existing structure to be checked to ensure it is able to support the loads applied by the scaffold (such scaffolding must be designed and checked by a structural engineer)

Figure 2.12   Trussed-out scaffold detail.

When working at heights or from platforms it is often necessary to have a secondary mechanism to prevent people falling from the structure or platform. Safety harnesses and belts, clipped to sound anchorage, with personnel using shock absorbers and arrest devices should be used. Working at heights always

ANALYSE FRAME (handwritten)

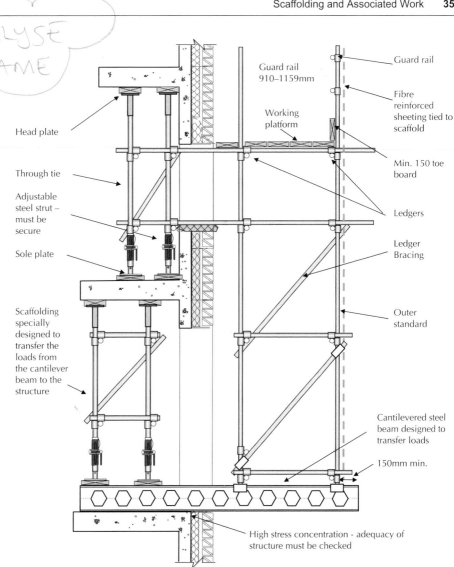

Guard rail
910–1159mm

Working
platform

Head plate

Through tie

Adjustable
steel strut –
must be
secure

Sole plate

Scaffolding
specially
designed to
transfer the
loads from
the cantilever
beam to the
structure

Guard rail

Fibre
reinforced
sheeting tied to
scaffold

Min. 150 toe
board

Ledgers

Ledger
Bracing

Outer
standard

Cantilevered steel
beam designed to
transfer loads

150mm min.

High stress concentration - adequacy of
structure must be checked

Note: Adequacy of existing structure to be checked to ensure it is able to
support the loads applied by the scaffold and cantilever beam

Figure 2.13   Cantilever scaffolding.

risk assessment (handwritten)
train staff (handwritten)
s.pension (handwritten)

presents a hazard; a full risk assessment of the task should be undertaken and all of the processes and equipment used to reduce the risks must be clearly identified. In the majority of cases personnel will need to be trained in the use of the equipment, and regular supervision is necessary to ensure that the equipment is used properly.

Trussed-out scaffolding

Cantilevered
scaffolding

Photograph 2.7   Trussed-out and cantilevered scaffolding.

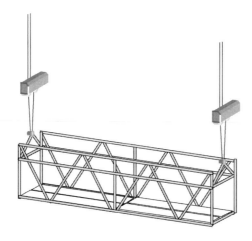

Figure 2.14 Slung scaffold – suspended cradle.

2.2 Refurbishment and facade retention

Existing buildings, both on the site to be developed and also those on neighbouring sites, affect the development of many sites in urban and semi-urban areas. Abutting buildings may need to be supported and protected for the duration of the project, during which time structures are removed and the new structure assembled. Temporary supporting works to existing structures may need to be provided to ensure that work can be undertaken safely while restoring and renovating properties, demolishing structures, retaining facades and refurbishing buildings.

Refurbishment

The limited amount of land in the UK that is available to build on has resulted in high land prices meaning that there is often greater economic advantage for refurbishing and converting existing properties. The stock of existing buildings available that no longer serve a viable or useful function is considerable. Many industrial buildings including mills, warehouses and breweries have been converted into high quality flats, or mixed use buildings with shops, restaurants, bars and residential apartments all housed in one building. New uses for a redundant building require a complete understanding of the building's construction, structural system, material content and service provision, as well as the cultural and historic context in which it is set. A checklist would need to cover issues such as:

❑ Historical and social context (including town planning restrictions)
❑ Economic constraints and potential (Life Cycle Analysis – LCA)

- ❑ Condition of fabric
- ❑ Condition of services
- ❑ Stability of the structure and foundations (especially the loading capacity)
- ❑ Acoustic and thermal properties
- ❑ Fire protection and escape provision
- ❑ Contaminants within the existing building, e.g. asbestos
- ❑ Health and safety
- ❑ Potential for re-use and recovery of materials (embodied energy etc.)
- ❑ Access limitations
- ❑ Scope for new use

There are a number of options available when deciding to renovate, refurbish or demolish a building. Table 2.2 identifies factors that may be considered and some of the options available.

Facade retention

Refurbishment makes use of the original structure and building fabric, and involves restoring, repairing and upgrading to the required standards. However, not all existing buildings have sufficient structural properties for the proposed new use and a considerable amount of structural work may need to be undertaken to ensure that the structure is made good. In many cases the foundations may need to be strengthened and underpinned and the structure reinforced. In some cases the structural work is so extensive that the only part of the original structure retained is the facade. Facade retention (sometimes referred to as facadism) involves retaining only the external building envelope. In some cases it may be as little as one elevation of the building only. The structure and majority of the building fabric is demolished to make way for a new structure behind the retained (historic) facade. Removing the main structural and lateral

Table 2.2   Refurbishment and demolition: factors to be considered

Factors to be considered when altering, refurbishing or demolishing buildings	
Factors considered	**Options**
Building function	No action
Adaptability and flexibility of building	Alteration
In-use cost: service and maintenance	Alteration and extension
Value of land and property	Refurbishment
Physical condition – structural soundness and stability	Partial demolition and new build
Historical or aesthetic interest	Demolish, remove historical or valuable materials and use on another development
	Facade retention and new build
	Demolition and rebuild
Green and sustainable issues	Demolition and redevelopment
Life cycle costing	

support (walls and floors) from the facade will render it unstable. A temporary support system must be put in place to hold the facade firmly in place while the existing structure is removed and the new structure installed. The temporary support system must be able to provide the necessary lateral stability and resist wind loads. The support systems may be located:

❏ Outside the curtilage of the existing building – external support
❏ Inside the curtilage of the existing building (behind the facade) – internal support
❏ Both external and internal to the existing building – part internal and part external

Figures 2.15 and 2.16 provide examples of external, internal and part internal–part external facade retention systems. Each support system is designed specifically to suit the facade that is being supported and the process used to construct the new building.

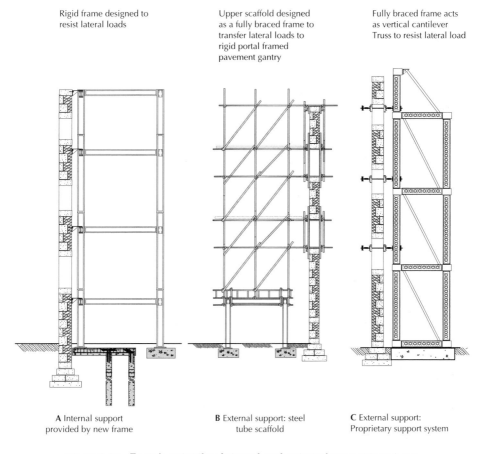

Rigid frame designed to resist lateral loads

Upper scaffold designed as a fully braced frame to transfer lateral loads to rigid portal framed pavement gantry

Fully braced frame acts as vertical cantilever Truss to resist lateral load

A Internal support provided by new frame

B External support: steel tube scaffold

C External support: Proprietary support system

Figure 2.15   Facade retention internal and external support systems.

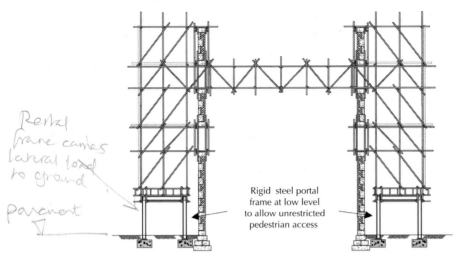

Portal frame carries lateral load to ground

Pavement

Rigid steel portal
frame at low level
to allow unrestricted
pedestrian access

A Tubular steel scaffold with flying truss

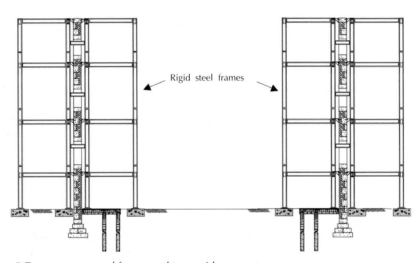

Rigid steel frames

B Temporary external frame used to provide support

Figure 2.16   Part internal–part external facade retention.

Various methods of retaining the facade and constructing the new works are
used. This is a specialist field, covered extensively by Highfield (1991, 2000).
A summary of the principal issues that must be addressed in facade retention
schemes include:

❑ Temporary support to the facade – throughout the works
   ○ Must retain the facade, prevent unwanted movement, allow for differ-
     ential movement and resist wind loads

❏ Permanently tying back the facade to the new structure
  ○ Facade ties must restrain the facade and prevent outward movement away from the new structure
  ○ The ties must not transmit any vertical loads from the existing structure to the facade
❏ Allowance for differential settlement between the new structure and the retained facade
  ○ Ties to the new structure must be capable of accommodating such movement
❏ Ensure the new foundations do not impair the stability of the retained facade
  ○ Underpinning may be necessary to ensure that settlement is controlled

Temporary support
Temporary support to the facade can be provided by steel tubular scaffolds constructed to hold the facade firmly in place until the new structural frame is built (Figure 2.17). Where possible, the scaffolding facade ties are taken through the window openings to avoid the need for breaking through the facade or drilling into the masonry to fix resin or mechanical anchors. Drilling and other potentially damaging operations should be avoided where feasible, especially if the facade is of architectural merit and/or town planning restrictions apply.

Where the wall is clamped with a through tie, timber packing either side of the tie is used to provide a good contact with the surface. Surfaces of a facade are often irregular and the thickness of the remaining structure may vary. Timber packing, felt and other slightly resilient and compressible materials should be used to secure the facade surface and protect it by preventing direct contact with metal supports.

When the temporary structural frame is in position the demolition operations to the main structure can commence. Once the new frame is constructed the facade can be tied to the new building. The lateral forces applied by the wind may mean that the scaffolding needs to be trussed out, with kentledge applied (load to hold the support down), or flying shores may be needed to transfer the loads (Figure 2.18).

The design of the shoring system is dependent on the position and number of walls and floors retained and the position of the new structure and its floors. The installation of shoring should be coordinated with the demolition and installation of the new frame. During the demolition operations it is essential that the integrity of the remaining structure be maintained. Only when the final structure is erected and the facade fully tied to it can the shoring be removed completely.

Flying shores may be used to brace the structure against wind loads. These can only be used where there is an adequate return, e.g. the opposite face of the building. Depending on the direction of the force, the loads are transferred across the truss, down the opposing scaffold and to the ground. Foundations

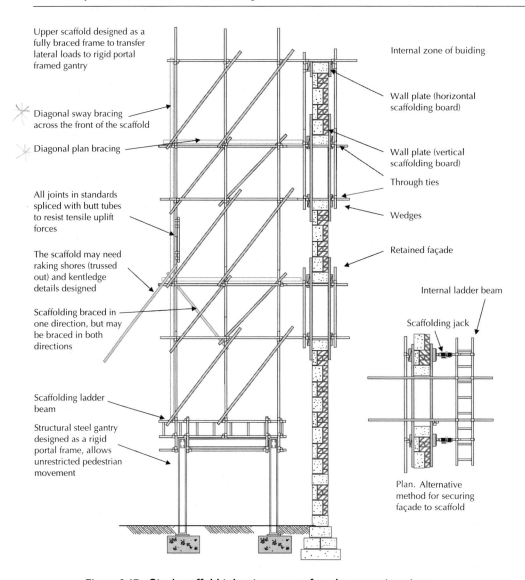

Upper scaffold designed as a fully braced frame to transfer lateral loads to rigid portal framed gantry

Diagonal sway bracing across the front of the scaffold

Diagonal plan bracing

All joints in standards spliced with butt tubes to resist tensile uplift forces

The scaffold may need raking shores (trussed out) and kentledge details designed

Scaffolding braced in one direction, but may be braced in both directions

Scaffolding ladder beam

Structural steel gantry designed as a rigid portal frame, allows unrestricted pedestrian movement

Internal zone of buiding

Wall plate (horizontal scaffolding board)

Wall plate (vertical scaffolding board)

Through ties

Wedges

Retained façade

Internal ladder beam

Scaffolding jack

Plan. Alternative method for securing façade to scaffold

Figure 2.17   Steel scaffold tube: temporary facade support system.

to the scaffold are used to ensure that the loads are adequately transferred. The scaffolding and the shores may be designed to resist both compression and tension so the foundations must be capable of resisting uplift and compression, acting as kentledge and a thrust block.

Deflection must be limited to preserve the integrity of the retained structure. Flying trusses are constructed with a camber to reduce the impact of sagging. Intermediated scaffold towers can be used to reduce sagging. Unfortunately

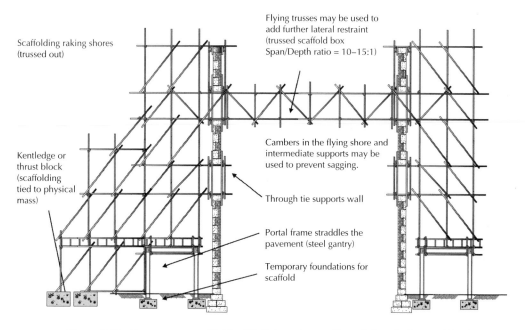

Scaffolding raking shores
(trussed out)

Flying trusses may be used to
add further lateral restraint
(trussed scaffold box
Span/Depth ratio = 10–15:1)

Kentledge or
thrust block
(scaffolding
tied to physical
mass)

Cambers in the flying shore and
intermediate supports may be
used to prevent sagging.

Through tie supports wall

Portal frame straddles the
pavement (steel gantry)

Temporary foundations for
scaffold

Figure 2.18   Temporary scaffold with flying shores (truss), kentledge and basing.

the use of intermediate supports may impede the new construction work. The construction of reinforced concrete lift shafts and service towers can be used to transfer the loads to permanent structures at an early phase in the construction process.

Where there is sufficient room outside the structure, raking shores will be used to provide the main structural support to the facade. Temporary works outside the building reduces the impact on the new build, which is carried out inside the structure.

Hybrid and proprietary facade support systems

There are various structural formwork or falsework systems available that can be used, in conjunction with steel tubular scaffolding, as a hybrid system. Alternatively the tubular, manufactured or patented systems can be used on their own to provide the required support (Photograph 2.8). The system illustrated in Figure 2.19 provides a schematic of the RMD support system; this can be used totally externally as shown in the diagram or internally or as a part internal–part external frame. The components fix together to provide a strong rigid frame. Often the areas between each lift of the supporting structure are used to house temporary site accommodation. In larger structures, multiple bays are used which are fully braced to provide the required support.

Photograph 2.8   Proprietary facade retention systems.

Fabricated support systems and use of new structure

It is also possible to limit the amount of temporary scaffolding and support systems by making use of the new structural frame (Figures 2.20 and 2.21). Although logistically complicated, in some buildings it is possible to bore through ground floors to construct new foundations, puncture holes through upper floors and walls, and erect part of the new structural frame before removing

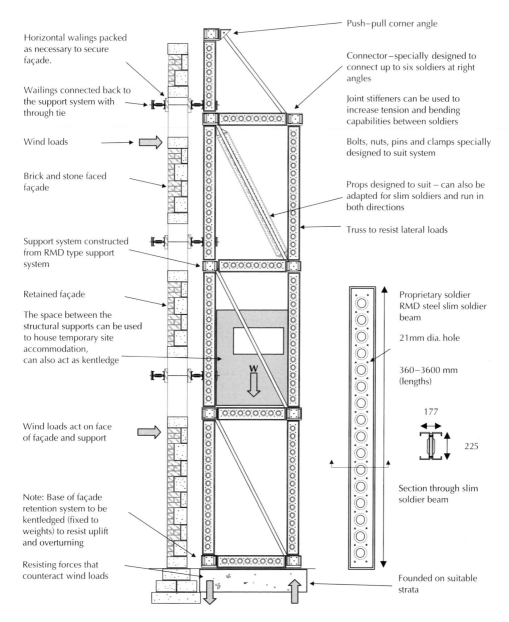

Horizontal walings packed as necessary to secure façade.

Wailings connected back to the support system with through tie

Wind loads

Brick and stone faced façade

Support system constructed from RMD type support system

Retained façade

The space between the structural supports can be used to house temporary site accommodation, can also act as kentledge

Wind loads act on face of façade and support

Note: Base of façade retention system to be kentledged (fixed to weights) to resist uplift and overturning

Resisting forces that counteract wind loads

Push–pull corner angle

Connector – specially designed to connect up to six soldiers at right angles

Joint stiffeners can be used to increase tension and bending capabilities between soldiers

Bolts, nuts, pins and clamps specially designed to suit system

Props designed to suit – can also be adapted for slim soldiers and run in both directions

Truss to resist lateral loads

Proprietary soldier RMD steel slim soldier beam

21mm dia. hole

360–3600 mm (lengths)

177

225

Section through slim soldier beam

Founded on suitable strata

Figure 2.19   Facade retention using proprietary RMD support system: fully braced frame (www.rmdkwikform.net adapted from Highfield, 2000).

the main supports of the existing structure. In order for such operations to be undertaken a thorough structural survey is required, and careful planning of the demolition sequence is necessary. Figures 2.20 and 2.21 show the facade tied to part of the new structure. All facade retentions schemes are expensive;

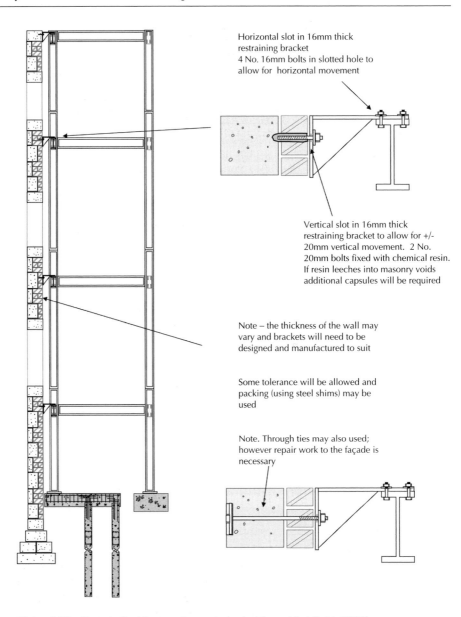

Horizontal slot in 16mm thick
restraining bracket
4 No. 16mm bolts in slotted hole to
allow for horizontal movement

Vertical slot in 16mm thick
restraining bracket to allow for +/-
20mm vertical movement. 2 No.
20mm bolts fixed with chemical resin.
If resin leeches into masonry voids
additional capsules will be required

Note – the thickness of the wall may
vary and brackets will need to be
designed and manufactured to suit

Some tolerance will be allowed and
packing (using steel shims) may be
used

Note. Through ties may also used;
however repair work to the façade is
necessary

Figure 2.20   Facade tied to new frame (adapted from Highfield, 2000).

rather than using a scaffolding system, some retention schemes make use of
specifically designed and fabricated rolled steel beams and columns to provide
support to the external face of the facade. While the structure surrounding the
exterior of the building is only temporary, each of the beams and columns can
be recycled and reused and the system may not be as expensive as first appears.

e/s

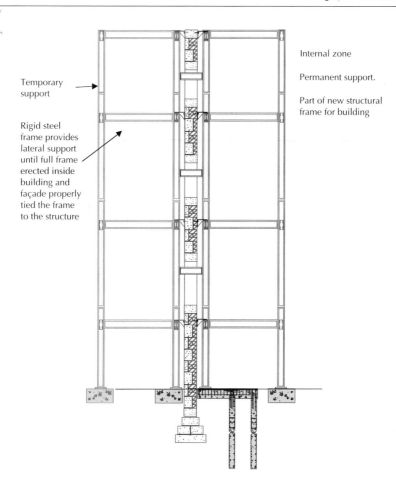

Temporary support

Rigid steel frame provides lateral support until full frame erected inside building and façade properly tied the frame to the structure

Internal zone

Permanent support.

Part of new structural frame for building

Figure 2.21   Steel frame: facade retention with temporary steel frame.

Inspections and maintenance

rusting

Appropriate safety inspections must be carried out, similar to those outlined for scaffolding systems. In situations where the works are prolonged, metal support systems are susceptible to rusting and may lose their loadbearing properties. The integrity of the system must be maintained at all times; any structural members that are damaged must be replaced. Inspections must be carried out after exposure to severe weather. The condition of the scaffold must be recorded and permission to use the temporary structure given to site personnel.

Slight movement of the facade is to be expected, thus the retained facade must be monitored for the duration of the works. If movement of the wall is detected and/or cracks develop in the fabric, investigation should be carried out to ensure that the wall is still structurally stable. Where cracking patterns suggest that de-lamination of the wall is occurring or the facade is losing its structural integrity, remedial works will be necessary. It is essential that monitoring and maintenance is continuous throughout the construction process and any operations necessary to ensure the facade and temporary works remain structurally sound are undertaken.

2.3 Demolition

[handwritten annotation: town planning, risk asses—, method statement, structural survey, planning/sequence, dem. survey]

The reasons for demolition vary. Some buildings may simply have outlived their functional use, some may have become derelict and uneconomical to repair and some may still be perfectly functional but need to be removed to make way for a new development. Buildings that have become structurally unsound through neglect may need to be demolished so that they do not pose a threat to the safety of those passing in close proximity to the building. Local authorities have the power to issue a dangerous structures notice on the building owners, requiring immediate action. In all cases the appropriate town planning office should be contacted to discuss the proposed demolition and then the appropriate consents applied for prior to any demolition work commencing.

Demolition of any property carries risks. It is essential that a full survey is undertaken so that a detailed method statement can be prepared and the appropriate demolition techniques used. Special measures are required for the controlled removal of some materials, such as asbestos, prior to demolition commencing. Demolition operations must be carefully planned and each stage monitored so that the structure can be taken down without any risk to those working on the site and those in the local vicinity of the building. Information that should be collected on a demolition survey includes:

- ❑ Existing services – live and unused services
- ❑ Natural and manmade water courses
- ❑ Presence of asbestos and other hazardous materials
- ❑ Distribution of loads
- ❑ Building structure, form and condition
- ❑ Evidence of movement and weaknesses in the structure
- ❑ Identification of hazards
- ❑ Distribution and position of reinforcement – especially post-tensioned beams
- ❑ Allowable loading of each floor (for demolition plant)
- ❑ Stability of the structure
- ❑ Survey of adjacent and adjoining structures
- ❑ Loads transferred through adjoining structures
- ❑ Loads transferred from adjoining structures
- ❑ Access to structure – allowable bearing strength of access routes

Photograph 2.9   Flying shores used to support existing structures during demolition.

Prior to demolition the following tasks should be undertaken:

- ❑ Conduct site survey
- ❑ Contact neighbours and relevant authorities (local authority, police) to discuss the options
- ❑ Identify structural hazards and reduce risks
- ❑ Select appropriate demolition technique
- ❑ Identify demolition phases and operations
- ❑ Identify communication and supervision procedures
- ❑ Organise logistics, and identify safe working areas and exclusion zones
- ❑ Identify need for temporary structures and controlled operations to avoid unplanned structural collapse
- ❑ Select material handling method
- ❑ Identify procedure for decommission services and plant
- ❑ Identify recycling and disposal methods and process
- ❑ Ensure health and safety processes are not compromised during the process

Photograph 2.9 shows the level of temporary support that may be required to existing structures during demolition.

Recycling demolition waste

A high proportion of demolition waste can be recycled and/or reused. Material recovery from the demolition makes environmental and economic sense, and the amount of material being recovered and reused is steadily increasing. The design and construction of new buildings should consider the whole life cycle of the building, which includes demolition (disassembly) and materials recovery. This requires clear decisions to be taken at the design and detailing phases. Method statements should clearly describe how the disassembly strategy works.

3 Foundations and Substructures

This chapter develops further the description of ground and foundations in *Barry's Introduction to Construction of Buildings* and introduces substructure and basement construction. The foundation of a building is that part of the substructure which is in direct contact with, and transmits loads to, the ground. The substructure is that part of a building or structure that is below natural or artificial ground level and which supports the superstructure. In practice the concrete that runs underneath walls, piers and columns and steel reinforced concrete rafts, which are spread under the whole building, are described as foundations. Steel and concrete columns, known as pile foundations, can be inserted or bored into the ground, transferring the building loads to load-bearing strata, which may be a considerable depth below the surface of the ground.

3.1 Ground stability

Ground is the term used for the earth's surface, which varies in composition within the following five groups: rocks; non-cohesive soils; cohesive soils; peat and organic soils; made-up ground and fill. Rocks include the hard, rigid, strongly cemented geological deposits such as granite, sandstone and limestone, and soils include the comparatively soft, loose, uncemented geological deposits such as gravel, sand and clay. Unlike rocks, soils, made-up ground and fill are compacted under the compression of the loads of buildings on foundations. The foundation of a building is designed to transmit loads to the ground so that any movements of the foundation are limited and thus will not adversely affect the functional requirements of the building or neighbouring buildings/ground. Movement of the foundations may be caused by the load of the building on the ground and/or by movements of the ground that are independent of the load applied to the building.

The applied load of buildings on foundations may cause settlement either through the compression of soil below foundations or because of shear failure due to overloading. Settlement movements on non-cohesive soils, such as gravel and sand, take place as the building is erected and this settlement is described as 'immediate settlement'. On cohesive soils, such as clay, settlement is a gradual process as water, or water and air, are expelled from pores in the soil. This settlement, which is described as 'consolidation settlement', may continue for several years after completion of the building. Anticipated ground movements and potential settlement will be accommodated in the design of the foundation system, which should also include for relative movement between different parts of the foundation.

If the building loads are not properly distributed and foundations not designed and constructed correctly, differential settlement may occur. Differential settlement occurs when different parts of the building settle into the ground at different rates. Figure 3.1 illustrates some of the causes of differential settlement. Movements of the foundation independent of the applied loads of buildings are due to seasonal changes or the effects of vegetation, which lead to shrinking or swelling of clay soils, frost heave, changes in ground water level and changes in the ground due to natural or artificial causes. The expansion of water in soils with low permeability due to freezing was described in *Barry's Introduction to Construction of Buildings*. The expansion and consequent heaving of the soil occur at the surface and for a depth of some 600 mm. The NHBC (2000) Standards recommend a minimum depth of 450 mm for all excavations to avoid frost action; however, most foundations are excavated to a minimum distance of 750 mm to avoid volume changes due to seasonal movement. The foundations of large buildings are generally some metres below the surface, at which level frost heave will have no effect in the UK. Correctly designed and constructed, the foundations will provide a firm and durable base, helping to prevent distortion of the structure and damage to underground services.

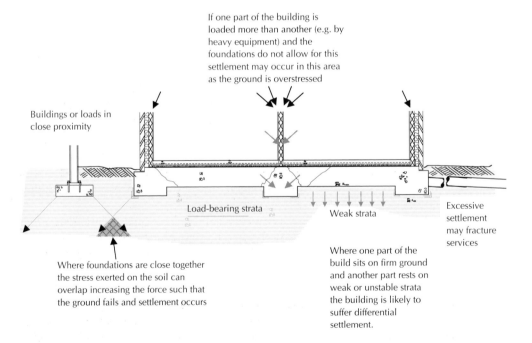

If one part of the building is loaded more than another (e.g. by heavy equipment) and the foundations do not allow for this settlement may occur in this area as the ground is overstressed

Buildings or loads in close proximity

Load-bearing strata

Weak strata

Excessive settlement may fracture services

Where foundations are close together the stress exerted on the soil can overlap increasing the force such that the ground fails and settlement occurs

Where one part of the build sits on firm ground and another part rests on weak or unstable strata the building is likely to suffer differential settlement.

Figure 3.1   Causes of differential settlement.

Rocks

Rocks may be classified as sedimentary, metamorphic and igneous according to their geological formation as shown in Table 3.1 or by reference to their presumed bearing value (Table 3.2), which is the net loading intensity considered appropriate to the particular type of ground for preliminary design purposes. The presumed bearing values are based on the assumption that foundations are carried down to unweathered rock. Hard igneous and gneissic rocks, in sound condition, have so high an allowable bearing pressure that there is little likelihood of foundation failure.

Table 3.1   Rocks

Group	Rock type
Sedimentary	Sandstones (including conglomerates) Some hard shales Limestones
Metamorphic	Some hard shales Slates Schists Gneisses
Igneous	Granite Dolerite Basalt

Hard limestones and hard sandstones are, when massively bedded, stronger than good quality concrete and it is rare that their full bearing capacity is utilised. Where water containing dissolved carbon dioxide runs over the face of limestone, the limestone may also dissolve into the solution. Water containing carbon dioxide, which flows along cracks or joints in the limestone, may further erode the limestone and reduce the soundness of the rock.

Schists and slates are rocks with pronounced cleavage. If the beds are shattered or steeply inclined a reduction in bearing values is made. Hard shales and hard mudstones, formed from clayey or silty deposits by intense natural compaction, have a fairly high allowable bearing pressure.

Soft sandstones have a very variable allowable bearing pressure depending on the cementing material. Soft shales and soft mudstones are intermediate between hard cohesive soils and rocks. They are liable to swell on exposure to water and soften.

Chalk and soft limestone include a variety of materials composed mainly of calcium carbonate and the allowable bearing pressure may vary widely. When exposed to water or frost these rocks deteriorate and should, therefore, be protected with a layer of concrete as soon as the final excavation level is reached.

Thinly bedded limestones and sandstones, which are stratified rocks, often separated by clays or soft shales, have a variable allowable bearing pressure depending on the nature of the separating material. Heavily shattered rocks have been cracked and broken up by natural processes. The allowable bearing pressure is determined by examination of loading tests.

Soils

Soils are commonly classified as non-cohesive or cohesive as the grains in the former show a marked tendency to be separate whereas the grains of the latter have a marked tendency to adhere to each other. These characteristics affect the behaviour of the soils under the load of buildings.

Characteristics of soils

The characteristics of a soil that affect its behaviour as a foundation are compressibility, cohesion of particles, internal friction and permeability. It is convenient to compare the characteristics and behaviour of clean sand, which is a coarse-grained non-cohesive soil, with clay, which is a fine-grained cohesive soil, as foundations to buildings.

❑ Compressibility.   Under load, sand is only slightly compressed due to the expulsion of water and some rearrangement of the particles. Because of its high permeability water is quickly removed and sand is rapidly compressed as building loads are applied. Compression of sand subsoils keeps pace with the erection of buildings so that once the building is completed no further compression takes place. Clay is very compressible, but due to its impermeability compression takes place slowly because of the very gradual expulsion of water through the narrow capillary channels in the clay. The compression of clay subsoil under the foundation of a building may continue for some years after the building is completed, with consequent gradual settlement.

❑ Cohesion of particles (plasticity).   Where there is negligible cohesion between particles of sand the soil is not plastic. If there is marked cohesion between particles of clay (which can be moulded, particularly when wet), this is plastic soil. The different properties of compressibility and plasticity of sand and clay are commonly illustrated when walking over these soils. A foot makes a quick indent in dry sand with little disturbance of the soil around the imprint, whereas a foot sinks gradually into clay with appreciable heaving of the soil around the imprint. Similarly, the weight of a building on sand or gravel causes rapid compression of the soil by a rearrangement of the particles with little disturbance of the surrounding soil, as illustrated in Figure 3.2. Under the load of a building, plastic clay is slowly compressed due to the gradual expulsion of air and water through the narrow capillary channels with some heave of the surrounding surface, as illustrated in

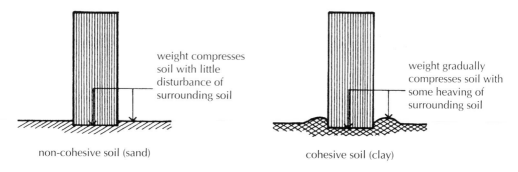

non-cohesive soil (sand)   cohesive soil (clay)

Figure 3.2   Cohesion of particles.

Figure 3.2. The surrounding surface heave may be pronounced as the load increases and the shear resistance of the clay is overcome, as explained later.

❑ Internal friction.   There is internal friction between the particles of dry sand and dry gravel. This friction is least with small particles of sand and greatest with the larger particles of gravel. The shape of the particles also plays a part in the friction, being least where particles are smooth faced and greatest where they are coarse faced. When internal friction is overcome, for example by too great a load from the foundation of a building, the soil shears and suddenly gives way.

There is little friction between fine particles of clay. Owing to the plastic nature of clay, shear failure under the load of a building may take place along several strata simultaneously with consequent heaving of the surrounding soil, as illustrated in Figure 3.3. The shaded wedge of soil below the building is pressed down and displaces soil at both sides, which moves along the slip surfaces indicated. In practice, the load on a foundation and the characteristics of the soil may not be uniform over the width of a building. In consequence the internal friction of the subsoil under the buildings may vary so that the shear of the soil may occur more pronouncedly on one side, as illustrated in Figure 3.4. This is an extreme, theoretical type of failure of a clay subsoil which is commonly used by engineers to calculate the resistance to shear of clay subsoils and presumes that the half cylinder of soil ABC rotates about centre O of slip plane ABC and causes heave of the surface on one side.

❑ Permeability.   When water can pass rapidly through the pores or voids of a soil, the soil is said to be permeable. Coarse-grained soils such as gravel and sand are permeable, and because water can drain rapidly through them they consolidate rapidly under load. Fine-grained soils such as clay have low permeability and because water passes very slowly through the pores, they consolidate slowly.

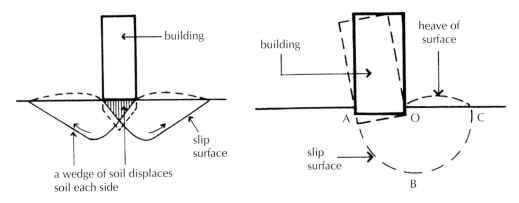

Figure 3.3   Shear failure.   **Figure 3.4   Plastic failure.**

Non-cohesive coarse-grained soils – gravel and sand

Non-cohesive coarse-grained soils such as gravels and sands consist of coarse grained, largely siliceous unaltered products of rock weathering. Gravels and sands composed of hard mineral particles have no plasticity and tend to lack cohesion, especially when dry. Under pressure from the loads on foundations the soils in this group compress and consolidate rapidly by some rearrangement of the particles and the expulsion of water. The three factors that principally affect the allowable bearing pressures on gravels and sands are density of packing of particles, grading of particles and the size of particles. The denser the packing, the more widely graded the particles of different sizes, and the larger the particles the greater the allowable bearing pressure. Ground water level and the flow of water may adversely affect allowable bearing pressures in non-cohesive soils. Where ground water level is near to the foundation level this will affect the density of packing, and the flow of water may wash out finer particles and so affect grading, both reducing bearing pressure. Non-cohesive soils should be laterally confined to prevent spread of the soil under pressure.

Cohesive fine-grained soils – clay and silt

Cohesive fine-grained soils such as clays and silts are a natural deposit of the finer siliceous and aluminous products of rock weathering. Clay is smooth and greasy to the touch, shows high plasticity, dries slowly and shrinks appreciably on drying. The principal characteristic of cohesive soils as a foundation is their susceptibility to slow volume changes. Under the pressure of the load on foundations, clay soils are very gradually compressed by the expulsion of water or water and air through the very many fine capillary paths so that buildings settle gradually during erection. The settlement will continue for some years after the building is completed. Seasonal variations in ground water and vigorous growth of trees and shrubs will cause appreciable shrinkage, drying and wetting expansion of cohesive soils. Shrinkage and expansion due to seasonal variations will extend to 1 m or more in periods of severe drought below the surface in Great Britain. Seasonal variation in moisture content will extend much deeper in soils below trees and can be up to 4 m or more below large trees. When shrubs and trees are removed to clear a site for building on cohesive soils, for some years after the clearance there will be ground recovery as the soil gradually recovers water previously taken out by trees and shrubs. This gradual recovery of water by cohesive soils and consequent expansion may take several years.

Peat and organic soil

Peat and organic soils contain a high proportion of fibrous or spongy vegetable matter from the decay of plants mixed with varying proportions of fine sand, silt or clay. These soils are highly compressible and will not serve as a stable foundation for buildings.

Made-up ground and fill

Made-up ground is the term used to describe where the ground level has been raised by spreading and sometimes compacting material from excavations or waste. Depending on the materials used, the way that they have been levelled, layered and compacted will affect the ability of the made-up ground to be used as a load bearing material.

Fill is the term used to describe the tipping of material into pits or holes, which are left after excavation, mining or quarrying, to raise the level to that of the surrounding ground. Made-up ground and fill will not usually serve as a stable foundation for buildings due to the extreme variability of the materials used to make up ground and the variability of the compaction or natural settlement of these materials. Where it is anticipated that the area to be filled will be used for further development, careful attention needs to be given to the materials used to fill the site, the depth of the layers in which the fill material is placed, and the compaction method(s) to be used.

The characteristics of the individual constituents of ground, rocks, soils and organic soils will provide an indication of the likely behaviour of a particular ground under the load of foundations. In practice, soils often consist of combinations of gravel, sand and clay in varying proportions, which combine the characteristics of the constituents.

Ground Instability

There are areas of ground that are unstable due to natural processes, such as landslip of sloping strata of rocks or soils, or due to human activities such as mining and surface excavation. Under the load of foundations the unstable ground may be subject to ground movement, which should be anticipated in the design of foundation. Land instability may be broadly grouped under the headings:

- ❑ Landslip
- ❑ Surface flooding and soil erosion
- ❑ Natural caves and fissures
- ❑ Mining and quarrying
- ❑ Landfill

The Environment Agency (www.environment-agency.gov.uk) has a considerable database of useful information on landfill sites, floodplains, subsidence and contaminated land. Some particularly striking images of coastal erosion can be found on the following Web pages: www.member.aol.com shows examples of erosion caused by the sea in East Anglia; www.digitalworlds.co.uk uses historic maps and images together with current photographs to demonstrates the loss of land to the sea. The Hadley Centre provides information on global warming and environmental changes (www.met-office.gov.uk/research/hadleycentre), which may be of use when thinking about how to detail and construct buildings to better cope with climate change. It is also necesssary to determine the nature of the subsoil through physical investigation, as described in *Barry's Introduction to Construction of Buildings*.

Landslip

Landslip may occur under natural slopes where weak strata of clay, clay over sand or weak rock strata may slip down a slope, particularly under steep slopes and where water acts as a lubricant to the slip movement. Landslides of superficial strata nearest to the surface, which will be most noticeable and therefore recorded, are those that will in the main cause land instability that may affect the foundations of buildings. Landslides of deeper strata that have occurred, or may occur, generally go unnoticed and will only affect deep excavations and foundations. The most noticeable landslides occur in cliff faces where the continuous erosion of the base of the cliff face by tidal movements of the sea undermines the cliff and causes collapse of the cliff face and subsidence of the supported ground. Similar landslip and subsidence may occur where an excavation is cut into a slope or hillside. The previously supported sloping strata are effectively undermined and may slip towards the excavation. Landslip is also common around excavations for deep coal mining, which may break through sloping strata and so encourage landslides. The Department of the Environment, Transport and the Regions (now DEFRA) has commissioned studies and prepared reports of areas liable to land instability in and around the coal mining areas of Great Britain. Similar studies have led to reports of areas liable to land instability due to landslip around areas of metal, stone, chalk and limestone quarrying.

This house was once over 500m away from the cliffs.

The soft clay and mud stone at the base of the cliff is removed and the sand stone at the top of the cliff falls away.

Numerous sea defences, including concrete, steel and timber piles, have been torn away offering little protection to the coast line.

New off shore rock piers are being tested. Whilst they may offer some protection they change the water movement causing sand to shift changing the shape of the coast line.

Photograph 3.1   Coastal erosion and sea defences.

Surface flooding

Surface flooding may affect the stability of surface ground, and the seasonal movement of water through permeable strata below the surface may cause gradual erosion of soils and permeable rocks that may lead to land instability. The persistent flow of water from fractured water mains and drains may cause gradual erosion of soil and lead to land instability. The incidence of surface flooding and erosion by below surface water is, by and large, known and recorded by the regional water authorities. Information on floodplains in the UK can be found on the following web sites: www.antiflood.com and www.homecheck.co.uk.

Natural caves and fissures

Natural caves and fissures occur generally in areas in the UK where soluble rock strata, such as limestone and chalk, have been eroded over time by the natural movement of subterranean water. Where there are caves or small cavities in these areas near the surface, land instability and subsidence may occur. The Department of the Environment, Transport and the Regions (now DEFRA) has prepared a review of information on the incidence of such cavities in the form of regional reports and maps showing the location and nature of known cavities and the likelihood of land instability due to the cavities.

Mining and quarrying

Mining and quarrying of mineral resources has been carried out for centuries over much of England and parts of Wales and Scotland. The majority of the mines and quarries have by now been abandoned and covered over. From time to time mining shafts collapse and the ground above may subside. Similarly, ground that has been filled over redundant quarries may also subside. There is potential for land instability and subsidence over those areas of the UK where mineral extraction has taken place. The Department of the Environment, Transport and the Regions has commissioned surveys and produced reports of those areas known or likely to be subject to land instability due to mining and quarrying activities. There are regional reports and atlases indicating the location of areas that may be subject to land instability subsidence. Coal mining areas have been comprehensively surveyed, mapped and reported. Other areas where comparatively extensive quarrying for stone, limestone, chalk and flint has taken place have been surveyed, and mapped and reported. Less extensive quarrying, for chalk for example in Norwich, has been included. The reports indicate those areas where subsidence is most likely to occur and the necessary action that should be taken preparatory to building works (www.environment -agency.gov.uk).

Landfill

Landfill is a general term to include the ground surface which has been raised artificially by the deposit of soil from excavations, backfilling, tipping, refuse disposal and any form of fill which may be poorly compacted, of uncertain composition and density and thus have indeterminate bearing capacity and be classified as unstable land. In recent years regional and local authorities have had some control and reasonably comprehensive details of landfill, which may give indication of the age, nature and depth of recent fill. The land over much of the area of the older cities and towns in the UK particularly on low-lying land, has been raised by excavation, demolition and fill. This overfill may extend some metres below the surface in and around older settlements and where soil excavated to form docks has been tipped to raise ground levels above flood water levels. Because of the variable and largely unknown nature of this fill, the surface is in effect unstable land and should be considered as such for foundations. There are no records of the extent and nature of this type of fill that has taken place over some considerable time. The only satisfactory method of assessing the suitability of such ground for foundations is by means of trial pits or boreholes to explore and identify the nature and depth of the fill.

3.2 Functional requirements

The primary functional requirements of foundations are strength and stability. To comply with Building Regulations the combined dead, imposed and wind loads of the building should be safely transmitted to the ground without causing movement of the ground that may impair the stability of any part of another building. Loading is concerned with the bearing strength of the ground relative to the loads imposed on it by the building. The foundation or foundations should be designed so that the combined loads from the building are spread over an area of the ground capable of sustaining the loads without undue movement. The pressure on the ground from the foundations of a new building increases the load on the ground under the foundations of an adjoining building and so increases the possibility of instability.

The building should also be constructed so that ground movement caused by swelling, shrinking or freezing of the subsoil or landslip or subsidence (other than subsidence arising from shrinkage) will not impair the stability of any part of the building. The swelling, shrinkage or freezing of subsoil is described in *Barry's Introduction to Construction of Buildings* and in this chapter relative to the general classification of soils.

Bearing capacity

The natural foundation of rock or soil on which a building is constructed should be capable of supporting the loads of the building without such settlement due to compression of the ground that may fracture connected services or impair the stability of the structure. For the majority of small buildings the bearing capacities for rocks and soils set out in Table 3.2 will provide an acceptable guide in the design of foundations. For heavy loads on foundations some depth below the surface, it may not be sufficient to accept the bearing capacities shown in Table 3.2 because of the uncertain nature of the subsoil. For example, the descriptions hard clays, stiff clays, firm clays and soft clays may not in practice give a sufficiently clear indication of an allowable bearing pressure to design an economical, safe foundation. For example clay soils, when overloaded, may be subject to shear failure due to the plastic nature of the soil. It is necessary, therefore, to have some indication of the nature of subsoils under a foundation by soil exploration.

Table 3.2   Bearing capacities

Group	Types of rocks and soils	Bearing capacity (kN/m^2)
I	1 Strong igneous and gneissic rocks in sound condition	10 000
Rocks	2 Strong limestones and strong sandstones	4 000
	3 Schists and slates	3 000
	4 Strong shales, strong mudstones and strong siltstones	2 000
	5 Clay shales	1 000
II	6 Dense gravel or dense sand and gravel	>600
Non-cohesive soils	7 Medium dense gravel or medium dense gravel and sand	>200 to 600
	8 Loose gravel or loose sand and gravel	<200
	9 Compact sand	>300
	10 Medium dense sand	100 to 300
	11 Loose sand	<100
III	12 Very stiff boulder clays and hard clays	300 to 600 150 to 300
	13 Stiff clays	150 to 300
Cohesive soils	14 Firm clays	75 to 150
	15 Soft clays and silts	75
IV	16 Peat and organic soils	Foundations carried down through peat to a reliable bearing stratum
V	17 Made ground or fill	Should be investigated with extreme care

Based on BS 8004:1986.

The difficulty with any system of subsoil exploration is that it is effectively impossible to expose or withdraw an undisturbed sample of soil in the condition it was underground. The operation of digging trial pits and boring to withdraw samples of soil will disturb and change the nature of the sample. In particular the compaction of the soil under the weight of the overburden of soil above will be appreciably reduced in the withdrawn sample of soil and so affect its property in bearing loads. The particular advantage of samples of subsoil withdrawn by boring is in providing an indication of the varying nature of subsoils at various levels and the means of making an analysis of the characteristics of the various samples as a guide to the behaviour of the subsoil under load to make an assumption of allowable bearing pressure.

Allowable bearing pressure is the maximum pressure that should be allowed by applying a factor of safety to the ultimate bearing capacity of a soil. The ultimate bearing capacity is the pressure at which a soil will fail and settlement of a foundation would occur. Allowable bearing pressures are determined by the application of a factor of safety of as much as 3 for cohesive soils, such as clay, and appreciably less for non-cohesive soils.

Foundation design

Failure of the foundation of a building may be due to excessive settlement by compaction of subsoil, collapse of subsoil by failure in shear or differential settlement of different parts of the foundation. The allowable bearing pressure intensity at the base of foundations is the maximum allowable net loading taking into account the ultimate bearing capacity of the subsoil, the amount and type of settlement expected and the ability of the structure to take up the settlement. Designing a foundation is a combined function of both the site conditions and the characteristics of the particular structure.

Bearing pressures

The intensity of pressure on subsoil is not uniform across the width or length of a foundation and decreases with depth below the foundation. In order to determine the probable behaviour of a soil under foundations, the engineer needs to know the intensity of pressure on the subsoil at various depths. This is determined by Boussinesq's equation for the stress at any point below the surface of an elastic body, and in practice is a reasonable approximation to the actual stress in soil.

By applying the equation, the vertical stress on planes at various depths below a point can be calculated and plotted as shown in Figure 3.5. The vertical ordinates at each level d_1, d_2 etc. represent graphically unit stress at points at that level. If points of equal stress A, B and C are joined, the result is a bulb of unit pressure extending down from L. If this operation is repeated for unit area under a foundation, the result is a series of bulbs of equal unit pressure, as illustrated in Figure 3.6. Thus the bulb of pressure gives an indication of likely stress in subsoils at various points below a foundation. The practical use of these bulbs of pressure diagrams is to check that at any point in the subsoil, under a foundation, the unit pressure does not exceed the allowable bearing pressure of the soil. From samples taken from the subsoil to sufficient depth below the proposed foundation, it is possible to verify that the unit pressure at any point below foundations does not exceed allowable bearing pressure.

Where there are separate foundations close together or a group of piles, then the bulbs of pressure of each closely spaced foundation effectively combine to act as a bulb of pressure that would be produced by one foundation of the same overall width. Where it is known from soil sampling that there is a layer or stratum of soil with low allowable bearing pressure at a given depth below

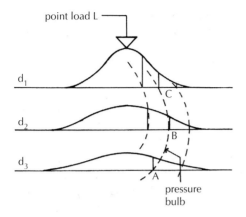

point load L

d_1

C

d_2

B

d_3

A

pressure bulb

Figure 3.5 Vertical stress distribution.

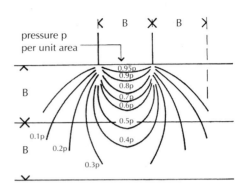

pressure p per unit area

B B

0.95p
0.9p
0.8p
0.7p
0.6p

0.5p

0.1p 0.4p

B 0.2p

0.3p

Figure 3.6 Bulbs of pressure under a strip foundation.

surface, a combined bulb of unit pressure greater than the allowable bearing pressure of the weak strata may be used to check that it does not cross the weak strata. The combined pressure bulb representing a quarter of the unit pressure on the foundations, shown in Figure 3.7, is used to check whether it intersects a layer or stratum of subsoil whose allowable bearing pressure is less than a quarter of unit pressure. If it does it is necessary to redesign the foundation. In practice it would be tedious to construct a bulb of pressure

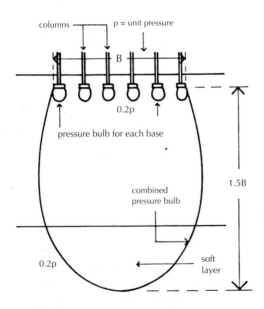

columns   p = unit pressure

B

0.2p

pressure bulb for each base

combined pressure bulb

1.5B

0.2p   soft layer

Figure 3.7 Combined pressure bulb.

diagram each time a foundation were designed and engineers today generally employ ready prepared diagrams or charts to determine pressure intensities below foundations.

Contact pressure

A perfectly flexible foundation uniformly loaded will cause uniform contact pressure with all types of soil. A perfectly flexible foundation supposes a perfectly flexible structure supporting flexible floors, roof and cladding. The CLASP system of building that was used for schools uses a flexible frame and was originally designed to accommodate movement in the foundation of buildings on land subject to mining subsidence. Most large buildings, however, have rigid foundations designed to support a rigid or semi-rigid frame.

The theoretical contact pressures between a perfectly rigid foundation and a cohesive and a cohesionless soil are illustrated in Figure 3.8, which shows the vertical ordinate (vertical lines) representing intensity of contact pressure at points below the foundation. In practice the contact pressure on a cohesive soil such as clay is reduced at the edges of the foundation by yielding of the clay, and as the load on the foundation increases more yielding of the clay takes place so that the stresses at the edges decrease and those at the centre of the foundation increase, as illustrated in Figure 3.9.

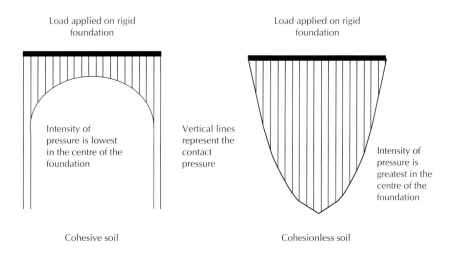

Figure 3.8 **Theoretical contact pressure.**

The contact pressure on a cohesionless soil such as dry sand remains parabolic, as illustrated in Figure 3.9, and the maximum intensity of pressure increases with increased load. If the foundation is below ground the edges stresses are no longer zero, as illustrated in Figure 3.9, and increase with increase of depth below ground.

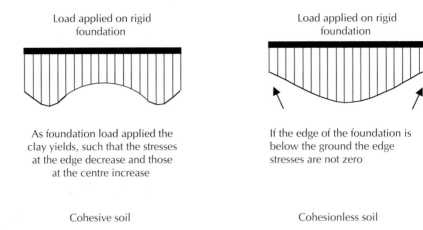

Load applied on rigid
foundation

Load applied on rigid
foundation

As foundation load applied the
clay yields, such that the stresses
at the edge decrease and those
at the centre increase

If the edge of the foundation is
below the ground the edge
stresses are not zero

Cohesive soil

Cohesionless soil

Figure 3.9   Contact pressure in practice.

For footings the assumption is made that contact pressure is distributed uniformly over the effective area of the foundation as differences in contact pressure are usually covered by the margin of safety used in design. For large spread foundations and raft foundations it may be necessary to calculate the intensity of pressure at various depths. An understanding of the distribution of contact pressures between foundation and soil will guide the engineer in his choice of foundation. For example, the foundation of a building on a cohesionless soil such as sand could be designed so that the more heavily loaded columns would be towards the edge of the foundation where contact pressure is least, and the lightly loaded columns towards the centre to allow uniformity of settlement over the whole area of the building, as illustrated in Figure 3.10. Conversely, a foundation on a cohesive soil such as clay would be arranged with the major loads towards the centre of the foundation where pressure intensity is least, as illustrated in Figure 3.11.

Differential settlement (relative settlement)

Parts of the foundation of a building may suffer different magnitudes of settlement due to variations in load on the foundations or variations in the subsoil, and different rates of settlement due to variations in the subsoil. These variations may cause distortion of a rigid or semi-rigid frame and consequent damage to rigid infill panels and cracking of loadbearing walls, rigid floors and applied finishes such as plaster and render. Some degree of differential settlement is inevitable in the foundation of most buildings, but so long as this is not pronounced or can be accommodated in the design of the building the performance of the building will not suffer.

The degree to which differential settlement will adversely affect a building depends on the structural system employed. Solid loadbearing brick and masonry walls can accommodate small differential settlement through small hair

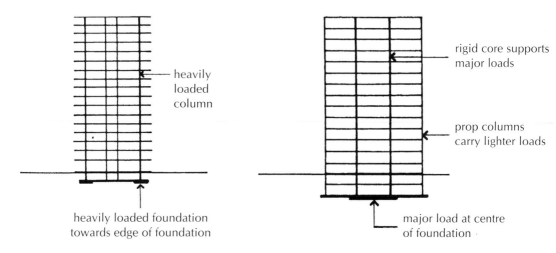

Figure 3.10   Foundation on a cohensionless subsoil.

Figure 3.11   Foundation on a cohesive subsoil.

cracks opening in mortar joints between the small units of brick or stone. These cracks, which are not visible, do not weaken the structure or encourage the penetration of rain. More pronounced differential settlement, such as is common between the main walls of a house and the less heavily loaded bay window bonded to it, may cause visible cracks in the brickwork at the junction of the bay window and the wall. Such cracks will allow rain to penetrate the thickness of the wall. To avoid this, either the foundation should be strengthened or some form of slip joint be formed at the junction of the bay and the main wall. High, framed buildings are generally designed as rigid or semi-rigid structures and any appreciable differential settlement should be avoided. Differential settlement of more than 25 mm between adjacent columns of a rigid or semi-rigid framed structure may cause such serious racking of the frame that local stress at the junction of vertical and horizontal members of the frame may endanger the stability of the structure and also crack solid panels within the frame. An empirical rule employed by engineers in the design of foundations is to limit differential settlement between adjacent columns to 1/500 of the distance between them.

Differential settlement can be reduced by a stiff structure or substructure or a combination of both. A deep hollow box raft (see later in Figure 3.34) has the advantage of reducing net loading intensity and producing more uniform settlement. A common settlement problem occurs in modern buildings where a tower or slab block is linked to a smaller building or low podium. Plainly there will tend to be a more pronounced settlement of the foundations of the tower or slab block than experienced in the smaller structure (podium). At the junction of the two structures there must be structural discontinuity and some form of flexible joint that will accommodate the differences in settlement. Figure 3.12 illustrates two examples of this arrangement. Although the tall and

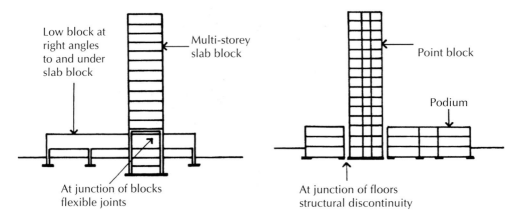

Figure 3.12   Relative settlement.

the low-rise building will appear to be one structure, they will in fact be separate structures that may move (settle) independently.

Shrinkable soils and vegetation

The type of soil and position of vegetation and trees will affect the depth of foundation (Figure 3.13). Precautions should be taken when constructing foundations on ground that is susceptible to expansion and shrinkage (Figures 3.14 and 3.15).

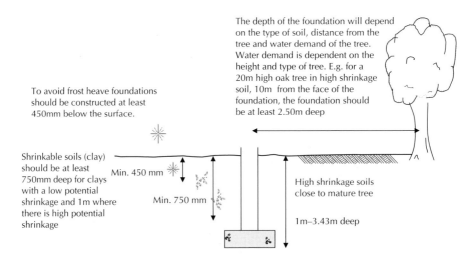

Figure 3.13   Depth of foundations and stability (information adapted from NHBC 2000).

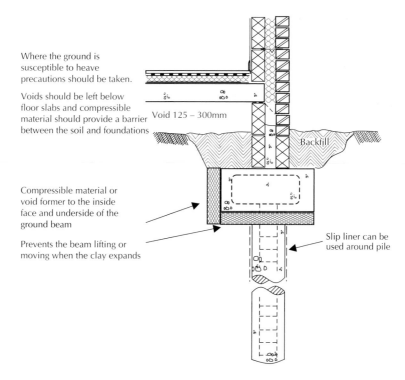

Where the ground is susceptible to heave precautions should be taken.

Voids should be left below floor slabs and compressible material should provide a barrier between the soil and foundations

Void 125 – 300mm

Backfill

Compressible material or void former to the inside face and underside of the ground beam

Prevents the beam lifting or moving when the clay expands

Slip liner can be used around pile

Figure 3.14   Precautions against heave: pile and ground beam.

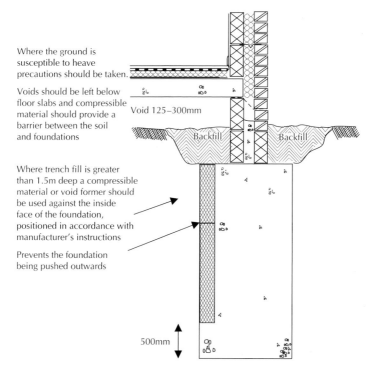

Where the ground is susceptible to heave precautions should be taken.

Voids should be left below floor slabs and compressible material should provide a barrier between the soil and foundations

Void 125–300mm

Backfill

Backfill

Where trench fill is greater than 1.5m deep a compressible material or void former should be used against the inside face of the foundation, positioned in accordance with manufacturer's instructions

Prevents the foundation being pushed outwards

500mm

Figure 3.15   Precautions against heave: trench fill.

3.3 Foundation types

Foundations may be classified as:

- ❏ Strip foundations
- ❏ Pad foundations
- ❏ Raft foundations
- ❏ Pile foundations

These are illustrated in Figure 3.16. See *Barry's Introduction to Construction of Buildings* for additional information.

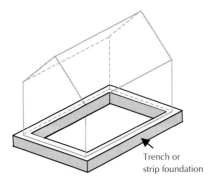

Trench or
strip foundation

A Strip foundations

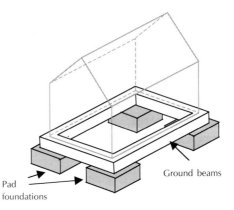

Pad
foundations

Ground beams

B Pad foundations

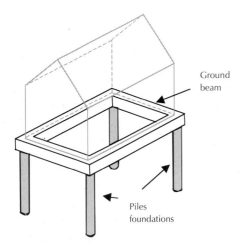

Ground
beam

Piles
foundations

C Pile foundations

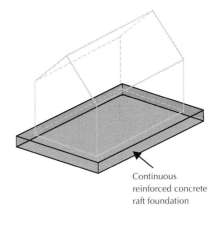

Continuous
reinforced concrete
raft foundation

D Raft foundations

Figure 3.16   Foundation types.

Strip foundations

Strip foundations consist of a continuous, longitudinal strip of concrete designed to spread the load from uniformly loaded walls of brick, masonry or concrete to a sufficient area of subsoil. The spread of the strip depends on foundation loads and the bearing capacity and shear strength of the subsoil. The thickness of the foundation depends on the strength of the foundation material.

A strip foundation, illustrated in Figure 3.17 (strip foundation), consists of a continuous, longitudinal strip of concrete to provide a firm, level base on which walls may be built. The foundation transfers the loads of the building to the loadbearing strata.

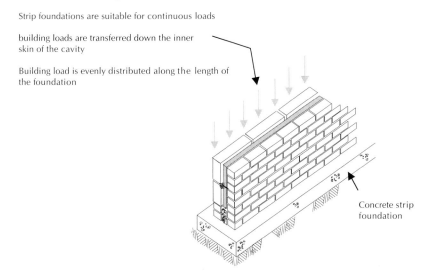

Strip foundations are suitable for continuous loads

building loads are transferred down the inner skin of the cavity

Building load is evenly distributed along the length of the foundation

Concrete strip foundation

Figure 3.17   Strip foundation.

Deep strip foundation or mass fill foundations

If the subsoil directly below the surface of the ground is weak or susceptible to moisture movement the foundation can be taken to a depth where the strata is stronger or the moisture content does not vary. If foundations are built on clay soils it is necessary to take the foundation to a depth of at least 750 mm (Figure 3.18). At such depths seasonal changes (wet winter conditions and dry summer), which result in changes to soil moisture content, are unlikely. Foundations on clay soils that are not taken to sufficient depths suffer considerable seasonal movement. During dry summer months as the ground dries and vegetation close to the building takes up any remaining moisture, the clay contracts and the foundations settle. If the settlement is not consistent across the whole of the building, cracks will form. In the winter months as the clay becomes saturated, the clay swells and the foundations lift. In many cases cracks that developed during the dry season will close up, but if the cracks have been filled or dislodged slightly, making it impossible for them to close, the building will lift and further cracks may develop.

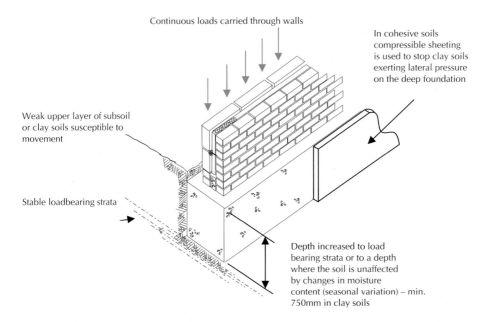

Continuous loads carried through walls

In cohesive soils compressible sheeting is used to stop clay soils exerting lateral pressure on the deep foundation

Weak upper layer of subsoil or clay soils susceptible to movement

Stable loadbearing strata

Depth increased to load bearing strata or to a depth where the soil is unaffected by changes in moisture content (seasonal variation) – min. 750mm in clay soils

Figure 3.18   Deep strip or mass fill strip foundation.

Width of foundations

The width of the foundation strip of concrete is determined by the need for room to lay the walling material below ground level and the requirement for the width of the strip to be adequate to spread the foundation loads to an area of soil capable of bearing the loads. The depth or thickness of the strip is determined by the shear strength of the concrete. A general rule is that the projection of the concrete strip each side of the wall should be no greater than the thickness of the concrete (see Figure 3.19).

Strip foundation

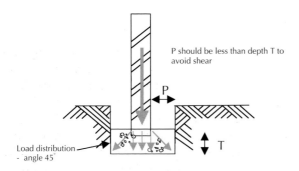

P should be less than depth T to avoid shear

P

Load distribution - angle 45°

T

Figure 3.19   Projections of strip foundations.

The tensile reinforcement allows
the width of the foundation to be increased

loads distributed over a greater area

load per unit area reduced

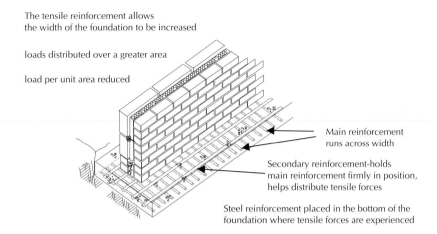

Main reinforcement
runs across width

Secondary reinforcement-holds
main reinforcement firmly in position,
helps distribute tensile forces

Steel reinforcement placed in the bottom of the
foundation where tensile forces are experienced

Figure 3.20   Steel reinforced wide strip foundation.

Wide strip foundations

Where the allowable bearing pressure on a subsoil is low, it may be necessary to use a comparatively wide and therefore thick strip of concrete to provide sufficient spread of foundation loads. As an alternative to a wide, thick strip of mass concrete, it may be economic to consider the use of a strip of reinforced concrete, illustrated in Figure 3.20 (wide strip foundation). The main reinforcement of mild steel rods is cast in the bottom, across the width of the strip, to provide additional tensile strength against the tendency to upward bending, with smaller secondary rods cast in along the length of the strip. The reinforcing rods are wired together and laid on and wired to either concrete or plastic spacers to provide sufficient concrete cover below the reinforcement to prevent destructive rusting. The cost of the reinforcement and extra labour have to be taken into account in considering the relative advantage of a reinforced concrete strip over a mass concrete strip. Concrete is spread and consolidated around the reinforcement and finished level ready for the walling.

Pad foundations

The foundation to piers of brick, masonry and reinforced concrete and steel columns is often in the form of a square or rectangular isolated pad of concrete to spread a concentrated load. The area of this type of foundation depends on the load on the foundation and the bearing and shear strength of the subsoil, and its thickness on the strength of the foundation material. The simplest form of pad foundation consists of a pad of mass concrete, as shown in Figures 3.21 and 3.22, illustrating a pier and foundation beam base for a small building. The heavily loaded pad foundations to the columns of framed buildings are generally taken down to a layer of compact subsoil at such a level that the

Photograph 3.2   Wide strip foundation – reinforcement cage.

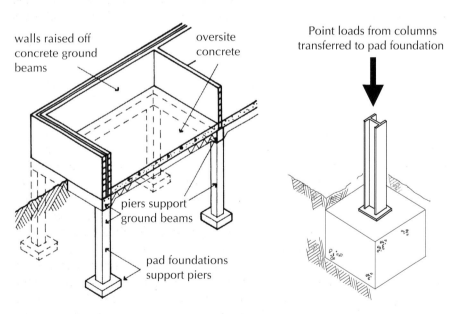

Figure 3.21   Pad foundation.   **Figure 3.22   Mass fill pad foundations.**

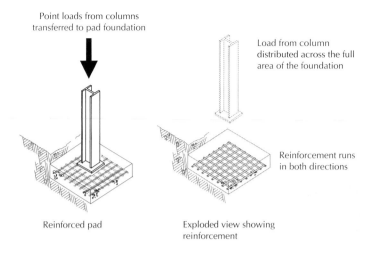

Point loads from columns
transferred to pad foundation

Load from column
distributed across the full
area of the foundation

Reinforcement runs
in both directions

Reinforced pad

Exploded view showing
reinforcement

Figure 3.23   Reinforced pad foundations.

excavation and necessary temporary support for the sides of the excavation
are justified by the allowable bearing pressure of the subsoil. For all but the
more heavily loaded columns, a mass concrete pad foundation may be used.
In this construction, the thickness of the concrete pad should be the same as the
projection of the pad around the column to resist the punching shear stresses
of the slender column on the wide spread base of the pad.

Pad foundations are often used in combination with ground beams. The
ground beams can be used to transfer continuous loads imposed by brickwork
or infill cladding panels to the reinforced or mass filled concrete pad foundations (Figure 3.24).

Where the subsoil for some depth below the surface has poor allowable
bearing pressure and the loads at the base of piers and columns are moderate,
it may be economical to use reinforced concrete pad foundations to spread
the loads over a sufficient area of soil. A reinforced concrete pad foundation
may be chosen where the depth of the necessary excavation is no more than a
few metres below the surface, to limit the necessary excavation and temporary
support for the sides of the excavation and allow for ease of access. A blinding
layer of weak concrete some 50 mm thick is spread and levelled in the bed of the
excavation to provide a firm working base. Two-way spanning, steel reinforcing rods are placed on and wired to spacers to provide cover of concrete below
reinforcement (Photograph 3.2). Concrete is spread over the reinforcement,
consolidated around and below reinforcement and consolidated and levelled
to the required thickness. Figure 3.23 is an illustration of a typical square reinforced concrete pad foundation. As an alternative to reinforced concrete pad
bases, pile foundations are used. Where the columns of a framed structure

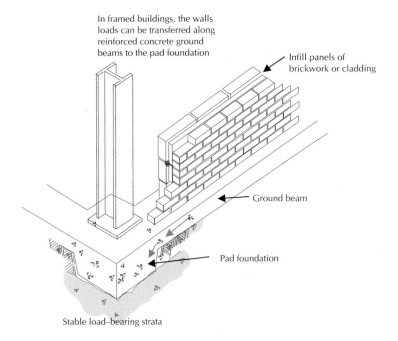

In framed buildings, the walls loads can be transferred along reinforced concrete ground beams to the pad foundation

Infill panels of brickwork or cladding

Ground beam

Pad foundation

Stable load–bearing strata

Figure 3.24   Pad and ground beam.

are comparatively closely spaced and pad foundations some few metres below the surface would be close to each other, it may be economical to form one continuous combined column foundation, as illustrated in Figure 3.25.

A mass concrete combined foundation may be used where the subsoil is sound and the allowable bearing pressure is comparatively high. The width of the continuous strip of concrete should be wide enough to spread the loads over an adequate area of subsoil and the thickness of the concrete at least equal to the projection of the strip each side of the columns. More usually, the continuous strip of concrete would be reinforced to limit the thickness and width of concrete. Figure 3.25 is an illustration of this type of foundation showing the top and bottom reinforcing bars wired to stirrups. The stirrups serve to position reinforcement and provide some resistance to shear.

Combined foundations

The foundations of adjacent columns are combined when a column is so close to the boundary of the site that a separate foundation would be eccentrically loaded. Combined foundations may also be used to resist uplift, overturning and opposing forces. Where a framed building is to be erected next to an existing building, it is usually necessary to use some form of cantilever or asymmetrical combined base foundation. A cantilever system would be used to transfer the loads from the columns, which are to be erected next to the

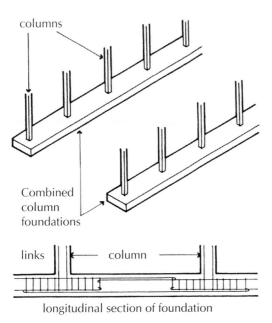

columns

Combined
column
foundations

links   column

longitudinal section of foundation

Figure 3.25   Combined column foundations.

existing building, away from the existing foundations. If the existing and new foundations were erected next to each other the combined force of the existing building and new structure may over stress the ground. The purpose of this arrangement is to ensure that the pressure on the subsoil under the base of the existing building is not so heavily surcharged by the weight on the foundation of the new building as to cause appreciable settlement. With the cantilever beam foundation illustrated in Figure 3.26 a beam supports the columns next to the existing building. The beam is cantilevered over a stub column so that the foundation is distant from the existing wall and unlikely to surcharge the soil under its foundation.

Where the subsoil under the wall of an adjoining building is comparatively sound and the load on columns next to the existing wall is moderate, it may be acceptable to use an asymmetrical combined base foundation such as that illustrated in Figure 3.27. Some of the load on the column base, next to the existing wall, will be transferred to the wider part of the reinforced concrete combined base to reduce the surcharge of load on the foundation of the existing wall. Where the boundary of a site limits the spread of the bases of columns next to the boundary line a system of rectangular or trapezoidal combined reinforced concrete bases may be used.

A rectangular, combined base is used where the columns next to the boundary are less heavily loaded than those distant from the boundary, as illustrated in Figure 3.28A. The load from heavily loaded columns next to the boundary

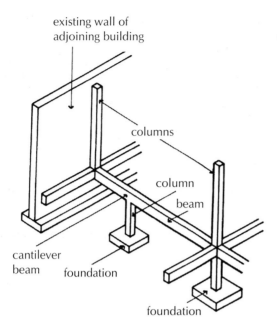

Figure 3.26   Cantilever beam foundation.

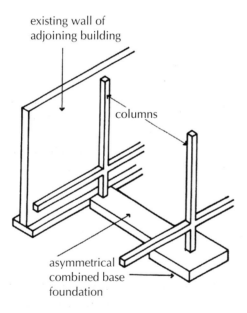

Figure 3.27   Asymmetrical combined base foundation.

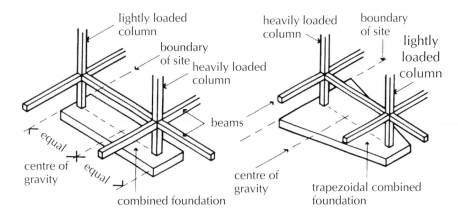

Figure 3.28   A, Rectangular combined base foundation. B, Trapezoidal combined base foundation.

can be more widely spread than that from less heavily loaded internal columns by the use of a trapezoidal, reinforced concrete, combined base illustrated in Figure 3.28B.

Raft foundations

A raft foundation is a continuous slab of concrete usually covering an area equal to or greater than the base of a building or structure to provide support for walls or lightly loaded columns and serve as a base for the ground floor. The word raft is used in the sense that the slab of concrete floats on the surface as a raft does on water. Raft foundations are used for lightly loaded structures on soils with poor bearing capacity and where variations in soil conditions necessitate a considerable spread of loads. Beam and raft and cellular raft foundations are used for more heavily loaded structures where the beams or cells of a raft are used to provide wide spread of loads. The three types of reinforced concrete raft foundations are:

❑ Solid slab raft
❑ Beam and slab raft
❑ Cellular raft (buoyant raft)

Solid slab raft foundation

Solid slab raft foundation is a solid reinforced concrete slab generally of uniform thickness, cast on subsoils of poor or variable bearing capacity, so that the loads from walls or columns of lightly loaded structures are spread over the whole area of the building. Concrete rafts are reinforced with mild steel rods to provide tensile strength against the upward or negative bending and resistance to shear stress due to the loads from walls or columns that are raised off the raft. For additional strength under the load of walls, rafts are commonly cast with downstand edge beams, as illustrated in Figure 3.29, and downstand beams under loadbearing internal walls. The solid slab raft foundation illustrated in Figure 3.29 is cast with a wide toe to the beam under external walls so that the concrete does not show above ground solely for appearance sake.

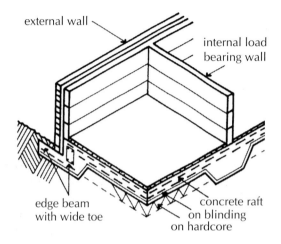

external wall

internal load bearing wall

edge beam with wide toe

concrete raft on blinding on hardcore

Figure 3.29   Solid slab raft foundation.

To provide a level bed for a concrete raft, it is necessary to remove vegetable topsoil and roughly level the surface. A bed of hardcore may be spread and compacted over the site to raise the ground floor level to or just above ground level. To provide a level bed for the concrete, a layer of blinding is spread either on a hardcore bed or directly on levelled soil to a thickness of 50 mm. The purpose of the blinding is to provide a level bed, which will prevent wet concrete running through it. The necessary reinforcement is placed and supported (with concrete spacers) to provide the necessary cover and the concrete spread, consolidated and finished level. A solid reinforced concrete slab raft foundation for lightly loaded piers or columns is illustrated in Figure 3.30. The columns are cast with haunched bases that are designed to provide a larger area at the base than that of the pier or column itself. This increased area of contact with the slab will reduce the punching shear tendency of the columns to force their way down through the slab. Alternatively, additional reinforcement is placed around the base of the column, preventing it punching through the base of the raft. This type of raft is best suited to lightly loaded columns or piers of one or two storey structures. The slab is cast on a bed of blinding. The slab is reinforced with two-way spanning rods and additional reinforcement around the column bases.

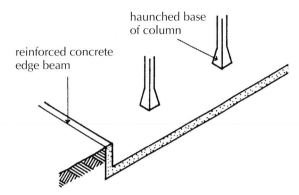

Figure 3.30   Solid slab reinforced concrete raft.

Beam and slab raft foundation

As a foundation to support the heavier loads of walls or columns a solid slab raft would require considerable thickness. To make the most economical use of reinforced concrete in a raft foundation supporting heavier loads it is practice to form a beam and slab raft. This raft consists of upstand or downstand beams that take the loads of walls or columns and spread them to the monolithically cast slab, which bears on natural subsoil. Figure 3.31 is an illustration of an upstand beam raft and Figure 3.32 shows a section through a downstand beam raft. Downstand beam raft foundations are often used where the soil is easily excavated and can support itself without the need for support. The area of the foundation would be stripped to formation level; this would be the underside of the raft slab construction. The downstand beams would then be excavated to the required depth. The reinforcing cages to the beams are lowered into the excavations with spacers attached to provide an adequate cover of concrete to the steel to inhibit rusting. Concrete is placed and consolidated up to the level of the underside of the slab.

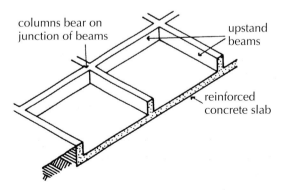

Figure 3.31   Reinforced concrete beam and slab raft.

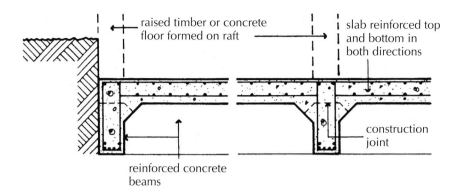

Figure 3.32   **Beam and slab raft with downstand beams.**

The reinforcement to the slab is set in place with spacers to provide concrete cover and chairs to hold the upper reinforcement in place. Concrete for the slab is spread and consolidated around the reinforcement and then levelled. Loadbearing walls are built off the slab over the beams and columns over the intersections of beams to spread loads over the area of the slab.

On granular soils, upstand beam rafts may be used. Temporary support is necessary to uphold the sides of the upstand beams. The slab is cast around the necessary reinforcement and the upstand beams inside temporary timber supports, around reinforcement to produce the upstand beam raft illustrated in Figure 3.33. With an upstand beam raft, it is plainly necessary to form a raised timber or concrete ground floor.

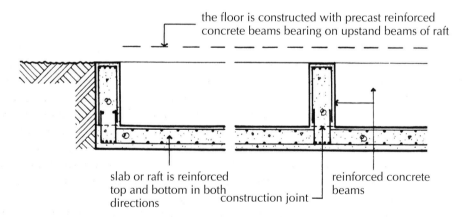

Figure 3.33   **Beam and slab raft with upstand beams.**

Cellular raft foundation (also called buoyant raft)

Where differential settlements are likely to be significant and the foundations have to support considerable loads, the great rigidity of the monolithically cast reinforced concrete cellular raft is an advantage. This type of raft consists of top and bottom slabs separated by and reinforced with vertical cross ribs in both directions, as illustrated in Figure 3.34. The monolithically cast reinforced concrete cellular raft has great rigidity and spreads foundation loads over the whole area of the substructure to reduce consolidation settlement and avoid differential settlement. A cellular raft may be the full depth of a basement storey, and the cells of the raft may be used for mechanical plant, storage or car parking space. The cells of a raft foundation are less suitable for habitable space as no natural light enters the rooms below ground.

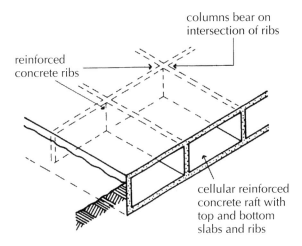

columns bear on intersection of ribs

reinforced concrete ribs

cellular reinforced concrete raft with top and bottom slabs and ribs

Figure 3.34   Cellular raft.

A cellular raft is also used when deep basements are constructed to reduce settlement by utilising the overburden pressure that occurs in deep excavations. This negative or upward pressure occurs in the bed of deep excavations in the form of an upward heave of the subsoil caused by the removal of the overburden, which is taken out by excavation. This often quite considerable upward heave can be utilised to counteract consolidation settlement caused by the load of the building and so reduce overall settlement.

When material (soil, rocks etc.) is excavated to make room for the cellular raft a large load is removed from the ground. As the load is removed the ground wants to lift (heave), as illustrated in Figure 3.35A.

If the weight of the foundation and building is similar to that of the excavated material, the stresses placed on the ground are similar to the loads exerted by the material that was previously excavated (Figure 3.35B). Or if the building is heavier than the materials removed, the stresses imposed are often only slightly more than those that the excavated ground previously imposed on the strata below it. Thus the chances of settlement occurring are reduced. Cellular

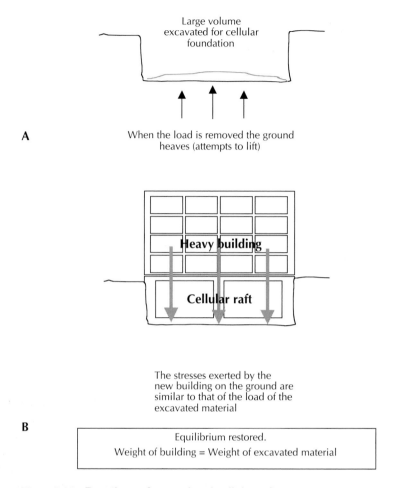

Figure 3.35 Reactions of ground and cellular rafts.

basements, when used in skyscrapers and other tall or heavy structures, may be up to three to four storeys deep (Figure 3.36). Deep cellular basements help reduce the additional stress imposed on the substrata.

Pile foundations

The word 'pile' is used to describe columns, usually of reinforced concrete, driven into or cast in the ground in order to carry foundation loads to some deep underlying firm stratum or to transmit loads to the subsoil by the friction of their surfaces in contact with the subsoil. The main function of a pile is to transmit loads to lower levels of ground by a combination of friction along their sides and end bearing at the pile point or base. Piles that transfer loads mainly by friction to clays and silts are termed friction piles and those

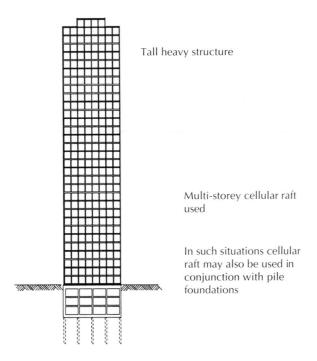

Tall heavy structure

Multi-storey cellular raft used

In such situations cellular raft may also be used in conjunction with pile foundations

Figure 3.36   Multi-storey cellular raft.

that mainly transfer loads by end bearing to compact gravel, hard clay or rock are termed end-bearing piles (Figure 3.37). Four or more piles may be used to support columns of framed structures. The columns are connected to a reinforced concrete pile cap connected to the pile, as illustrated in Figure 3.38. Piles may be classified by their effect on the subsoil as displacement piles or non-displacement piles. Displacement piles are driven, forced or cut (by an auger) into the ground to displace subsoil. The strata are penetrated. No soil is removed during the operation. Solid concrete or steel piles and piles formed inside tubes which are driven into the ground and which are closed at their lower end by a shoe or plug, which may either be left in place or extruded to form an enlarged toe, are all forms of displacement pile. Non-displacement piles are formed by boring or other methods of excavation that do not substantially displace subsoil. Sometimes the borehole is lined with a casing or tube that is either left in place or extracted as the hole is filled.

Driven piles are those formed by driving a precast pile and those made by casting concrete in a hole formed by driving. Bored piles are those formed by casting concrete in a hole previously bored or drilled in the subsoil.

Driven piles – concrete
Square, polygonal or round section reinforced concrete piles are cast in moulds in the manufacturer's yard and are cured to develop maximum strength.

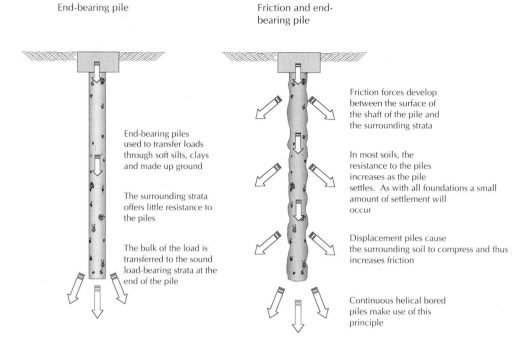

End-bearing pile

Friction and end-
bearing pile

End-bearing piles
used to transfer loads
through soft silts, clays
and made up ground

The surrounding strata
offers little resistance to
the piles

The bulk of the load is
transferred to the sound
load-bearing strata at the
end of the pile

Friction forces develop
between the surface of
the shaft of the pile and
the surrounding strata

In most soils, the
resistance to the piles
increases as the pile
settles. As with all foundations a small
amount of settlement will
occur

Displacement piles cause
the surrounding soil to compress and thus
increases friction

Continuous helical bored
piles make use of this
principle

Figure 3.37   End-bearing and friction piles.

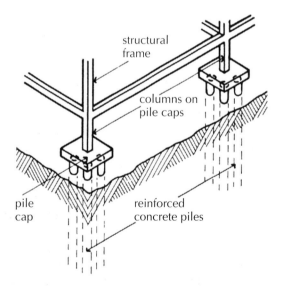

structural
frame

columns on
pile caps

pile
cap

reinforced
concrete piles

Figure 3.38   Pile foundation.

The placing of the reinforcement and the mixing, placing, compaction and curing of the concrete can be accurately controlled to produce piles of uniform strength and cross-section. The precast piles are often square section with chamfered edges, as illustrated in Figure 3.39. The head of the pile is reinforced with helical binding wire; this helps prevent damage that would otherwise be caused by driving the pile into the ground. Once the pile is in place the concrete at the top of the pile is removed to expose the main reinforcement. The helical reinforcement can be removed once the main reinforcement bars, which will be tied into the pile cap, are exposed.

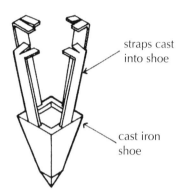

straps cast into shoe

cast iron shoe

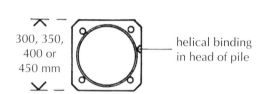

300, 350, 400 or 450 mm

helical binding in head of pile

Figure 3.39   Head of pile.

Figure 3.40   Shoe of pile.

To assist driving, the foot of the pile may be finished with a cast iron shoe, as illustrated in Figure 3.40. Figure 3.41 is an illustration of a typical pile. The piles are lifted into position and driven into the ground by means of a mechanically operated drop hammer attached to a mobile piling rig. To increase the length of the pile to the required depth, additional segments are added (Figure 3.42). The pile is driven in until a predetermined 'set' is reached. The word 'set' is used to describe the distance that a pile is driven into the ground by the force of the hammer falling a measurable distance. From the weight of the hammer and the distance it falls, the resistance of the ground can be calculated and the bearing capacity of the pile calculated. To connect the top of the precast pile to the reinforced concrete foundation the top 300 mm of the length of the pile is broken to expose reinforcement to which the reinforcement of the foundation is connected.

Precast driven piles are not in general used on sites in built-up areas. Difficulties are often experienced when attempting to move large precast piles through narrow streets; however, using smaller sections (Figure 3.42) can overcome this problem. The logistics of moving precast and prefabricated objects should always be considered when selecting construction methods. The noise, vibration and general disturbance caused by driving (hammering) piles into the ground can be a nuisance. Where vibration is excessive, or buildings and structures are

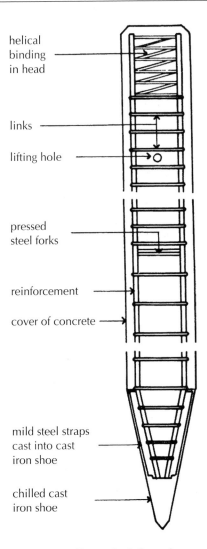

helical
binding
in head

links

lifting hole

pressed
steel forks

reinforcement

cover of concrete

mild steel straps
cast into cast
iron shoe

chilled cast
iron shoe

Figure 3.41   Precast reinforced concrete pile.

sensitive to vibration, damage may be caused to adjacent buildings, structures and services. Driven piles are used as end-bearing piles in weak subsoils where they are driven to a firm underlying stratum. Driven piles give little strength in bearing due to friction of their sides in contact with soil, particularly when the surrounding soil is clay. This is due to the fact that the operation of driving moulds the clay around the pile and so reduces frictional resistance between the pile and the surrounding clay. In coarse-grained cohesionless soils where the piles do not reach a firm stratum, driven piles act as friction-bearing piles due to the action of pile driving, which compacts the coarse particles around the sides of the pile and so increases frictional resistance and in compacting the soil

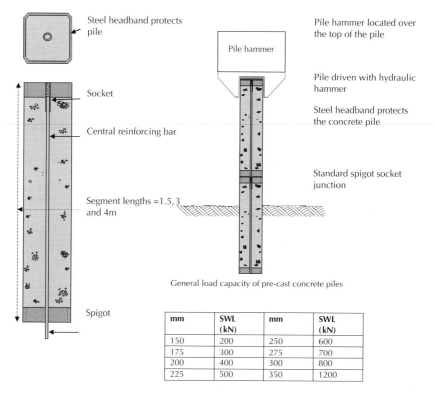

General load capacity of pre-cast concrete piles

mm	SWL (kN)	mm	SWL (kN)
150	200	250	600
175	300	275	700
200	400	300	800
225	500	350	1200

Figure 3.42   Precast concrete pile segments (adapted from www.roger-bullivant.co.uk).

increases its strength. This type of piled foundation is sometimes described as a floating foundation, as is a cast-in-place piled foundation, as bearing is mainly by friction and in effect the piles are floating in the subsoil rather than bearing on firm soil.

Driven tubular steel piles

Tubular steel piles are very similar in principle to precast concrete piles. Typically the piles are 6 m long, although shorter 3 m segmental piles can be used in areas where access and headroom is restricted (Figure 3.43 and Photograph 3.3). The piles are particularly suitable for driving in difficult or uncertain ground conditions up to 50 m deep. Hard driving conditions caused by fill, obstructions and boulders can also be dealt with. The piles are capable of taking large axial loads. Tubular steel piles can also accommodate horizontal loads resulting from bending moments and horizontal reaction, and can resist vertical tension loads that are a result of uplift and heave reactions.

Normally the piles are driven with an open end so that soil fills and plugs the void; in exceptional circumstances the end of the pile can be closed by welding

Piles are lifted into position, checked for plumb and driven into the ground until the required set is reached.

The exposed pile is then broken away and the reinforcement tied into the pile cap.

Photograph 3.3   Driven precast concrete piles (www.roger-bullivant.co.uk).

and end plate. The void at the top of the pile is filled with concrete. Reinforcement can be positioned in the pile, allowing it to be tied into the pile cap's reinforcement cage. The piles are top driven using rigs with hydraulic hammers. The site should be firm, dry and level ready to receive the piling rigs, which can weigh up to 35 tonnes.

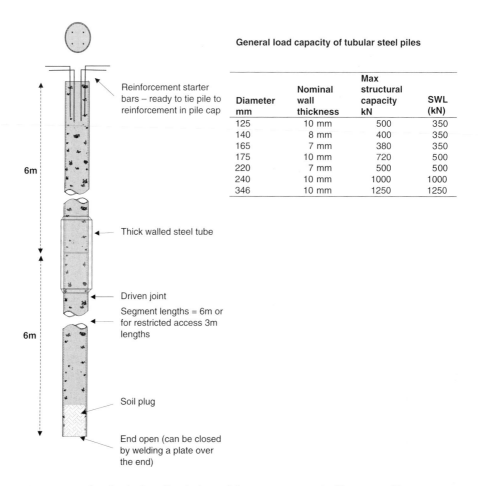

General load capacity of tubular steel piles

Diameter mm	Nominal wall thickness	Max structural capacity kN	SWL (kN)
125	10 mm	500	350
140	8 mm	400	350
165	7 mm	380	350
175	10 mm	720	500
220	7 mm	500	500
240	10 mm	1000	1000
346	10 mm	1250	1250

Reinforcement starter bars – ready to tie pile to reinforcement in pile cap

6m

Thick walled steel tube

Driven joint

Segment lengths = 6m or for restricted access 3m lengths

6m

Soil plug

End open (can be closed by welding a plate over the end)

Figure 3.43   Steel tubular piles (adapted from www.roger-bullivant.co.uk).

Driven cast-in-place piles

Driven cast-in-place piles are of two types; the first has a permanent steel or concrete casing and the second uses a temporary casing. The purpose of driving and maintaining a permanent casing is to consolidate the subsoil around the pile casing by the action of driving. The lining is left in place to protect the concrete cast inside the lining against weak strata of subsoil that might otherwise fall into the pile excavation. Permanent casings also protect the green concrete (concrete which has not set) of the pile against static or running water that may erode the concrete. The lining also protects the concrete against contamination. Figure 3.44 is an illustration of a driven cast-in-place pile with a permanent reinforced concrete casing. Precast reinforced concrete shells are threaded on a steel mandrel. Metal bands and bitumen seal joints between shells. The mandrel and shells are lifted on to the piling rig and then driven

A Steel pile placed in position

B Next section is lifted into place

C Pile sections fixed together
and driven to required set

D Exposed piles will be cut
ready to receive the pile cap

Photograph 3.4   Driven segmental steel piles (www.roger-bullivant.co.uk).

into the ground. At the required depth the mandrel is removed, a reinforcing cage is lowered into the shells and the pile completed by casting concrete inside the shells. This type of pile is used principally in soils of poor bearing capacity and saturated soils where the concrete shells protect the green concrete cast inside them, from static or running water.

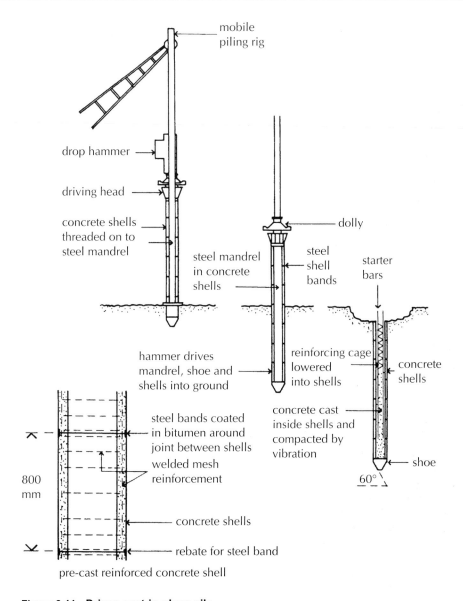

Figure 3.44   Driven cast-in-place pile.

A driven cast-in-place pile without permanent casing is illustrated in Figure 3.45. The base of a steel lining tube, supported on a piling rig, is filled with ballast. A drop hammer rams the ballast and the tube into the ground. At the required depth, the tube is restrained and the ballast is hammered in to form an enlarged toe as shown (in Figure 3.45). Concrete is placed by hammering it inside a lining tube; the tube is gradually withdrawn. The effect of driving the tube and the ballast into the ground is to compact the soil around the pile, and

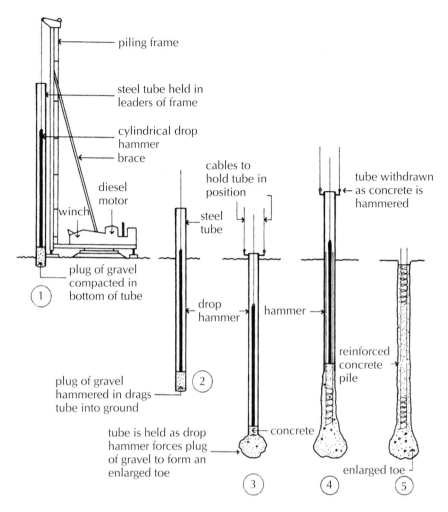

Figure 3.45   Driven cast-in-place concrete pile.

the subsequent hammering of the concrete consolidates it into pockets (voids) and weak strata. The enlarged toe provides additional bearing area at the base of the pile. This type of pile acts mainly as a friction pile.

Another type of driven cast-in-place pile without permanent casing is formed by driving a lining tube with cast iron shoe into the ground with a piling hammer operating from a piling rig, as illustrated in Figure 3.46. Concrete is placed and consolidated by the hammer as the lining tube is withdrawn. The particular application of this type of pile is for piles formed through a substratum so compact as to be incapable of being taken out by drilling. The purpose of the cast iron shoe, which is left in the ground, is to penetrate the compact stratum through which the pile is formed.

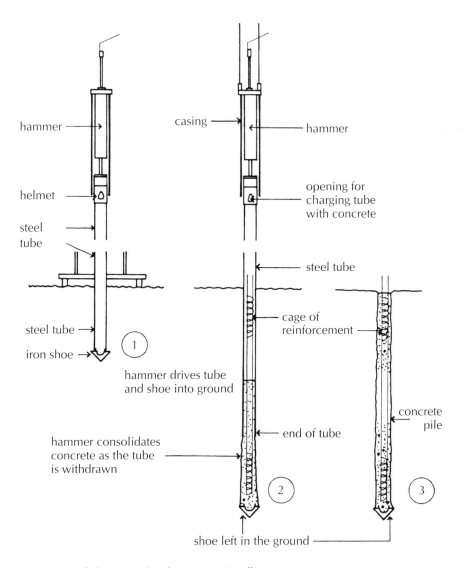

hammer

casing

hammer

helmet

opening for
charging tube
with concrete

steel
tube

steel tube

steel tube

cage of
reinforcement

iron shoe

1

hammer drives tube
and shoe into ground

concrete
pile

hammer consolidates
concrete as the tube
is withdrawn

end of tube

2

3

shoe left in the ground

Figure 3.46   Driven cast-in-place concrete pile.

Jacked piles

Figure 3.47 illustrates a system of jacked piles that are designed for use in
cramped working conditions, as for example where an existing wall is to
be underpinned and headroom is restricted by floors and in situations
where the vibration caused by pile driving might damage existing buildings.
Where the wall to be underpinned has a sound concrete base a small area below
the foundation is excavated. This provides sufficient space for small beams, the
pile jack and the first section of pile to be inserted.

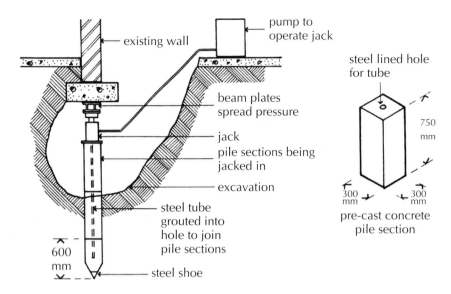

Figure 3.47 Pre-cast concrete jacked pile.

First the pile cap with the steel shoe is inserted and driven into the ground. The jack then retracts and the next section is inserted between the driven pile section and the jack. The pile sections are then repeatedly inserted between the jack and then driven into the ground, as illustrated in Figure 3.47. The precast concrete sections are jacked into the ground, as illustrated. Some systems use hollow precast concrete pile sections. Where hollow sections are used, reinforcement can be inserted into the void and concreted in position; alternatively lengths of steel tube are often inserted and grouted in position, making a strong connection between all of the sections. Once the jack is removed a concrete cap is cast on top of the pile and up to the underside of the concrete base.

When the wall to be underpinned has a poor base and the wall or structure above might be disturbed by either the area excavated for underpinning or the jacking, then an alternative process must be used. One option is to insert pairs of piles each side of the wall. Steel or reinforced concrete beams (often called needles) are then inserted through the wall above the foundation, but below ground level. The needles will be used to support the wall and transfer the loads to the piles either side of the original foundation. When piles are formed on both sides of the wall they are jacked in against temporary units loaded with kentledge. As there is no building foundation to jack against, a temporary loaded structure (kentledge) must be used so that the jack can drive piles from the structure into the ground. Once the piles are jacked into position the jack and kentledge are removed. The piles and needles can then be tied together using a reinforced concrete pile cap.

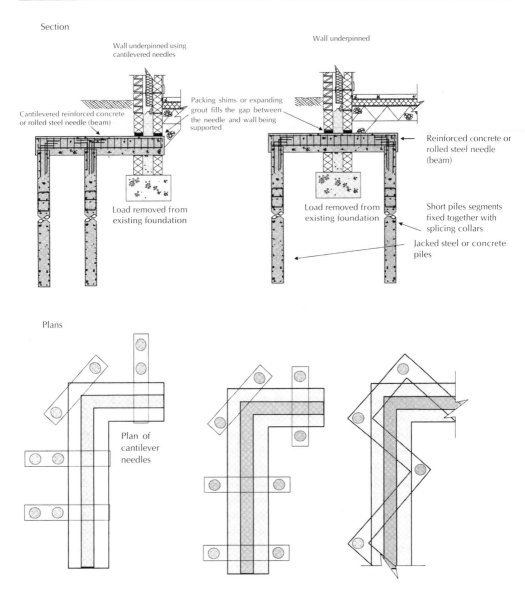

Section

Wall underpinned using
cantilevered needles

Wall underpinned

Cantilevered reinforced concrete
or rolled steel needle (beam)

Packing shims or expanding
grout fills the gap between
the needle and wall being
supported

Reinforced concrete or
rolled steel needle
(beam)

Load removed from
existing foundation

Load removed from
existing foundation

Short piles segments
fixed together with
splicing collars

Jacked steel or concrete
piles

Plans

Plan of
cantilever
needles

Figure 3.48   Various underpinning arrangements.

Bored piles

Auger bored piles

A hole is bored or drilled by means of earth drills (mechanically operated augers), which withdraw soil from the hole into which the pile is to be cast (Photograph 3.5). Occasionally, it is necessary to lower or drive in steel lining tubes as the soil is taken out, to maintain the sides of the drilled hole. As the pile is cast the lining tubes are gradually withdrawn. The mechanical rigs used to install the piles come in a range of sizes, from small units weighing just a few tonnes to large rigs exceeding 20 m and weighing in excess of 50 tonnes (Photograph 3.5).

(Note: the auger is not attached)

Photograph 3.5   Continuous flight auger rigs (www.roger-bullivant.co.uk).

Although not common nowadays, some boring can be fixed to tripods rather than the typical tracked rigs (Figure 3.49). Advantages of such equipment are that the piling rigs are light and easily manipulated. Because all of the arisings are brought to the surface, a precise analysis of the subsoil strata is obtained from the soil withdrawn. A disadvantage of piles cast in the ground is that it is not possible to check that the concrete is adequately compacted and whether there is adequate cover of concrete around the reinforcement.

Figure 3.49 illustrates the drilling and casting of a bored cast-in-place pile. Soil is withdrawn from inside the lining tubes with a cylindrical clay cutter that is dropped into the hole, which bites into and holds the cohesive soil. The cutter is then withdrawn and the soil knocked out of it. Coarse-grained soil is withdrawn by dropping a shell cutter (or bucket) into the hole. Soil, which is retained on the upward hinged flap, is emptied when the cutter is withdrawn. The operation of boring the hole is more rapid than might be supposed and a pile can be bored and cast in a matter of hours. Concrete is cast under pressure through a steel helmet, which is screwed to the top of the lining tubes. The application of air pressure at once compacts the concrete and simultaneously

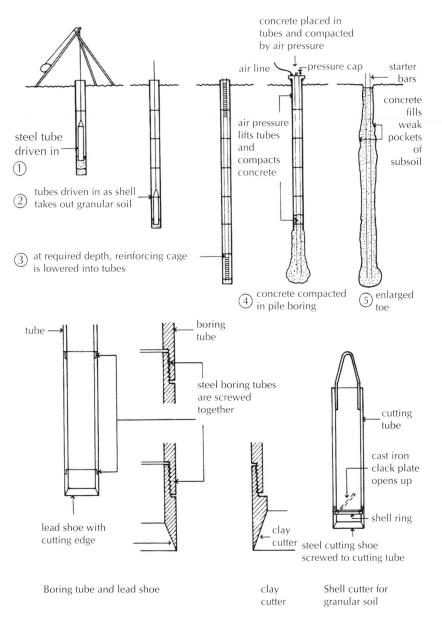

concrete placed in
tubes and compacted
by air pressure

air line — pressure cap   starter
bars

① steel tube
driven in

② tubes driven in as shell
takes out granular soil

③ at required depth, reinforcing cage
is lowered into tubes

air pressure
lifts tubes
and
compacts
concrete

concrete
fills
weak
pockets
of
subsoil

④ concrete compacted
in pile boring

⑤ enlarged
toe

tube →

boring
tube

steel boring tubes
are screwed
together

cutting
tube

cast iron
clack plate
opens up

shell ring

lead shoe with
cutting edge

clay
cutter

steel cutting shoe
screwed to cutting tube

Boring tube and lead shoe

clay
cutter

Shell cutter for
granular soil

Figure 3.49   Bored cast-in-place concrete pile.

lifts the helmet and lining tubes as the concrete is compacted. As the lining tubes
are withdrawn, protruding sections are unscrewed and the helmet refixed until
the pile is completed. As the concrete is cast under pressure it extends beyond
the circumference of the original drilling to fill and compact weak strata and
pockets in the subsoil, as illustrated in Figure 3.49. Because of the irregular

shape of the surface of the finished pile it acts mainly as a friction pile to form what is sometimes called a floating foundation. As the pile continues to settle into the soil the friction forces surrounding the pile increase.

Large diameter bored pile
Figure 3.50 illustrates the formation and casting of a large diameter bored pile. A tracked crane supports hydraulic rams and a diesel engine which operates a Kelly bar and rotary bucket drill. The diesel engine rotates the Kelly bar and bucket. In the bottom of the bucket are angled blades that rotate, excavating the strata and filling the bucket with soil. The hydraulic rams force the bucket into the ground. The filled bucket is raised and emptied and drilling proceeds. In non-cohesive soils the excavation is lined with steel lining tubes. To provide increased end bearing the drill can be belled out to twice the diameter of the pile.

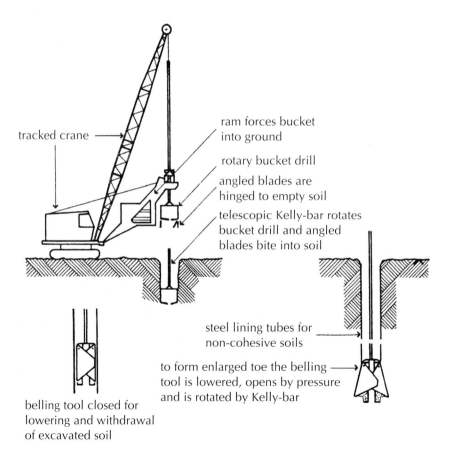

tracked crane

ram forces bucket into ground

rotary bucket drill

angled blades are hinged to empty soil

telescopic Kelly-bar rotates bucket drill and angled blades bite into soil

steel lining tubes for non-cohesive soils

to form enlarged toe the belling tool is lowered, opens by pressure and is rotated by Kelly-bar

belling tool closed for lowering and withdrawal of excavated soil

Figure 3.50   Bored cast-in-place reinforced concrete pile.

Rotary drilling equipment is commonly used for piles to be cast in cohesive soils. A tractored based rig supports a diesel engine and crane jib. A cable run from the motor up the jib supports a large, square drilling rod or Kelly bar that passes through a turntable, which rotates the bar to which is attached a drilling auger. The weight of the rotating Kelly bar causes the augur to drill into the soil. The augur is withdrawn from time to time to clear it of excavated soil. Where the subsoil is reasonably compact, the reinforced concrete pile is cast in the pile hole and consolidated around the reinforcing cage. In granular subsoil the excavation may be lined with steel lining tubes that are withdrawn as the pile is cast in place. This type of pile is often used on urban sites where a number of piles are to be cast, because it will cause the least vibration to disturb adjacent buildings and create the least noise disturbance.

Continuous flight auger (CFA) piles

CFA piles are formed using hollow stem auger boring techniques. The auger has a hollow central tube surrounded by a continuous thread. The helical cutting edge is continuous along the full length of the auger. The hollow tube that runs down the centre of the shaft is used for pumping concrete into the hole as the cutting device is withdrawn (Figure 3.51 and Photograph 3.6). As the

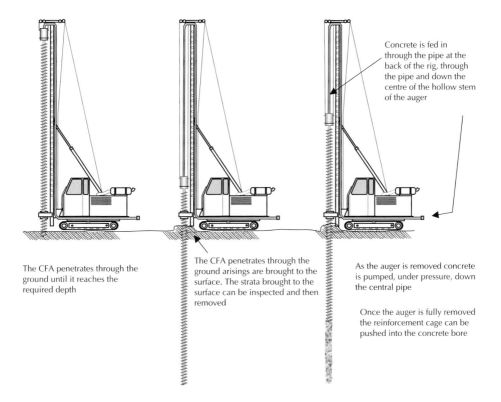

Concrete is fed in through the pipe at the back of the rig, through the pipe and down the centre of the hollow stem of the auger

The CFA penetrates through the ground until it reaches the required depth

The CFA penetrates through the ground arisings are brought to the surface. The strata brought to the surface can be inspected and then removed

As the auger is removed concrete is pumped, under pressure, down the central pipe

Once the auger is fully removed the reinforcement cage can be pushed into the concrete bore

Figure 3.51   Continuous flight auger (CFA) piles.

Continuous flight auger

Arisings
brought to the
surface

Photograph 3.6   Continuous flight auger (www.roger-bullivant.co.uk).

CFA rig cuts into the ground, arisings are brought to the surface. The arisings allow the soil to be inspected at regular intervals, giving an indication of the strata below the surface. Once the rig has produced a bore to a calculated depth, through a known strata, the auger is steadily withdrawn as concrete is pumped under pressure through the hollow stem. The concrete simultaneously fills the void left by the auger as it is extracted. The concrete reinforcement cage is pushed into the bore after the pile has been concreted. Spacers are fixed to the side of the reinforcement so that the cage is positioned centrally in the pile and adequate concrete cover is maintained around the reinforcement. CFA piles are suitable for a range of ground conditions, are relatively quiet and cause less noise and vibration when compared to hammer driven piles; they can accommodate large working loads and are quick to install.

Bored displacement piles
Continuous helical displacement piles
Continuous helical displacement (CHD) piles are becoming a popular alternative to CFA piles. The displacement piles have the advantage that they do not

produce arisings; this is particularly useful on contaminated sites. In most ground conditions the continuous helical displacement piles have enhanced load carrying capacity, compared with a CFA pile of similar dimensions. Due to the compaction of the soil, friction between the pile and the strata is increased. The increased strength gained in some ground conditions enables the pile length to be shortened, resulting in shorter installation times and more economical foundations.

The piles are formed using a multi-flight, bullet ended shaft, which is driven by a high torque rotary head (Figure 3.52 and Photograph 3.7). This enables the ground to be penetrated without bringing any material to the surface

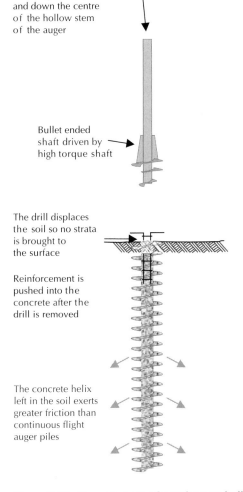

Concrete is fed in through the pipe and down the centre of the hollow stem of the auger

Bullet ended shaft driven by high torque shaft

The drill displaces the soil so no strata is brought to the surface

Reinforcement is pushed into the concrete after the drill is removed

The concrete helix left in the soil exerts greater friction than continuous flight auger piles

Figure 3.52   Components of continuous helical displacements (CHD) piles.

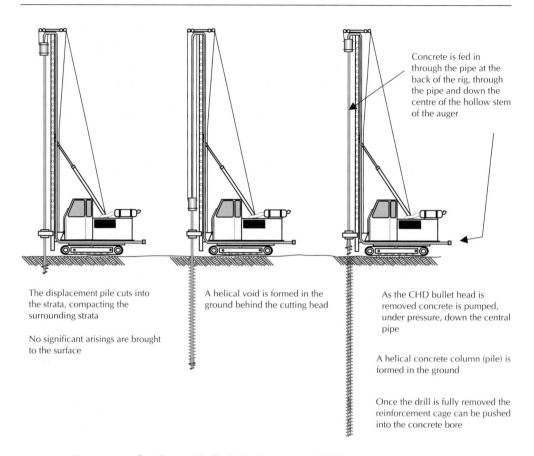

Concrete is fed in through the pipe at the back of the rig, through the pipe and down the centre of the hollow stem of the auger

The displacement pile cuts into the strata, compacting the surrounding strata

No significant arisings are brought to the surface

A helical void is formed in the ground behind the cutting head

As the CHD bullet head is removed concrete is pumped, under pressure, down the central pipe

A helical concrete column (pile) is formed in the ground

Once the drill is fully removed the reinforcement cage can be pushed into the concrete bore

Figure 3.53   Continuous helical displacements (CHD) piles.

(Figure 3.53). Some slight heaving of the surface may occur as the ground is compressed; however, this is normally negligible. The pile is drilled to the calculated or proven depth, the shaft is then reversed and extracted whilst concrete is simultaneously pumped under pressure into the helical void that remains. Once the auger is totally extracted, reinforcement can be pushed into the concrete as a single bar or cage (Photograph 3.7).

Testing piles
Piles are often tested using dynamic load, static load or sonic integrity methods. These three methods are described briefly here.

Dynamic load methods
Dynamic load methods of testing are suitable for most types of pile but are more frequently used on precast concrete or tubular steel piles. The test is usually used on small piling works where the cost of static load testing cannot be justified. The test determines the load bearing capacity of the pile, skin friction

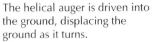

The helical auger is driven into the ground, displacing the ground as it turns.

The drive checks the resistance and set. As the ground is displaced little arisings are brought to the surface.

Once the required depth and set is achieved the auger is reversed and withdrawn, pumping concrete into the helical pile void as it rotates.

A reinforcing cage with spacers, which ensure correct cover is maintained, is simply pushed into the wet concrete.

Photograph 3.7   Continuous helical flight displacement piles (www.roger-bullivant.co.uk).

and end bearing. Other characteristics such as hammer energy transfer, pile integrity, pile stresses, driving and load displacement behaviour can also be determined.

To dynamically test a pile the pile is struck by a hammer using the piling rig. Two strain transducers and accelerometers (measures speed and acceleration) are firmly attached to the face of the pile near to the head of the pile (Photograph 3.8A). As the pile is struck the equipment measures the force and acceleration of the pile. The information is relayed to the monitoring equipment. Once analysed, the data provides models of shaft friction distribution, bearing capacity and load settlement behaviour.

Sonic integrity testing
Sonic integrity testing is normally used on continuous flight auger (CFA) piles, continuous helical displacement (CHD) or other piles foundations formed using insitu concrete. The integrity method is fast and reliable. A large number of piles can be tested in a single visit. The pile determines the reliability, morphology (form and composition) and quality of construction of the pile.

Before the pile can be tested it must be sufficiently cured, free of latence and trimmed back to sound concrete. It is preferable to carry out the test at the final cut-off level of the pile. A small hand-held hammer is used to strike the pile. A series of low strain acoustic shock waves are sent through the piles. As the waves pass down the pile, the sound waves rebound where changes in impedance occur. The rebound (echo) is then recorded by a small accelerometer (instrument for measuring speed and acceleration), which is held against the pile head. The response is monitored and stored, and a graphical representation produced for immediate inspection (Photograph 3.8B).

Static load testing
Static load testing is used to determine the displacement characteristics of a pile. All piles are suited to static load testing. Static load testing frames are assembled specifically for the test. Major piling contractors assemble frames capable of accommodating loads up to 4000 kN.

A known load has to be applied in the form of kentledge (loaded test frame) or tension pile reaction. Load can also be applied by fixing a frame to piles already installed in the ground. Other piles can then be tested against the frame load. Once an adequate reaction has been provided (load), the test is carried out using a hydraulic jack and calibrated digital load cell. Time, load, temperature and displacement data are recorded.

Pile caps and spacing of piles
Piles may be used to support pad, strip or raft foundations. Commonly a group of piles is used to support a column or pier base. The load from the column or

A Dynamic load testing   **B** Sonic integrity testing

Photograph 3.8   Dynamic load testing method and sonic integrity testing (www.roger-bullivant.co.uk).

pier is transmitted to the piles through a reinforced concrete pile cap, which is cast over the piles. To provide structural continuity the reinforcement of the piles is linked to the reinforcement of the pile caps through starter bars protruding from the top of the cast-in-place piles or through reinforcement exposed by breaking off the top concrete from precast piles. The exposed reinforcement of the top of the piles is wired to the reinforcement of the pile caps. Similarly, starter bars cast in and linked to the reinforcement of the pile caps protrude from the top of the pile caps for linking to the reinforcement of columns. Figure 3.54 illustrates typical arrangements of pile caps. The spacing of piles should be wide enough to allow for the necessary number of piles to be driven or bored to the required depth of penetration without damage to adjacent construction or to other piles in the group. Piles are generally formed in comparatively close groups for economy in the size of the pile caps to which they are connected. As a general rule the spacing, centre to centre of friction piles, should be not less than the perimeter of the pile, and the spacing of end-bearing piles not less than twice the least width of the pile.

A Kentledge

B Tensile pile reaction

C Static load testing-jack

D Time, load, temperature and displacement recorded

Photograph 3.9   Piles – static load testing equipment (www.roger-bullivant.co.uk).

Ground stabilisation

There are a number of different methods that can be used for improving the general ground condition of a site. In many cases improving the ground reduces the cost of foundations. Where the ground has been improved by compaction and consolidated, traditional foundation methods may be used rather than an expensive system of piles. Some sites which have been built up or are unstable may need improving just to provide a sound hard standing so that heavy plant can operate on the site safely. Other sites require more permanent

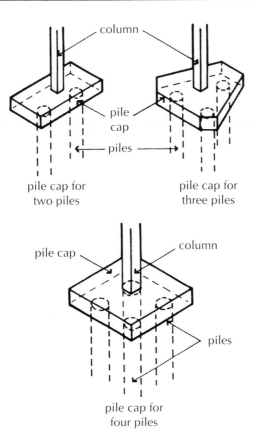

pile cap for
two piles

pile cap for
three piles

pile cap for
four piles

Figure 3.54   Concrete pile caps.

improvement, ensuring that the new building or structure and access to and around the structure remain stable.

Dynamic compaction

Dynamic compaction and consolidation using tamping systems can enhance the ground conditions up to considerable depths. The ground is consolidated by repeatedly dropping specially designed tampers into the ground. Two systems are commonly used. The first uses a flat-bottomed tamper, the alternative, more modern, method uses cone-shaped tampers. Flat bottom tampers can be slower and tend to create more noise and vibration than the cone system of tamping. Modern methods tend to use vertical guiders (or leaders) to control the fall and rise of the tamper (Photograph 3.10); with traditional methods the load (tamper) was simply attached to a cable. When lifting and lowering the weight, time had to be allowed for the tamper to stabilise.

Ground conditions suitable for dynamic compaction include natural granular soils, made ground and land-filled refuse sites. The technique can also be used as part of a more significant earthworks operation, where the ground is built up in layers, compacted and consolidated. Where fill is built up in layers, the fill may take the form of unmodified material (as previously excavated) or soils which are modified or stabilised using additives, such as quicklime, pulverised fuel ash (PFA), cement etc.

To achieve the desired effect, several passes may be required. Careful monitoring and testing is required; grid levels may need to be taken before and after each pass. Trial drops should be taken to determine the optimum treatment regime, monitor the imprint and depths, and measure pore water pressures, as necessary. To determine the allowable bearing capacity, accurate measurements are taken of the penetration achieved by application of particular energy (known load from a known height). Analysis of the levels can be used to calculate the amount of void closure and the degree of densification. Using dynamic compaction bearing capacities of 50 to 150 kN/m^2 can be achieved. Greater bearing capacities may be achievable, depending on the ground condition.

Different shaped tamper heads are available with a variety of weights, depending of the degree of consolidation and compaction required (Figure 3.55). Figure 3.56 shows a typical pattern of work. Three passes are used in this example to achieve the required compaction and consolidation. Initial tamping is undertaken using a single pointed tamper; the tamper is up to 2.5 m long with a mass of 10 tonnes. The ground is tamped on a grid with the objective of densifying deep layers. Subsequent tamping is then undertaken using multipointed tampers (Photograph 3.10). The multipoint tamper improves ground consolidation at shallower depths (Figures 3.55 and 3.56).

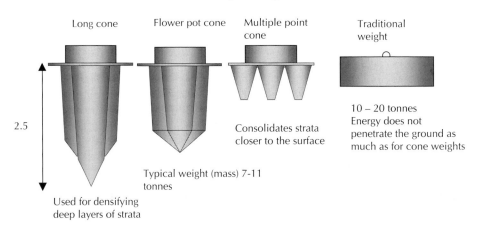

Figure 3.55   Typical cone type tampers (adapted from www.roger-bullivant.co.uk).

Vibro compaction

Vibro compaction (also called vibro displacement or vibro replacement) uses large vibrating mandrels (vibrating shafts or rods) to penetrate, displace and compact the soil (Figure 3.57 and Photograph 3.11). When the mandrel is

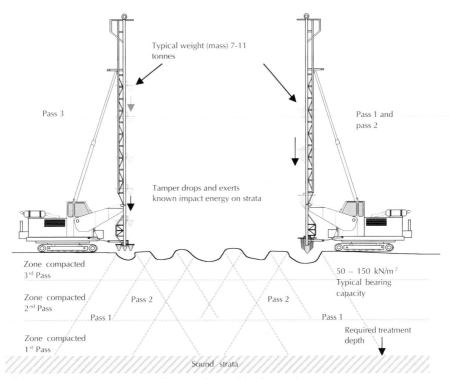

Typical weight (mass) 7-11 tonnes

Pass 3

Pass 1 and pass 2

Tamper drops and exerts known impact energy on strata

Zone compacted
3rd Pass

50 – 150 kN/m^2
Typical bearing
capacity

Zone compacted
2nd Pass

Pass 2

Pass 2

Pass 1

Pass 1

Required treatment
depth

Zone compacted
1st Pass

Sound strata

A Section: dynamic compaction – schematic of cone tampers

First drop

Second drop

Plan of first two passes
for compaction

B Plan – compaction pattern

Figure 3.56   Dynamic compaction (adapted from www.roger-bullivant.co.uk).

removed from the ground the subsequent void is filled with stone. The mandrel is then forced back through the stone, further displacing and compacting the ground and stone. The method that produces stone columns in the ground compacts the surrounding strata, enhancing the ground bearing capacity and limiting settlement. Typical applications include support of foundations, slabs, hard standings, pavements, tanks or embankments. Soft soils, man-made and other strata can be reinforced to achieve improved specification requirements whilst

Densifying cone

Photograph 3.10   Dynamic compaction (www.roger-bullivant.co.uk).

slopes can be treated to limit the risk of slip failure. The allowable ground bearing capacities for low to medium rise buildings and industrial developments is in the region of $100 \, \text{kN/m}^2$ to $200 \, \text{kN/m}^2$. Beneath tanks, on embankments or slopes $100 \, \text{kN/m}^2$ can be achieved. Ground conditions may allow heavier loads to be supported.

The high powered mandrel penetrates the ground, in a vertical plane to the designed depth. When the mandrel is extracted the resulting bore is filled with suitable aggregate or stabilised solid; these are compacted in layers of 200–600 mm increments. The shape of the mandrel (poker) tip is designed to ensure high compaction of the stone or stabilised soil column and surrounding strata. The taper on the mandrel causes increased densification of the strata.

Vibro displacement and compaction can be used in granular and cohesive soils. In granular soils the ground bearing capacity is improved by the introduction of columns of compacted stone or stabilised soil, and the compaction and desification of the granular soil that surrounds each column.

Cohesive soils are not compacted in the same way that granular soils are. In clays the stone columns help to share and distribute the loads. The columns of stone carry the loads down the pile and distribute them through the strata; however, they are not end bearing. Where the density and consistency of the ground varies, the installation of stone columns helps to stabilise the ground, enhances load bearing performance and makes the conditions more uniform, thus limiting differential settlement. The benefits of vibro compaction include:

❏ Buildings can be supported on conventional foundations (normally reinforced and shallow foundations)
❏ Work can commence immediately following the vibro displacement. Foundations can be installed straight away
❏ The soil is displaced. No soil is produced
❏ Contaminants remain in the ground – reducing disposal and remediation fees
❏ Economical, when compared with piling or deep excavation works
❏ Can be used to regenerate brownfield sites
❏ Can use reclaimed aggregates and soils

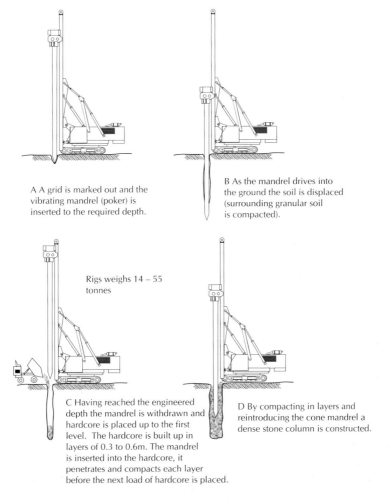

A A grid is marked out and the vibrating mandrel (poker) is inserted to the required depth.

B As the mandrel drives into the ground the soil is displaced (surrounding granular soil is compacted).

Rigs weighs 14 – 55 tonnes

C Having reached the engineered depth the mandrel is withdrawn and hardcore is placed up to the first level. The hardcore is built up in layers of 0.3 to 0.6m. The mandrel is inserted into the hardcore, it penetrates and compacts each layer before the next load of hardcore is placed.

D By compacting in layers and reintroducing the cone mandrel a dense stone column is constructed.

Figure 3.57   Vibro displacment – typical sequence.

The vibrating poker is driven into the ground, displacing and compacting the ground.

Once the poker (high powered vibrating mandrel) is removed a void is left.

Hardcore is then poured into the hole. The poker is then reinserted through the hardcore forcing the hardcore into the surrounding ground.

The hardcore pile, which has been forced into the ground, compacts the surrounding area.

Photograph 3.11   Vibro displacement and compaction (www.roger-bullivant.co.uk).

Vibro flotation

Vibro floatation uses a similar process to vibro compaction, except that the vibrating poker has high-pressure water jets at the tip of the poker. The water jets help achieve the initial penetration into the ground. The advantage of this system is that the water jets help the vibrator penetrate hard layers of ground. A major disadvantage is that the system is messy and imprecise, thus rarely used.

Pressure grouting

In permeable soils, or soils where it is known that small cavities may be within the ground, pressure grouting may be used to fill the voids. Holes are drilled into the ground using mechanically driven augers. As the auger is withdrawn cement slurry is forced down a central tube into the bore under pressure. Pressures of up to $70\,000\,\mathrm{N/mm^2}$ can be exerted by the grout on the surrounding soil. The slurry contains cementious additives, such as PFA, microsilica, chemical grout, cement or a mixture. PFA is cheap and often used as a bulk filler to improve the bearing capacity of the ground. As the grout enters the void it forces itself into the voids, cavities and fissures in the soils and rock. In weak soils it will displace and compact the ground as it fills the voids. As the voids are filled the ground becomes stiffer, more stable and water resistant. Pressure grouting can also be used around basements and coffer dams to reduce the hydrostatic pressure on the structure. To create a water resistant barrier the bores and subsequent columns of grout are placed at close centres. PFA is generally used as the ground modification and stabilisation material, whereas the expensive chemical mixes (resin or epoxy mixes) or those containing microsilica are used to fill small voids and improve the ground's water resistance.

Soil modification and recycling

With the increased use of brownfield, reclaimed and landfill sites for construction purposes there is a need for faster, more effective and economical methods of improving the ground conditions. Plant has been developed that is capable of cutting into site soils (including contaminated material), breaking up the strata, then grading and crushing the material before mixing it with cementious additives and relaying it to provide a compacted modified or stabilised hard standing (Figure 3.58).

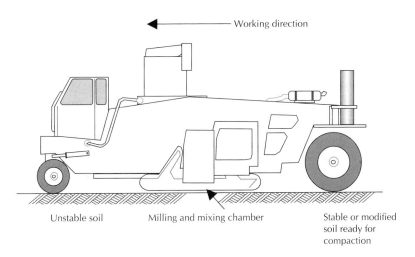

Working direction

Unstable soil   Milling and mixing chamber   Stable or modified
                                                           soil ready for
                                                           compaction

Figure 3.58   Soil modification, stabilisation and recycling machine.

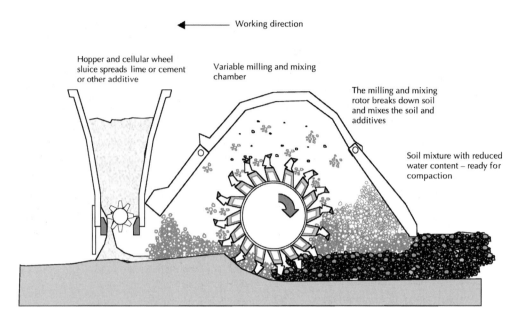

Figure 3.59   Schematic diagram: soil modification, stabilisation and recycling (adapted from www.roger-bullivant.co.uk).

Roger Bullivant has developed a soil stabilisation and modification system. The system uses a large cutting wheel that breaks down and pulverises the soil, and in a mixing chamber mixes the additives with the graded soil (Figure 3.59 and Photograph 3.12). In cohesive soils, quicklime is most commonly used as the additive, whereas in more granular soils the additives are usually cement, microsilica or pulverised fuel ash. Other additives may also be used to enhance the properties of the stabilised soil further. Where the soil needs improved strength and stability, and only small doses of additives are needed, the stabilised soil will immediately provide a workable strata. The cementious additives will normally continue to enhance the soil as they continue to mature and react over time. Some soils that are normally considered unsuitable may be modified, with further additives, to provide workable material; however, this process would be regarded as soil modification rather than stabilisation.

The additives used in soil stabilisation increase the strength of the soil, providing more workable materials which can be better compacted to maximise bearing capacity and minimise settlement. The technique can be used to provide stabilised or modified materials for earthworks, or may be used to provide permanent load transfer platforms or hard standings. Soil stabilisation can be used to treat and neutralise certain contaminants or encapsulate the contaminants, removing the need for expensive removal and disposal.

The soil stabilisation rig shown has a large cutting wheel that breaks down and pulverises the soil, this is then mixed with either quicklime for cohesive soils or cement, microsilica or pulverised fuel ash in granular soils. The ground is then levelled and rolled.

Photograph 3.12   Soil modification and stabilisation (www.roger-bullivant.co.uk).

3.4 Substructures and basements

The foundation substructure of multi-storey buildings is often constructed below natural or artificial ground level. In towns and cities the ground for some metres below ground level has often been filled, over the centuries, to an artificial level. Filled ground is generally of poor and variable bearing capacity and is not good material on which to place foundations. It is generally necessary and expedient, therefore, to remove the artificial ground and construct a substructure or basement of one or more floors below ground. Similarly, where there is a top layer of natural ground of poor and variable bearing capacity, it is often removed and a substructure formed. Where there are appreciable differences of level on a building site, a part or the whole of the building may be below ground level as a substructure.

The natural or artificial ground around the substructure is often permeable to water and may retain water to a level above that of the lower level or floor of a substructure. Ground water in soil around a substructure will impose pressure on both the walls and floor of a substructure (hydrostatic pressure). The pressure of water is often considerable and may penetrate small cracks. The cracks in the construction may be due to construction joints that are not watertight, shrinkage or movement of dense concrete walls and floors, and even dense, solidly-built brick walls may allow water to penetrate. To limit the penetration of ground water under pressure it is practice to build in water stops across construction and movement joints in concrete walls and floors, and to line brick walls and concrete floors with a layer of impermeable material in the form of a waterproof lining like a tank, hence the term 'tanking to basements' (Figure 3.60 and 3.61).

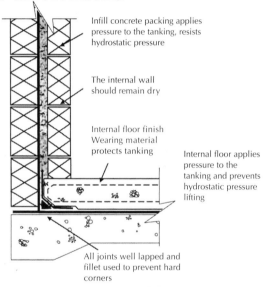

Infill concrete packing applies pressure to the tanking, resists hydrostatic pressure

The internal wall should remain dry

Internal floor finish Wearing material protects tanking

Internal floor applies pressure to the tanking and prevents hydrostatic pressure lifting

All joints well lapped and fillet used to prevent hard corners

Figure 3.60   Type A Internal basement tanking system.

Another approach is to accept that there will be some penetration of ground water through the external concrete wall, but not to allow this water into the useable part of the basement (Figure 3.62). Cavity walls are constructed with an external wall that retains the soil (the structural wall), a clear cavity that is drained at the bottom and an internal wall that provides a dry surface. Water that manages to penetrate the external structure runs down the external face of the cavity and is guided through channels to a sump where it is pumped out of the building. The external structural wall can be constructed of dense reinforced concrete with water stops, or could be formed using contiguous, secant or steel piles.

The design of a basement is dependent on use, site conditions, construction conditions and waterproofing system. Table 3.3 provides a brief summary of the types of construction that are suitable for different basements.

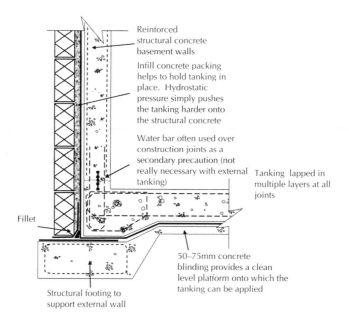

Reinforced structural concrete basement walls

Infill concrete packing helps to hold tanking in place. Hydrostatic pressure simply pushes the tanking harder onto the structural concrete

Water bar often used over construction joints as a secondary precaution (not really necessary with external tanking)

Tanking lapped in multiple layers at all joints

Fillet

50–75mm concrete blinding provides a clean level platform onto which the tanking can be applied

Structural footing to support external wall

Figure 3.61   Type A External basement tanking system – with blockwork protection.

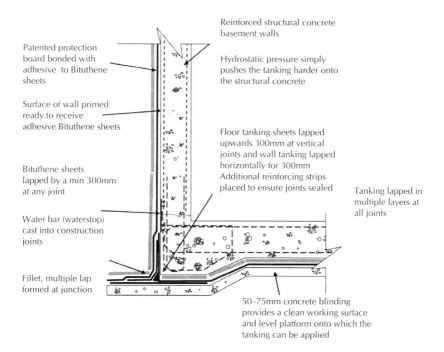

Patented protection board bonded with adhesive to Bituthene sheets

Reinforced structural concrete basement walls

Surface of wall primed ready to receive adhesive Bituthene sheets

Hydrostatic pressure simply pushes the tanking harder onto the structural concrete

Floor tanking sheets lapped upwards 300mm at vertical joints and wall tanking lapped horizontally for 300mm Additional reinforcing strips placed to ensure joints sealed

Bituthene sheets lapped by a min 300mm at any joint

Tanking lapped in multiple layers at all joints

Water bar (waterstop) cast into construction joints

Fillet, multiple lap formed at junction

50–75mm concrete blinding provides a clean working surface and level platform onto which the tanking can be applied

Figure 3.62   Type A External basement tanking system – with protective board (adapted from www.uk.graceconstruction.com).

Table 3.3 Type and level of protection required to suit use of basement

Level of protection required to suit the type of basement

Grade of construction	Use of basement	Level of performance and conditions required	Type of basement construction	Comment
Grade 1. Basic utility rooms	Plant rooms, car park	Some seepage and some damp patches may occur (tolerable) >65% relative humidity 15–32°C temp	Reinforced concrete to BS 8110 Type B structure	Check that the ground water does not contain chemicals. Some chemicals may degrade or have other deleterious effects on the structure and internal finishes.
Grade 2. Improved utility	Workshops and plant rooms, retail storage areas	No water penetration or seepage, but moisture vapour is tolerable 35–5% relative humidity <15°C storage up to 42°C for plant rooms	Type A Type B to BS 8007 (Watertight concrete)	Good supervision of all stages of construction is necessary to ensure watertight construction. Membranes should be applied in multiple layers and lapped joints.
Grade 3. Habitable	Ventilated residential and working areas, for example offices, restaurants and leisure centres	Dry environment, tightly controlled 40–60% relative humidity 18–29°C temp. depending on use	Type A Type B to BS 9007 Type C to BS 8110	Good supervision of all stages of construction is necessary to ensure watertight construction. Membranes should be applied in multiple layers and lapped joints.
Grade 4. Special	Archives, paper stores and computer rooms	Controlled environment that is totally dry 35–50% relative humidity 13–22°C temp	Type A Type B to BS 8007 Combined with vapour proof membrane Type C Ventilated wall cavity and vapour barrier to inner skin and floor protection	Good supervision of all stages of construction is necessary to ensure watertight construction. Membranes should be applied in multiple layers and lapped joints.

Brief description of basement types

Type A – Tanking Membrane. Waterproof membrane formed out of mastic asphalt, bitumen, rubber/bitumen compound, bonded sheet membranes, Bituthene, or polymer modified bitumen is either applied externally, internally or sandwiched between the basement wall. Bentonite clay can also be used.
Type B – Monolithic Concrete. The structure itself provides the necessary waterproofing. The structure is formed with dense reinforced concrete, water bar is often used at construction joints and additives may be introduced to the concrete to make it denser and more water resistant.
Type C – Drained Cavity. It is anticipated that water will penetrate the external structure. An internal skin of blockwork or concrete (with drainage former) is used to form a cavity. Water that enters the structure is drained off and pumped out of the building.

Adapted from BS 8102:1990, BSI.

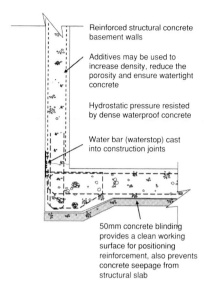

Reinforced structural concrete basement walls

Additives may be used to increase density, reduce the porosity and ensure watertight concrete

Hydrostatic pressure resisted by dense waterproof concrete

Water bar (waterstop) cast into construction joints

50mm concrete blinding provides a clean working surface for positioning reinforcement, also prevents concrete seepage from structural slab

Figure 3.63   Type B Waterproof concrete basement.

Waterstops to concrete walls and floors

Dense concrete, which is practically impermeable to water, would by itself effectively exclude ground water (Figure 3.63). However, in some situations it is difficult to prevent movement and the formation of cracks caused by shrinkage, structural, thermal and moisture movement. As concrete dries out and sets,

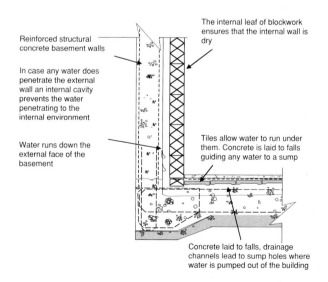

The internal leaf of blockwork ensures that the internal wall is dry

Reinforced structural concrete basement walls

In case any water does penetrate the external wall an internal cavity prevents the water penetrating to the internal environment

Water runs down the external face of the basement

Tiles allow water to run under them. Concrete is laid to falls guiding any water to a sump

Concrete laid to falls, drainage channels lead to sump holes where water is pumped out of the building

Figure 3.64   Type C Traditional drained cavity basement.

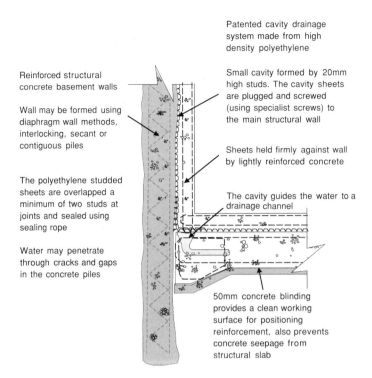

Patented cavity drainage system made from high density polyethylene

Reinforced structural concrete basement walls

Wall may be formed using diaphragm wall methods, interlocking, secant or contiguous piles

The polyethylene studded sheets are overlapped a minimum of two studs at joints and sealed using sealing rope

Water may penetrate through cracks and gaps in the concrete piles

Small cavity formed by 20mm high studs. The cavity sheets are plugged and screwed (using specialist screws) to the main structural wall

Sheets held firmly against wall by lightly reinforced concrete

The cavity guides the water to a drainage channel

50mm concrete blinding provides a clean working surface for positioning reinforcement, also prevents concrete seepage from structural slab

Figure 3.65   Type C Cavity drain formed using high density studded polyethlene sheets (adapetd from www.riw.co.uk).

it shrinks and this inevitable drying shrinkage causes cracks, particularly at construction joints, through which ground water will penetrate.

Waterstops

As a barrier to the penetration of water through construction joints and movement joints in concrete floors and walls underground, it is practice to either cast PVC waterstops against and across joints or to cast rubber waterstops into the thickness of concrete. The first method is generally used where water pressure is low, and the alternative, second method where water pressure is high. The first method is the most economical as it merely involves fixing the PVC stops to the formwork. Movement joints are formed right across and up the whole height of large buildings and filled with an elastic material that can accommodate the movement due to structural, thermal and moisture changes. These movement joints are formed at intervals of not more than 30 m.

Movement joints are formed in the main to accommodate thermal movement due to expansion and contraction of long lengths of solid structure and at angles and intersections right across the width and up the whole height of buildings including floors and roofs. In effect, movement joints create separate structures

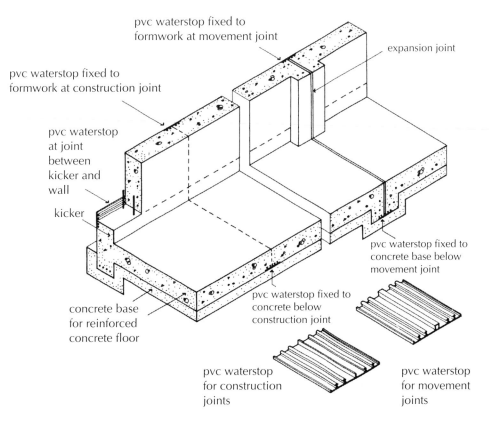

pvc waterstop fixed to
formwork at movement joint

expansion joint

pvc waterstop fixed to
formwork at construction joint

pvc waterstop
at joint
between
kicker and
wall

kicker

pvc waterstop fixed to
concrete base below
movement joint

concrete base
for reinforced
concrete floor

pvc waterstop fixed to
concrete below
construction joint

pvc waterstop
for construction
joints

pvc waterstop
for movement
joints

Figure 3.66   PVC waterstops.

each side of the joint. In framed structures, movement joints are usually formed between a pair of columns and pairs of associated beams.

PVC waterstops, illustrated in Figure 3.66, are fixed to the inside face of the timber formwork to the outside face of walls and to the concrete base under reinforced concrete floors so that the projecting dumbbells are cast into the concrete floors and walls. The large dumbbell in the centre of the waterstops for movement joints is designed to accommodate the larger movement likely at these joints. Provided the concrete is solidly consolidated up to the stops, this system will effectively act as a waterstop. At the right-angled joints of waterstops, preformed cross-over sections of stops are heat welded to the ends of straight lengths of stop. Rubber waterstops are cast into the thickness of concrete walls and floors, as illustrated in Figure 3.67. Plain web stops are cast in at construction joints and centre bulb stops at expansion joints. These stops must be firmly fixed in place and supported with timber edging to one side of the stop so that concrete can be placed and compacted around the other half of the stop without moving it out of place. At

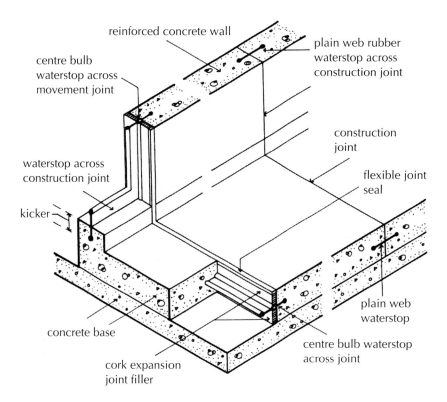

Figure 3.67   Rubber waterstops.

the junction of the joints hot vulcanising joins the stops. Hot vulcanising is where a hot iron heats the PVC and as the PVC melts the two ends merge together.

For waterstops to be effective, concrete must be placed and firmly compacted up to the stops, and the stops must be secured in place to avoid them being displaced during placing and compacting of concrete. Waterstops will be effective in preventing penetration of water through joints provided they are solidly cast up to or inside sound concrete and there is no gross contraction at construction joints or movement at expansion joints.

Tanking

The term tanking is used to describe a continuous waterproof lining to the walls and floors of substructures to act as a tank to exclude water.

Mastic asphalt

The traditional material for tanking is mastic asphalt, which is applied and spread hot in three coats to a thickness of 20 mm for vertical and 30 mm for horizontal work. Joints between each layer of asphalt in each coat should be

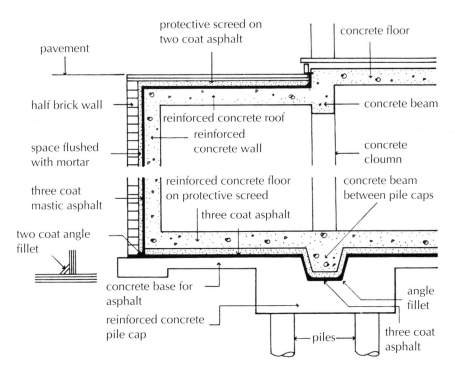

Figure 3.68   Mastic asphalt tanking.

staggered at least 75 mm for vertical and 150 mm for horizontal work with the joints in succeeding coats. Angles are reinforced with a two-coat fillet of asphalt. Asphalt tanking should be applied to the outside face of structural walls and under structural floors so that the walls and floors provide resistance against water pressure on the asphalt and the asphalt keeps water from the structure. Figure 3.68 is an illustration of asphalt tanking applied externally to the reinforced concrete walls and floor of a substructure or basement. The horizontal asphalt is spread in three coats on the concrete base and over pile caps and extended 150 mm outside of the junction of the horizontal and vertical asphalt and the angle fillet. The horizontal asphalt is then covered with a protective screed of cement and sand 50 mm thick. The reinforced concrete floor should be cast on the protective screed as soon as possible to act as a loading coat against water pressure under the asphalt below.

  When the reinforced concrete walls have been cast in place and have dried, the vertical asphalt is spread in three coats and fused to the projection of the horizontal asphalt with an angle fillet. A half brick protective skin of brickwork is then built, leaving a 40 mm gap between the wall and the asphalt. The gap is filled solidly with mortar, course by course, as the wall is built. The half brick wall provides protection against damage from backfilling and the mortar filled gap ensures that the asphalt is firmly sandwiched up to the structural wall. In

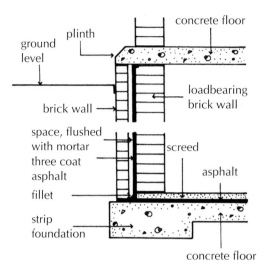

Figure 3.69   Mastic asphalt tanking.

Figure 3.68 the asphalt tanking is continued under a paved forecourt. Where vertical asphalt is carried up on the outside of external walls it should be carried up at least 150 mm above ground to join a damp proof course.

Figure 3.69 is an illustration of mastic asphalt tanking to a concrete floor and loadbearing brick wall to a substructure. The protective screed to the horizontal asphalt and protecting outer wall and mortar filled gap to the vertical asphalt serve the same functions as they do for a concrete substructure. As a key for the vertical asphalt, the horizontal joints in the external face of the loadbearing wall should be lightly raked out and well brushed when the mortar has hardened sufficiently.

Where the walls of substructures are on site boundaries and it is not possible to excavate to provide adequate working space to apply asphalt externally, a system of internal tanking may be used. The concrete base and structural walls are built, the horizontal asphalt is spread on the concrete base and a 50 mm protective screed spread over the asphalt. Asphalt is then spread up the inside of the structural walls and joined to the angle fillet reinforcement at the junction of horizontal and vertical asphalt. A loading and protective wall, usually of brick, is then built with a 40 mm mortar filled gap up to the internal vertical asphalt. The internal protective and loading wall, which has to be sufficiently thick to resist the pressure of water on the asphalt, is usually one brick thick. A concrete loading slab is then cast on the protective screed to act against water pressure (also called hydrostatic pressure) on the horizontal asphalt. An internal asphalt lining is rarely used for new buildings because of the additional floor and wall construction necessary to resist water pressure on the asphalt. Internal asphalt is sometimes used where a substructure to an existing building is to be waterproofed.

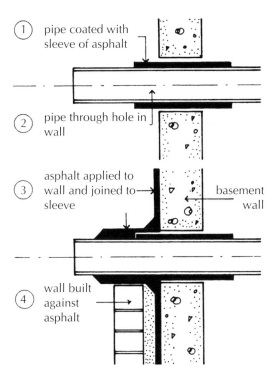

① pipe coated with
   sleeve of asphalt

② pipe through hole in
   wall

③ asphalt applied to
   wall and joined to→
   sleeve

   basement
   wall

④ wall built
   against
   asphalt

Figure 3.70   Four stages in forming asphalt collar around pipe.

Service pipes for water, gas and electricity and drain connections that are run through the walls of a substructure that is lined with asphalt tanking are run through a sleeve that provides a watertight seal to the perforation of the asphalt tanking and allows for some movement between the service pipe or drain and the sleeve. The sleeve is coated with asphalt that is joined to the vertical asphalt with a collar of asphalt, which runs around the sleeve, as illustrated in Figure 3.70.

Asphalt that is sandwiched in floors as tanking has adequate compressive strength to sustain the loads normal to buildings. The disadvantages of asphalt are that asphalting is a comparatively expensive labour intensive operation and that asphalt is a brittle material that will readily crack and let in water if there is differential settlement or appreciable movements of the substructure. In general, the use of asphalt tanking is limited to substructures with a length or width of not more than about 7.5 m to minimise the possibility of settlement or movement cracks fracturing the asphalt.

Bituminous membranes

As an alternative to asphalt, bituminous membranes are commonly used for waterproofing and tanking to substructures. The membrane is supplied as a

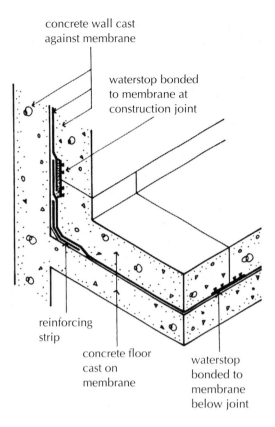

concrete wall cast
against membrane

waterstop bonded
to membrane at
construction joint

reinforcing
strip

concrete floor
cast on
membrane

waterstop
bonded to
membrane
below joint

Figure 3.71 Bituminous sheet membrance tanking.

sheet of polythene or polyester film, or sheet bonded to a self-adhesive rub-
ber/bitumen compound or a polymer modified bitumen. The heavier grades
of these membranes are reinforced with a meshed fabric sandwiched in the self-
adhesive bitumen. The membrane is supplied in rolls about 1 m wide and 12 to
18 m long, with the self-adhesive surface protected with a release paper backing. The particular advantage of these membranes is that their flexibility can
accommodate small shrinkage, structural, thermal and moisture movements
without damage to the membrane. Used in conjunction with waterstops to
concrete substructures these membranes may be used instead of tanking (see
Figures 3.60, 3.61 and 3.71).

The surface to which the membrane is applied by adhesion of the bitumen
coating must be dry, clean and free from any visible projections that would
puncture the membrane. The membrane is applied to a dry, clean float finished
screed for floors and to level concrete wall surfaces on which all projecting nibs
from formwork have been removed and cavities filled. The vertical surface to
which the membrane is to be applied is first primed. The rolls of sheet are

laid out, the paper backing removed and the membrane laid with the adhesive bitumen face down or against walls and spread out and firmly pressed on to the surface with a roller. Joints between long edges of the membrane are overlapped 75 mm and end joints 150 mm, and the overlap joints are firmly rolled in to compact the join. Laps on vertical wall surfaces are overlapped so that the sheet above overlaps the sheet below. At construction and movement joints the membrane is spread over the joint with a PVC or rubber waterstop cast against or in the concrete.

Bitumen membranes are formed outside structural walls and under structural floors, as illustrated in Figure 3.71, with an overlap and fillet at the junction of vertical and horizontal membranes. To protect the vertical membrane from damage by backfilling, a protective concrete or half brick skin should be built up to the membrane. At angles and edges a system of purposely cut and shaped cloaks and gussets of the membrane material is used over which the membrane is lapped, as illustrated in Figure 3.72. To be effective as a seal to the vulnerable angle joints, these overlapping cloaks and gussets must be carefully shaped and applied. The effectiveness of these membranes as waterproof tanking depends on dry, clean surfaces free from protrusions or cavities, and careful workmanship in spreading and lapping the sheets, cloaks and gussets.

Cavity tanking

Cavity drain structures make allowances for the small amount of water that may pass through the external wall. The basement is constructed with two walls forming a void between the external and internal leaf, and a cavity is formed in the walls around the basement and below the floor. Traditionally cavities were formed with floor tiles, which created a void, and two separate leafs of masonry walling. Nowadays rolls of studded 1 mm thick polyethylene are used, with raised studs (domes), which stand 20 mm from the surface. The sheets of polyethylene, which come in rolls 20 m long by 1.4 m wide, are applied to the structural walls and floors to form a very small but effective void. The sheets are temporarily fixed to the wall using an effective plug and screw system, and then permanently held and protected by a concrete screed floor and insitu concrete wall. Any water that penetrates through the external wall or floor is guided to drainage channels where it is then pumped out of the building. The cavity should be ventilated to help minimise the build up of water vapour.

Basement walls

There is often insufficient space to provide for the basement excavation and adequate batter so that the basement walls can be constructed out of the ground. It is becoming more common for deep basements to be constructed using diaphragm walls, with the aid of large clam grabs and bentonite slurry, and auger bored piled walls using secant, interlocking and contiguous piles.

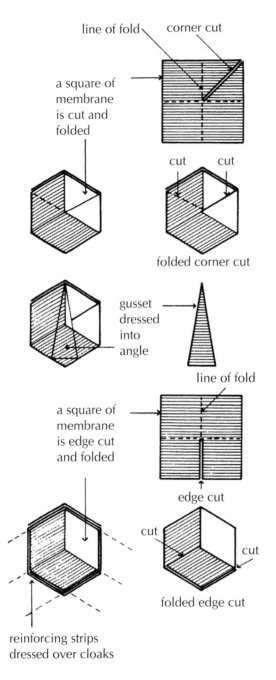

line of fold   corner cut

a square of
membrane
is cut and
folded

cut   cut

folded corner cut

gusset
dressed
into
angle

line of fold

a square of
membrane
is edge cut
and folded

edge cut

cut

cut

folded edge cut

reinforcing strips
dressed over cloaks

Figure 3.72   Internal angle cloaks to bituminuos membrane.

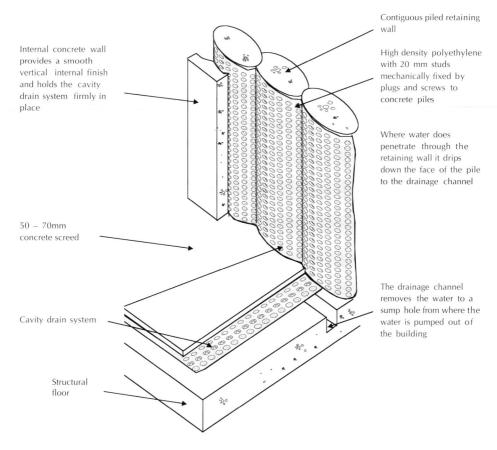

Internal concrete wall provides a smooth vertical internal finish and holds the cavity drain system firmly in place

Contiguous piled retaining wall

High density polyethylene with 20 mm studs mechanically fixed by plugs and screws to concrete piles

Where water does penetrate through the retaining wall it drips down the face of the pile to the drainage channel

50 – 70mm concrete screed

The drainage channel removes the water to a sump hole from where the water is pumped out of the building

Cavity drain system

Structural floor

Figure 3.73   Drained cavity system.

Contiguous piles

With the development of continuous flight auger piling (CFA) rigs, bored piles are becoming more common as a form of permanent wall and foundation. Contiguous piles (Figures 3.74 and 3.75A) are formed by drilling bored piles at close centres. Piles vary in diameter from 300 to 2400 mm although piles greater than 1200 mm are rarely used. A small gap is left between each pile; typically this ranges between 20 and 150 mm (Photograph 3.13). The size of the gap depends on the soil strength. Due to the gaps between each pile, the wall has a limited effect on ground water control. The gaps may be filled to provide a more water resistant structural concrete facing wall. The use of CFA rigs to form the piles limits the depth of pile to 30–55 m depending on the type of rig. In practice walls are usually constructed to a maximum of 25 m; some piles may go much deeper to provide vertical load capacity (Gaba, Simpson, Powrie, Beadman, 2003). A ring beam is cast along the top of the piles linking all of the piles together; this provides extra rigidity and strength and helps to distribute any loads placed on

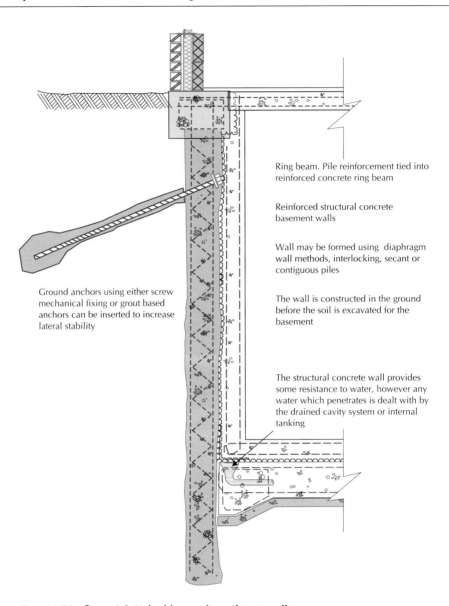

Ring beam. Pile reinforcement tied into reinforced concrete ring beam

Reinforced structural concrete basement walls

Wall may be formed using diaphragm wall methods, interlocking, secant or contiguous piles

The wall is constructed in the ground before the soil is excavated for the basement

The structural concrete wall provides some resistance to water, however any water which penetrates is dealt with by the drained cavity system or internal tanking

Ground anchors using either screw mechanical fixing or grout based anchors can be inserted to increase lateral stability

Figure 3.74   Secant, interlocking and contiguous piles.

top of the piles. Ground anchors can also be used to help resist the overturning forces caused by the surrounding strata and hydrostatic pressure (Photograph 3.14). Grout is normally forced through the anchor to tie it securely into the ground. Where the area surrounding the retaining structure accommodates roads, structures or other property, grout may be forced into the ground to produce a positive pressure on the ground, which counteracts the possibility of settlement caused by the excavation of the basement and movement of the retaining wall. The excavation of the basement may cause settlement

Reinforcement exposed ready to
receive ring beam cage

Continuous flight auger

Ground anchors

Small gap between each pile

Grout injected through ground
anchors

Surplus grout returns through the gaps
in the pile

Concrete ring beam connects all of the piles

Photograph 3.13   Contiguous piles and ground anchors (www.roger-bullivant.co.uk).

due to the vibration, which causes the surrounding ground to compact. Also, as the surrounding ground applies its load on the retaining wall some slight movement will occur. Vibration and movement of the retaining wall will result in settlement of the surrounding ground; the pressure exerted by pressurised grout can be used to remove the potential of settlement.

Secant piles
Secant piles consist of overlapping and interlocking piles (Figure 3.75B, C, D and E and Photograph 3.15). Female (primary) piles are bored using CFA rigs

The auger cuts through the retaining wall inserting a steel tube or sleeve. Reinforcing rods or tensile cables may also be installed. Finally grout is forced through the sleeve and penetrates into the surrounding ground. The grout forms a bulb around the anchor and ties it securely to the ground.

Photograph 3.14   Diaphragm walls (courtesy of D. Highfield).

and cast first. The secondary male piles are then drilled, secanting (cutting into) into the female pile. The system is often used in the construction of deep basement walls. Secant walls are often considered to be a more economical alternative to diaphragm walls.

Depending on the type of secant pile construction either one or both piles are reinforced to resist the lateral loads. When the secant wall is in place the excavated face can be covered with a layer of structural concrete. The concrete can be either sprayed or cast against the wall, providing a fair faced concrete finish. A reinforced concrete ring beam connects all of the piles together, improving the structural stability of the wall. The beam will also help to distribute any loads placed on top of the wall. The depth of the wall is usually limited to 25 m; however, it is possible to construct secant walls to a depth of 55 m. Multi-storey basements are usually stabilised by ground anchors and cross-bracing to resist lateral forces. Secant piles can be constructed as hard/soft, hard/firm or hard/hard walls.

Hard/soft secant piles
For hard/soft piles the female piles, which are cast first, are constructed of a soft pile mix; this usually takes the form of a cement and bentonite mix or cement, bentonite and sand mix. The mix has a weak characteristic strength of

Lightly reinforced concrete wall provides a smooth and clean surface for the bitumen tanking to be applied and helps to resist hydrostatic pressure.

Reinforcement

The interlocking secant piles help to resist hydrostatic pressure and retain the ground.

Sheet tanking is laid under the structural floor and turned up against the concrete wall.

Once all of the tanking is fixed to the concrete wall a further concrete wall will be cast against the tanking, trapping the tanking and preventing hydrostatic pressure pushing the tanking off the wall.

Photograph 3.15   Secant piling and internal tanking (courtesy of D. Highfield).

$1–3\,\text{N}/\text{mm}^2$. The female piles are used as a water retaining structure rather than a load-bearing column. Soft piles can retain up to 8 m head of ground water. The unreinforced soft pile is not usually used as a permanent wall material. As the bentonite and cement mix dries it will shrink and crack, losing its water-resisting properties. Some soft piles have been designed to retain their water-resisting properties for the life of the structure; often this necessitates the mix remaining hydrated throughout the life of the building.

Hard/firm secant piles
In the hard/firm pile arrangement the female pile has a characteristic strength of $10–20\,\text{N}/\text{mm}^2$; during the wall's construction the strength of the pile is held low by adding a retarding agent to the concrete mix (Figure 3.75C). Female piles

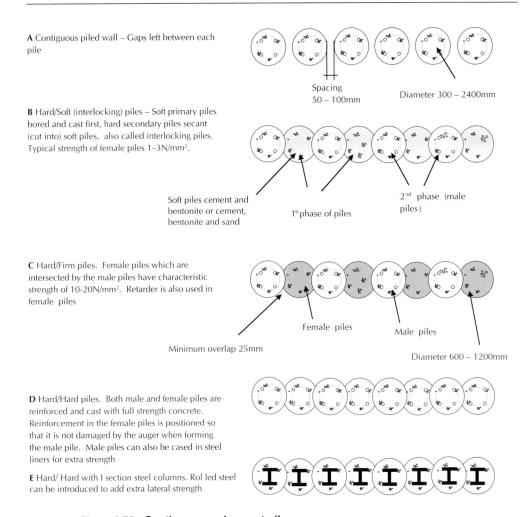

A Contiguous piled wall – Gaps left between each pile

Spacing
50 – 100mm

Diameter 300 – 2400mm

B Hard/Soft (interlocking) piles – Soft primary piles bored and cast first, hard secondary piles secant (cut into) soft piles, also called interlocking piles. Typical strength of female piles 1–3N/mm².

Soft piles cement and bentonite or cement, bentonite and sand

1st phase of piles

2nd phase (male piles)

C Hard/Firm piles. Female piles which are intersected by the male piles have characteristic strength of 10-20N/mm². Retarder is also used in female piles

Female piles

Male piles

Minimum overlap 25mm

Diameter 600 – 1200mm

D Hard/Hard piles. Both male and female piles are reinforced and cast with full strength concrete. Reinforcement in the female piles is positioned so that it is not damaged by the auger when forming the male pile. Male piles can also be cased in steel liners for extra strength

E Hard/ Hard with I section steel columns. Rol led steel can be introduced to add extra lateral strength

Figure 3.75   Contiguous and secant piles.

are usually designed to hit their target strength within 56 days rather than the more typical 28 days. Obviously, such practice ensures that the construction of the male pile is easier as the auger has to exert less force when secanting the female pile. Piles usually overlap a minimum of 25 mm.

Hard/hard secant piles

With hard/hard secant piles both male and female piles are cast with full strength concrete and both are fully reinforced. The female piles are cast first and a high torque-cutting casing is used to drill through the female pile. The reinforcement in the female pile is positioned so that the rig does not cut through it when boring the male pile. The depth of overlap is usually about 25 mm; considerable care is required to ensure that this is maintained along the full length

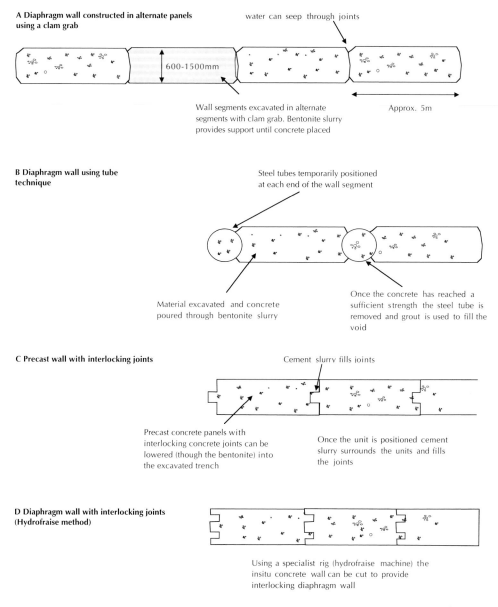

A Diaphragm wall constructed in alternate panels using a clam grab

water can seep through joints

600-1500mm

Wall segments excavated in alternate segments with clam grab. Bentonite slurry provides support until concrete placed

Approx. 5m

B Diaphragm wall using tube technique

Steel tubes temporarily positioned at each end of the wall segment

Material excavated and concrete poured through bentonite slurry

Once the concrete has reached a sufficient strength the steel tube is removed and grout is used to fill the void

C Precast wall with interlocking joints

Cement slurry fills joints

Precast concrete panels with interlocking concrete joints can be lowered (though the bentonite) into the excavated trench

Once the unit is positioned cement slurry surrounds the units and fills the joints

D Diaphragm wall with interlocking joints (Hydrofraise method)

Using a specialist rig (hydrofraise machine) the insitu concrete wall can be cut to provide interlocking diaphragm wall

Figure 3.76   Diaphragm walls.

of the pile. I section beams can also be added to the pile to further increase the lateral strength of the wall (Figure 3.75E).

Diaphragm walls
Diaphragm walls are formed by excavating a segmented trench, to form a continuous wall (Figure 3.76). While the deep trench is being excavated it is filled with bentonite slurry (supporting fluid). The slurry fills the trench and exerts

pressure on the sides of the excavation, thus preventing the excavation walls collapsing. The excavation is carried out using a clamshell grab (Figure 3.77 and Photograph 3.16), which digs the material out through the bentonite slurry. The width of the wall is determined by the width of the grab. The diaphragm wall is cast in the ground. Using a tremie tube about 200 mm diameter the concrete is fed into the trench and placed in position. As the concrete settles at the base of the excavation the bentonite slurry is displaced, drawn out of the excavation and cleaned for reuse. The disposal of bentonite at the end of the operation is expensive as the mixture is treated as a contaminated material. The diaphragm wall is constructed in alternate panels usually around 5 m long. The excavators are usually guided by shallow concrete beams that are cast so that the beam faces form the desired position of the wall. Diaphragm walls have been constructed to depths of 120 m; however, there are some practical difficulties when attempting to splice and link the reinforcement cages over such depths.

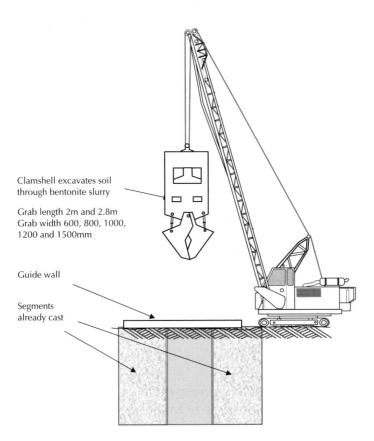

Clamshell excavates soil
through bentonite slurry

Grab length 2m and 2.8m
Grab width 600, 800, 1000,
1200 and 1500mm

Guide wall

Segments
already cast

Figure 3.77   Diaphragm walls – clamshell rig (adapted from Bachy Soletanche, www.bacsol.co.uk).

Clamshell attached to crane

Clamshell excavator

Bentonite pumped into the excavation to support the sides

Concrete guide walls help position the clamshell

Photograph 3.16   Clamshell excavating diaphragm wall (courtesy of D. Highfield).

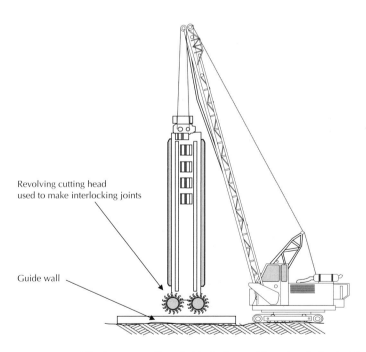

Revolving cutting head used to make interlocking joints

Guide wall

Figure 3.78   Diaphragm walls – hydrofraise cutting rigs (adapted from Bachy Soletanche, www.bacsol.co.uk).

The joints between each section can be cast using steel tubes or interlocking junctions to reduce the ingress of water through the joints (Figures 3.76B, C). The hydrofraise machine is used to cut an interlocking surface into the previously cast segment of wall (Figures 3.76D and 3.78). Interlocking precast concrete sections can also be used. Once the precast concrete sections are in place grout is used to fill any remaining joints.

4 Single-Storey Frames, Shells and Lightweight Coverings

This chapter describes the construction of single-storey buildings such as sheds, warehouses, factories and other buildings generally built on one floor. Over the past 60 years most single-storey buildings have been constructed with a structural frame of steel or reinforced concrete supporting lightweight roof and wall coverings. Thin, lightweight sheets fixed to comparatively slim structural frames to provide clear spans between supports can support the small imposed loads on roofs (see also Chapters 5, 6 and 7). A large proportion of the buildings in this category are constructed to serve a very specific purpose for a relatively short period of time, after which the market and hence the required performance of the building will have changed. It is not uncommon for sheds and warehouses to have a specified design life of between 15 and 30 years. After this time the building is demolished and materials recycled, or alternatively (and less likely) considerable works of repair and renewal are required to maintain minimum standards of comfort and appearance. As a consequence the materials used are selected primarily for economy of initial cost, tend to have limited durability and are prone to damage in use.

In traditional building forms one material could serve several functional requirements, for example, a solid loadbearing brick wall which provides strength, stability, exclusion of wind and rain, resistance to fire and to a small extent thermal and sound insulation. In contrast the materials used in the construction of lightweight structures are, in the main, selected to perform specific functions. Steel sheeting is used as a weather envelope and to support imposed loads, layers of insulation for thermal and sound resistance, thin plastic sheets for daylight, and a slender frame to support the envelope and imposed loads. The inclusion of one material for a specific purpose is likely to have a significant impact on the performance of adjacent materials, thus the designer needs to look at the performance of individual materials *and* the performance of the whole assembly.

4.1 Lattice truss, beam, portal frame and flat roof structures

Early single-storey framed structures for factories and warehouses consisted of brick side walls or steel columns supporting triangular frames (trusses) fabricated from small section steel members, pitched to support purlins, rafters and slate roofing. The minimum pitch for roof slates determined the pitch of the roof in this economical and functional building type. With developments in sheet roofing materials the minimum pitch has been reduced. To reduce the volume of unusable roof space that has to be heated and also to reduce the visual impact of the roof area it is common practice to construct single-storey buildings with low pitch roof frames, either as portal frames or as lattice beam or rafter frames (Figure 4.1). The pitch may be as low as 2.5°. Alternatively flat roof structures may be used.

Functional requirements

The functional requirements of framed structures are:

❑ Strength and stability
❑ Durability and freedom from maintenance
❑ Fire safety

Strength and stability

The strength of a structural frame depends on the strength of the material used in the fabrication of the members of the frame and also on the stability of the frame, which is dictated by the way in which the members are connected and on the bracing across and between frames. Steel is the material most used in framed structures because of its good compressive and tensile strength, and good strength to weight ratio. Hot rolled steel and cold-formed strip steel provide a wide range of sections suited to the economical fabrication of structural frames. Concrete has good compressive strength but poor tensile strength and so it is used as reinforced concrete in structural frames to benefit from the tensile strength of the steel and the compressive strength of the concrete. The concrete also provides protection against corrosion and damage by fire to the steel. Timber is often used in the fabrication of roof frames because it has adequate tensile and compressive strength to support the comparatively light loads. Timber tends to be used instead of steel to form lightweight roof frames because of its ease of handling and fixing.

Durability and freedom from maintenance

On exposure to air and moisture, unprotected steel corrodes to form an oxide coating, i.e. rust, which is permeable to moisture and thus encourages progressive corrosion, which may in time adversely affect the strength of the material. To inhibit rust, steel is painted, coated with zinc or encased in concrete. Painted surfaces will require periodic repainting. Any cutting and drilling operations will damage zinc or painted coatings. Reinforced concrete is highly durable and the surface will need little maintenance other than periodic cleaning. Seasoned, stress-graded timber treated against fungal and insect attack should require no maintenance during its useful life other than periodic staining or painting for appearance.

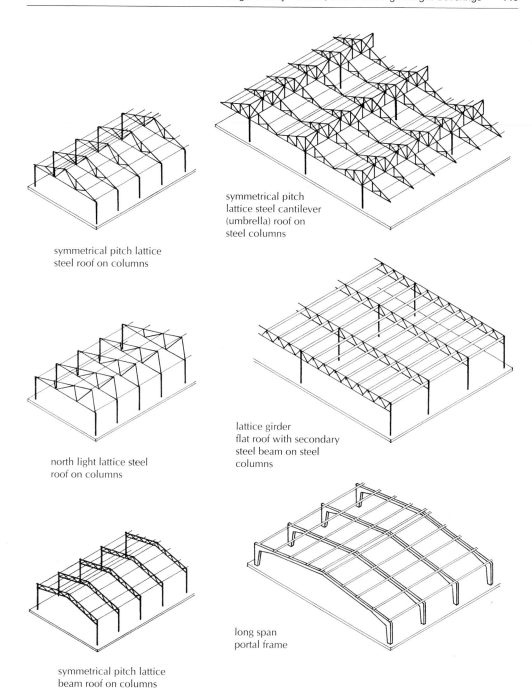

symmetrical pitch lattice
steel roof on columns

symmetrical pitch
lattice steel cantilever
(umbrella) roof on
steel columns

north light lattice steel
roof on columns

lattice girder
flat roof with secondary
steel beam on steel
columns

symmetrical pitch lattice
beam roof on columns

long span
portal frame

Figure 4.1   Typical lattice and portal frame construction.

Fire safety

All loadbearing structures (including roofs) should be designed so that they do not fail prematurely during a fire. Providing the structure with the necessary fire resistance helps to reduce the risk posed by falling debris to building users, pedestrians and fire fighters (ODPM, 2002, B3 section 8).

Elements of the structure that give support or stability to another element of the building must have no less fire resistance than the other supporting elements. Similarly, if a roof provides stability and support to columns then the roof must have at least the same fire resistance as the columns. All roofs should have sufficient fire resistance to resist exposure from the underside of the roof, remaining sound for a minimum of 30 minutes. The same provision also applies to roofs that form part of a fire escape (ODPM, 2002, B Appendix A, Table A1). Where the roof performs the function of a floor, the minimum period of fire resistance is dependent on the purpose of the building and the height of the building (see Table 4.1). If the building is constructed with a basement, this will also have an impact on the required fire resistance.

Table 4.1   Typical fire resistance periods for roofs that form floors (extract from ODPM, 2002, B Appendix A, Table A2)

MINIMUM PERIOD OF FIRE RESISTANCE (minutes)				
	Upper storey height (Height in m above ground)			
Purpose of building	not more than 5 m	not more than 18 m	not more than 30 m	more than 30 m
Residential	30	60	90	120
Flats and maisonettes				
Office				
Not sprinklered	30	60	90	Not permitted
Sprinklered	30	30	60	120
Shop and commercial				
Not sprinklered	60	60	90	Not permitted
Sprinklered	30	60	60	120
Assembly and recreation				
Not sprinklered	60	60	90	Not permitted
Sprinklered	30	60	60	120
Industrial				
Not sprinklered	60	90	120	Not permitted
Sprinklered	30	60	90	120

Note: Minimum resistance for compartment walls is 60 minutes

Lattice truss construction

'Lattice' is a term used to describe an open grid of slender members fixed across or between each other, usually in a regular pattern of cross-diagonals or as a rectilinear grid. 'Truss' is used to define the action of a triangular roof framework, where the spread under load of sloping rafters is resisted by the horizontal tie member that is secured to the feet of the rafters (which trusses or ties them against spreading).

Symmetrical pitch steel lattice truss construction

The simple, single bay shed frame illustrated in Figure 4.2 is a relatively economic structure. The small section, mild steel members of the truss can be cut and drilled with simple tools, assembled with bolted connections and speedily erected without the need for heavy lifting equipment. Similarly the structure can be readily dismantled and reused or recycled when no longer required.

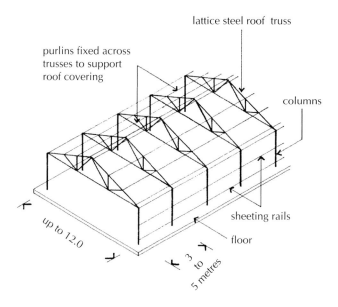

Figure 4.2   Single-bay symmetrical pitch lattice steel roof on steel columns.

The small section, steel angle members of the truss are bolted to columns and purlins, and sheeting rails are bolted to cleats, to support roof and side wall sheeting. These frames can be fabricated off site and quickly erected on comparatively slender mild steel I-section columns fixed to concrete pad foundations. The bolted, fixed base connection of the foot of the columns to the concrete foundation provides sufficient strength and stability against wind pressure on the side walls and roof. Wind bracing provides stability against wind pressure on the end walls and gable ends of the roof.

The depth of the roof frames at mid span provides adequate strength in supporting dead and imposed loads as well as rigidity to minimise deflection under load. For maximum structural efficiency the pitch of the rafters of the frames should be not less than 17° to the horizontal. This large volume of roof space cannot be used for anything other than housing services such as lighting and heating, and where the activity enclosed by the building needs to be heated it makes for an uneconomical solution. Trusses are often limited to spans of approximately 12 m and, for economy in the use of small section purlins and sheeting rails, the spacing of trusses is usually between 3.0 and 5.0 m. Much larger trusses can be and are fabricated to provide large clear spaces.

Rooflights are usually used to provide reasonable penetration of daylight to the interior of the building, as illustrated in Figure 4.3. The thin sheets of profiled steel sheets used to clad the walls have poor resistance to accidental damage and vandalism. As an alternative to steel columns and cladding, load-bearing brick walls may be used for single-bay buildings to provide support for the roof frames. The masonry walls provide better durability to accidental damage and vandalism. The roof frames are positioned on brick piers, which provide additional stiffness to the wall and transfer the loads of the roof to the foundations, as shown in Figure 4.4. As an alternative, a low brick upstand wall may be raised to a height of around 1500 mm as protection against knocks, with wall sheeting above.

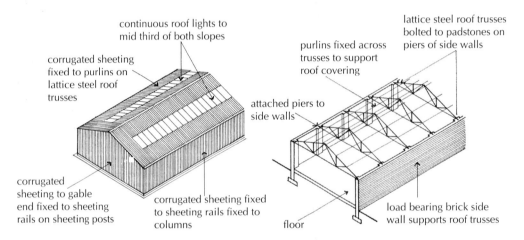

Figure 4.3   Single-bay symmetrical pitch lattice steel roof on columns with corrugated sheeting.

Figure 4.4   Single-bay symmetrical pitch lattice steel roof on brick side walls.

North light steel lattice truss construction
The north light roof has an asymmetrical profile with the south-facing slope pitched at 17° or more to the horizontal and the north-facing slope at around 60°, as illustrated in Figures 4.5 and 4.6. To limit the volume of the roof space that cannot be used (and has to be heated), most north light roofs are limited to spans of up to about 10 m.

Multi-bay lattice steel roof truss construction
To cover large areas it is common practice to use two or more bays of symmetrical pitch roofs to both limit the volume of roof space and the length of the

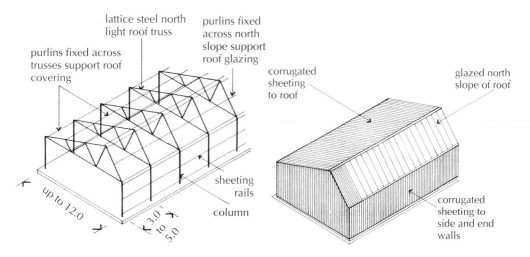

Figure 4.5   Single-bay north light lattice steel roof trusses on steel columns.

Figure 4.6   Single-bay lattice steel north light roof on columns with corrugated sheeting.

members of the trusses. To avoid the use of closely spaced internal columns (which may obstruct the working floor area) to support roof trusses, a valley beam is used. The valley beam supports the roof trusses between the internal columns, as illustrated in Figure 4.7. The greater the clear span between internal columns, the greater the depth of the valley beam and the greater the volume of unused roof space, as illustrated in Figure 4.8.

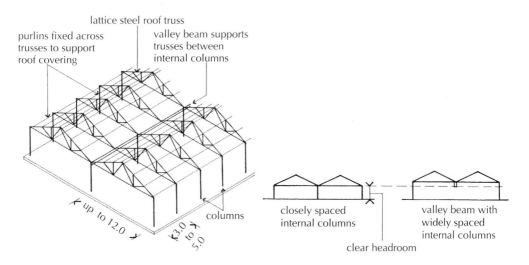

Figure 4.7   Two-bay symmetrical pitch lattice steel roof and columns with valley beam.

Figure 4.8   Increased volume of unused roof space with widely spaced internal columns.

Cantilever (umbrella) multi-bay lattice steel truss roof

Figure 4.9 shows a cantilever (or umbrella) roof with lattice steel girders constructed inside the depth of each bay of trusses at mid span. The lattice girder supports half of each truss with each half cantilevered each side of the truss, hence the name.

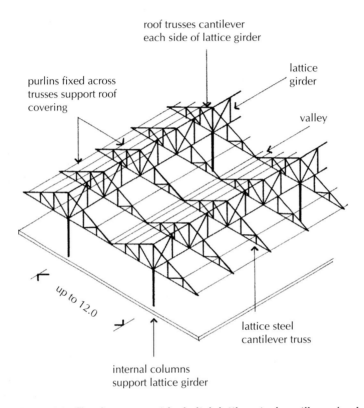

Figure 4.9   Two-bay symmetrical pitch lattice steel cantilever (umbrella) roof.

Lattice steel truss construction

Lattice steel trusses are often fabricated from one standard steel angle section with two angles positioned back to back for the rafters and main tie, and a similar angle for the internal struts and ties, as illustrated in Figure 4.10. The usual method of joining the members of a steel truss is with steel gusset plates, which are cut to shape to contain the required number of bolts at each connection. The flat gusset plates are fixed between the two angle sections of the rafters and main tie and to the intermediate ties and struts. Bearing plates fixed to the foot of each truss provide a fixing to the columns. The members of the truss are bolted together through the gusset plates.

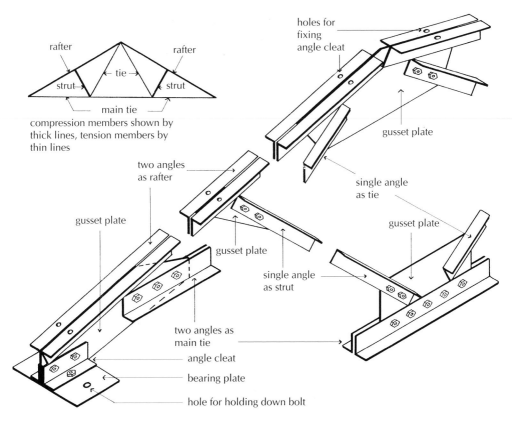

Figure 4.10   Lattice steel truss construction.

Standard I-section steel columns are used to support the roof trusses. A steel base plate is welded or fixed with bolted connections, with gusset plates and angle cleats, to the base of the columns. The column base plate is levelled with steel packing plates and then grouted in position with non-shrinkable cement. The base plate rests on the concrete pad foundation, to which it is rigidly fixed with four holding down bolts, cast or set into the foundation, as illustrated in Figure 4.11. The rigid fixing of the columns to the foundation bases provides stability to the columns, which act as vertical cantilevers in resisting lateral wind pressure on the side walls and the roof of the building. A cap is welded or fixed with bolted connections to the top of each column and the bearing plates of truss ends are bolted to the cap plate (Figure 4.12).

Lattice trusses can be fabricated from tubular steel sections that are cut, mitred and welded together as illustrated in Figure 4.13. Because of the labour involved in the cutting and welding of the members they tend to be more expensive than a similar sized angle section truss; however, they have greater structural efficiency and are visually more attractive. The truss illustrated in

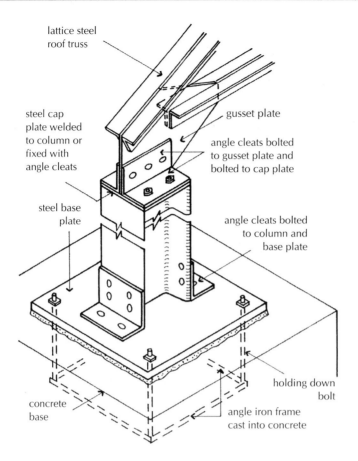

lattice steel
roof truss

steel cap
plate welded
to column or
fixed with
angle cleats

gusset plate

angle cleats bolted
to gusset plate and
bolted to cap plate

steel base
plate

angle cleats bolted
to column and
base plate

holding down
bolt

concrete
base

angle iron frame
cast into concrete

Figure 4.11   Cap and base of steel column support for lattice steel truss.

Figure 4.13 has a raised tie, which affords some increase in working height below the raised part of the tie.

Steel lattice beam roof construction

The two structural forms best suited to the use of deep profiled steel roof sheeting are lattice beam and portal frame. The simplest form of lattice beam roof is a single-bay symmetrical pitch roof constructed as a cranked lattice beam or rafter.

Symmetrical pitch lattice steel beam roof construction

The uniform depth lattice beam is cranked to form a symmetrical pitch roof with slopes of between 5 and 10°, as illustrated in Figure 4.14. Because of the low pitch of the roof there is little unused roof space and this form of

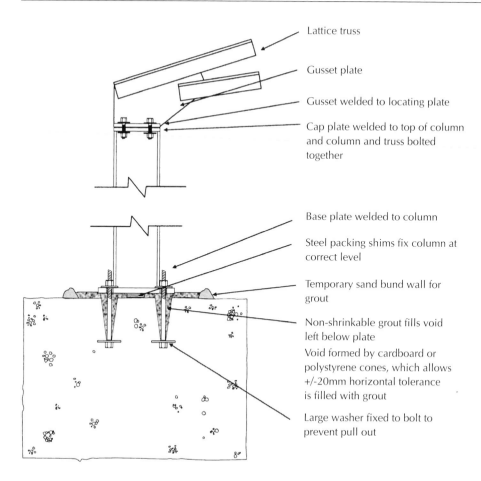

Lattice truss

Gusset plate

Gusset welded to locating plate

Cap plate welded to top of column and column and truss bolted together

Base plate welded to column

Steel packing shims fix column at correct level

Temporary sand bund wall for grout

Non-shrinkable grout fills void left below plate

Void formed by cardboard or polystyrene cones, which allows +/-20mm horizontal tolerance is filled with grout

Large washer fixed to bolt to prevent pull out

Figure 4.12   Cap and base of steel column support for lattice steel truss.

construction is preferred to lattice truss construction where the space is to be heated. The beams are fabricated from tubular and hollow rectangular section steel, which is cut and welded together with bolted site connections at mid span to facilitate the transportation of half-lengths to the site. The top and bottom chord of the beams are usually of hollow rectangular section for ease of fabrication. End plates, welded to the lattice beams, are bolted to the flanges of I-section columns. Service pipes and small ducts may be run through the lattice frames, and larger ducts suspended below the beams inside the roof space.

Multi-bay symmetrical pitch lattice steel beam roof construction

For multi-bay symmetrical pitch lattice beam roofs it is usual to fabricate a form of valley beam roof as illustrated in Figure 4.15. The valley beam is designed to

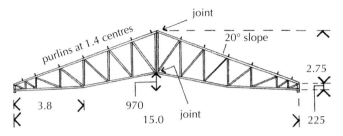

Raised tie tubular steel lattice truss

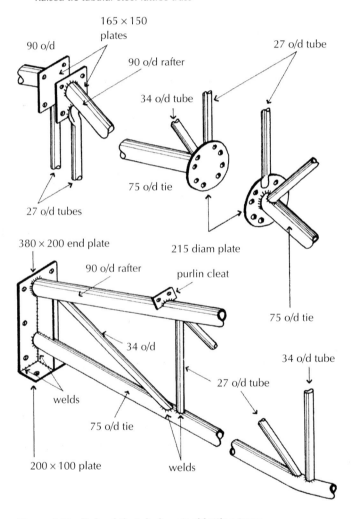

Figure 4.13   Raised tie tubular steel lattice truss.

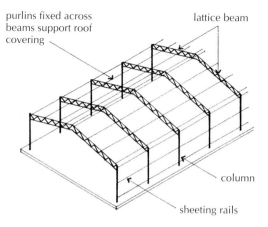

Figure 4.14   Single-bay symmetrical pitch lattice beam and column frame.

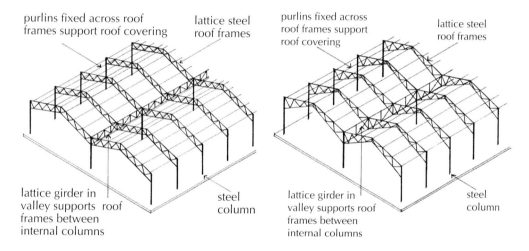

Figure 4.15   Two-bay lattice beam roof on steel columns.

Figure 4.16   Two-bay lattice steel butterfly roof.

be the same depth as the beams to prevent any increase in the unwanted volume of roof space. To provide the maximum free floor space a form of butterfly roof with deep valley beams is used, as illustrated in Figure 4.16. The deeper the valley beams, the greater the spacing between internal columns and the greater the unused roof space.

Steel portal frames

Rigid portal frames are an economic alternative to lattice truss and lattice beam roofs, especially for single-bay buildings. To be effective a pitched roof portal frame should have as low a pitch as practical to minimise spread at the knee of the portal frame (spread increases with the pitch of the rafters of a portal frame). The knee is the rigid connection of the rafter to the post of the portal. Portal frames with a span of up to 15 m are defined as short span, frames between 16 and 35 m as medium span and frames with a span of 36 to 60 m as long span. Because of the considerable clear spans afforded by the portal frame there is little advantage in using multi-bay steel portal systems, where the long span frame would be sufficient. For short- and medium-span frames the apex or ridge, where the rafters join, is usually made as an on-site, rigid bolted connection for convenience in transporting half portal frames to the site. Long-span portal frames may have a pin joint connection at the ridge to allow some flexure between the rafters of the frame, which are pin jointed to the foundation bases.

For economy short- and medium-span steel portal frames are often fabricated from one mild steel I-section for both rafters and posts, with the rafters welded to the posts without any increase in depth at the knee, as shown in Figure 4.17. Short-span portal frames may be fabricated off site as one frame, transported to site and craned into position. Larger span portals are assembled on site with bolted connections of the rafters at the ridge with high strength friction grip (hsfg) bolts (see Chapter 5).

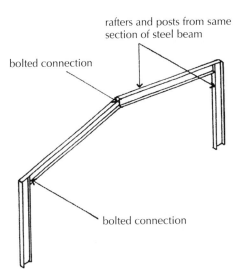

rafters and posts from same section of steel beam

bolted connection

bolted connection

Figure 4.17   Short span steel portal frame.

Many medium- and long-span steel portal frames have the connection of the rafters to the posts haunched at the knee to make the connection deeper and hence stiffer, as illustrated in Figure 4.18. The haunched connection of the rafters to the posts can be fabricated either by welding a cut I-section to the

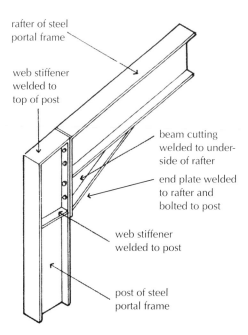

rafter of steel
portal frame

web stiffener
welded to
top of post

beam cutting
welded to under-
side of rafter

end plate welded
to rafter and
bolted to post

web stiffener
welded to post

post of steel
portal frame

Figure 4.18   Haunch to steel portal frame.

underside of the rafter (as illustrated in Figure 4.18) or by cutting and bending
the bottom flange of the rafter and welding in a steel gusset plate.

   In long-span steel portal frames the posts and lowest length of rafters, towards the knee, may often be fabricated from cut and welded I-sections so that
the post section and part of the rafter is wider at the knee than at the base and
ridge of the rafter (Figure 4.19).

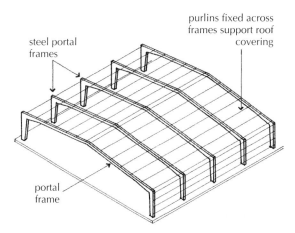

purlins fixed across
frames support roof
covering

steel portal
frames

portal
frame

Figure 4.19   Long span steel portal frames.

The junction of the rafters at the ridge is often stiffened by welding cut I-sections to the underside of the rafters at the bolted site connection as shown in Figure 4.20.

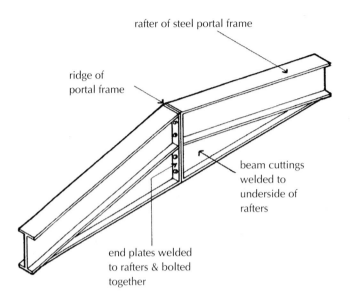

rafter of steel portal frame

ridge of
portal frame

beam cuttings
welded to
underside of
rafters

end plates welded
to rafters & bolted
together

Figure 4.20   Stiffening at ridge of steel portal frame.

Steel portal frames may be fixed or pinned to bases to foundations. For short-span portal frames, where there is relatively little spread at the knee or haunch, a fixed base is often used. The steel plate, which is welded through gusset plates to the post of the portal frame, is set level on a bed of cement grout on the concrete pad foundation and is secured by four holding-down bolts, which are set or cast into the concrete foundation (illustrated in Figure 4.21 and Photograph 4.1). A pinned base is made by sitting the portal base plate on a small steel packing piece on to a separate base plate, which bears on the concrete foundation. Two anchor bolts, either cast or set into the concrete pad foundation, act as holding-down bolts to the foot of the portal frame as illustrated in Figure 4.22. The packing between the plates allows some flexure of the portal post independent of the foundation.

Short-span portal frames are usually spaced between 3 and 5 m apart and medium-span frames at between 4 and 8 m apart to suit the use of angle or cold formed purlins and sheeting rails. Long-span portals are usually spaced at between 8 and 12 m apart to economise on the number of comparatively expensive frames. Channel, Zed, I-section or lattice purlins and sheeting rails support roof sheeting or decking and wall cladding. With flat and low pitch portal frames it is difficult to achieve a watertight system of roof glazing, therefore a system of monitor lights is sometimes used. These lights are formed by welded,

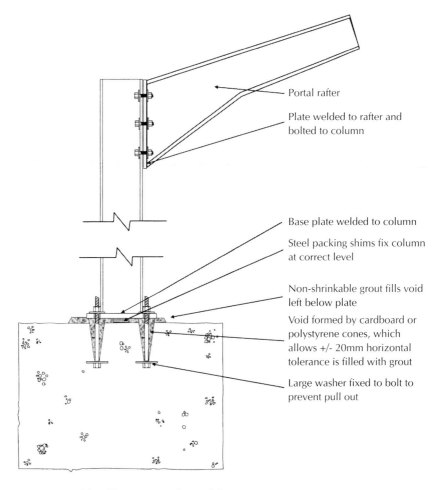

Portal rafter

Plate welded to rafter and bolted to column

Base plate welded to column

Steel packing shims fix column at correct level

Non-shrinkable grout fills void left below plate

Void formed by cardboard or polystyrene cones, which allows +/- 20mm horizontal tolerance is filled with grout

Large washer fixed to bolt to prevent pull out

Figure 4.21   Fixed base to steel portal frame.

Base plate welded to column

Holding down bolts

Steel packing shims fix column at correct level

Once positioned a temporary sand bund wall will be positioned around the column base and filled with grout. The grout is non-shrinkable and fills the void left below plate

Photograph 4.1   Fixed base with four holding-down bolts and base plate of portal packed to correct level (courtesy of D. Highfield).

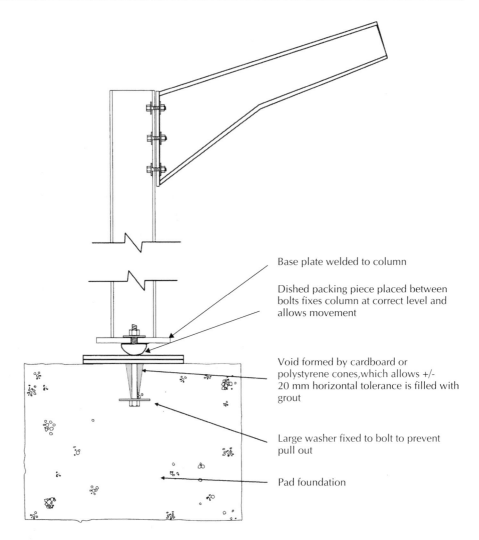

Base plate welded to column

Dished packing piece placed between bolts fixes column at correct level and allows movement

Void formed by cardboard or polystyrene cones,which allows +/- 20 mm horizontal tolerance is filled with grout

Large washer fixed to bolt to prevent pull out

Pad foundation

Figure 4.22 Pinned base to steel portal frame.

cranked I-section steel purlins fixed across the portal frames (Figure 4.23). The monitor lights finish short of the eaves to avoid any unnecessary complications that would otherwise occur. The monitor lights can be constructed to provide natural and controlled ventilation to the interior.

Wind bracing
The side wall columns (stanchions) and their fixed bases that support the roof frames are designed to act as vertical cantilevers to carry the loads in bending and shear that act on them from horizontal wind pressure on the roofing and cladding. The rigid knee joint between rafter and post will carry the loads

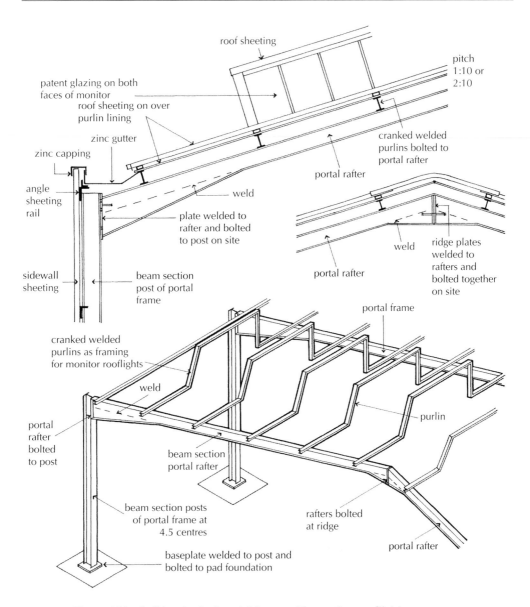

roof sheeting

pitch
1:10 or
2:10

patent glazing on both
faces of monitor
roof sheeting on over
purlin lining

cranked welded
purlins bolted to
portal rafter

zinc gutter

zinc capping

portal rafter

angle
sheeting
rail

weld

plate welded to
rafter and bolted
to post on site

weld   ridge plates
welded to
rafters and
bolted together
on site

portal rafter

sidewall
sheeting

beam section
post of portal
frame

portal rafter

portal frame

cranked welded
purlins as framing
for monitor rooflights

weld

purlin

portal
rafter
bolted
to post

beam section
portal rafter

beam section posts
of portal frame at
4.5 centres

rafters bolted
at ridge

portal rafter

baseplate welded to post and
bolted to pad foundation

Figure 4.23   Solid web steel portal frame with monitor rooflights.

from horizontal wind pressure on roof and side wall cladding. Where internal columns are comparatively widely spaced it is usually necessary to use a system of eaves bracing to assist in the distribution of horizontal loads. The system of eaves bracing shown in Figure 4.24 consists of steel sections fixed between the tie or bottom chord of roof frames and columns. To transfer the loads from wind pressure on the gable ends, a system of horizontal gable girders is formed at tie or bottom chord level.

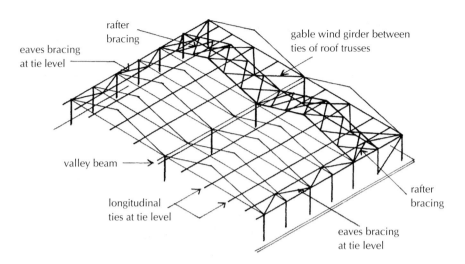

Figure 4.24   Wind bracing to steel truss roof on steel columns.

Structural bracing and wind bracing

Additional bracing is used to assist in setting out the building, to stabilise the roof frames, square up the ends of the building and offer additional resistance to the wind. The rafter bracing between the end frames, illustrated in Figure 4.24, serves to stabilise the rafters of the roof frames. Longitudinal ties between roof frames stabilise the frames against probable uplift due to wind pressure. The vertical bracing in adjacent wall frames at gable end corners hold the building square and serve as bracing against wind pressure on the gable ends of the building (Photograph 4.2).

Photograph 4.2   Wind bracing fixed to portal frame (courtesy of G. Throup).

Purlins and sheeting rails

Purlins are fixed across rafters and sheeting rails across the columns to provide support and fixing for roof and wall cladding and insulation (Figures 4.25 and 4.26). The spacing and size of the purlins and the sheeting rails are determined by the type of roof and wall cladding used. As a general rule, the

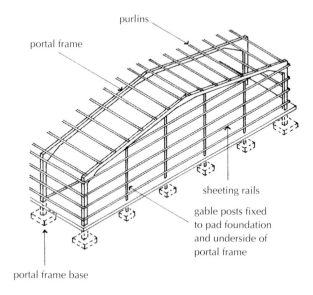

portal frame

purlins

sheeting rails

gable posts fixed
to pad foundation
and underside of
portal frame

portal frame base

Figure 4.25   Gable end framing.

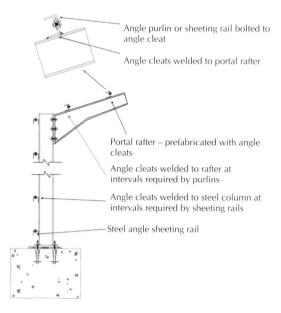

Angle purlin or sheeting rail bolted to
angle cleat

Angle cleats welded to portal rafter

Portal rafter – prefabricated with angle
cleats

Angle cleats welded to rafter at
intervals required by purlins

Angle cleats welded to steel column at
intervals required by sheeting rails

Steel angle sheeting rail

Figure 4.26   Connection of purlin to truss and sheeting rails to columns.

deeper the profile of the sheeting the greater its safe span and the further apart the purlins and sheeting rails may be fixed.

Mild steel angles and purlin rails are sometimes used, but these tend to have been replaced by a range of standard sections, purlins and rails in galvanised, cold formed steel strip. The sections most used are Zed and Sigma (Figure 4.27), with more complex sections with stiffening ribs also produced. These thin section purlins and rails help to facilitate direct fixing of the sheeting by self-tapping screws.

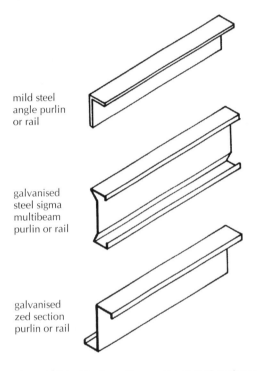

mild steel
angle purlin
or rail

galvanised
steel sigma
multibeam
purlin or rail

galvanised
zed section
purlin or rail

Figure 4.27   Steel section purlin and sheeting rails to support sheet metal and composite sheeting.

Purlins and sheeting rails are fixed to structural supports with cleats, washer plates and sleeves as illustrated in Figure 4.28. Anti-sag bars are fixed between cold formed purlins to stop them twisting during the fixing of roof sheeting and to provide lateral restraint to the bottom flange against uplift due to wind pressure. The purlins also derive a large degree of stiffness from the sheeting. Anti-sag bars and apex ties are made from galvanised steel rod that is either hooked or bolted between purlins, as illustrated in Figure 4.29. The apex ties provide continuity over the ridge. For the system to be effective there must also be some form of stiffening brace or strut at the eaves.

The secret fixing for standing seam roof sheeting for low pitched roofs does not provide lateral restraint for cold formed purlins, thus it is necessary to

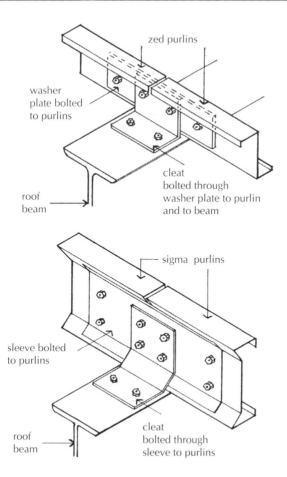

Figure 4.28   Washer plates and sleeves for continuity over supports.

use a system of braces between purlins. The braces are manufactured from galvanised steel sections and bolted between purlins with purpose-made apex braces, as illustrated in Figure 4.30.

To support the wall sheeting (cladding), sheeting rails are fixed across, or between, the steel columns and/or vertical frame members (Figures 4.27 and 4.28). Zed or Sigma section rails are bolted to cleats and then bolted to the structural frame. A system of side rail struts is fixed between rails to provide strength and stability against the weight of the sheeting. The side rails are fabricated from lengths of galvanised mild steel angle, with a fixing plate welded to each end, thus enabling the rails to be bolted to the sheeting rails. A system of tie wires is also used to provide additional restraint as shown in Figure 4.31.

Timber provides an alternative material for short- and medium-span purlins between structural frame members. The ease of cutting and simplicity of fixing makes treated timber a convenient and economic alternative to steel.

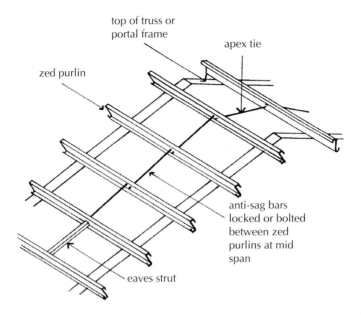

Figure 4.29   **Anti-sag bars to Zed purlins.**

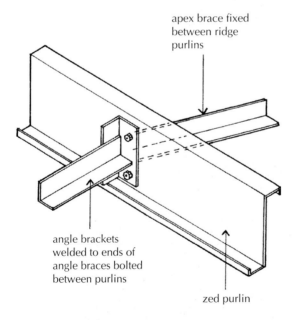

Figure 4.30   **Purlin braces.**

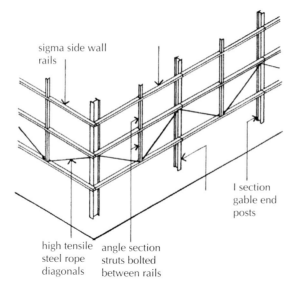

sigma side wall rails

I section gable end posts

high tensile steel rope diagonals

angle section struts bolted between rails

Figure 4.31 Struts and ties to side wall rails.

Sheeting rails for cladding

Steel rope prevents rails sagging

Photograph 4.3 Sheeting rails (courtesy of G. Throup).

Pre-cast reinforced concrete portal frames

Following the end of the Second World War (1945) there was a shortage of steel, which led to the widespread use of reinforced concrete portal frames for single storey structures, such as agricultural sheds, storage and factory buildings. A limited range of standard frames is cast in standard moulds under factory conditions. The comparatively small spans, limited sizes and bulky nature of the frames resulted in this method being used much less than steel. The advantage of concrete is its good fire resistance and relative freedom from maintenance.

For convenience of casting, transportation and erection on site, precast reinforced concrete portal frames are usually cast in two or more sections, which are bolted together on site at the point of contraflexure in the rafters and/or at the junction of post and rafter (Figure 4.32). The portal frames are typically spaced between 4.5 and 6 m apart to support pre-cast reinforced concrete purlin and sheeting rails. Alternatively timber or cold formed steel Zed purlins and sheeting rails may be used. The bases of the concrete portals are placed in mortices cast in concrete foundations and grouted in position. Alternatively base plates can be used in the same way that they are used in steel portal frames. The base plate is welded to the reinforcement and cast into the foot of the concrete frame at the same time as the rest of the precast frame. The clear span for standard single-bay structures may be up to 24 m, as shown in Figure 4.32.

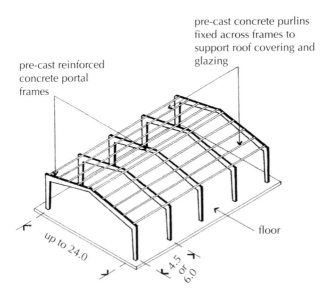

Figure 4.32   Single-bay symmetrical pitch portal frames.

Figure 4.33 is an illustration of a two-bay symmetrical pitch concrete portal frame. In this example the rafter is bolted to the post at the point of contraflexure. The internal posts are shaped to accommodate a precast reinforced concrete valley gutter, which is bolted to the rafters and laid to a fall. The concrete purlins are fixed by loops protruding from their ends, which fit over studs cast in the rafters, as shown in Figure 4.34.

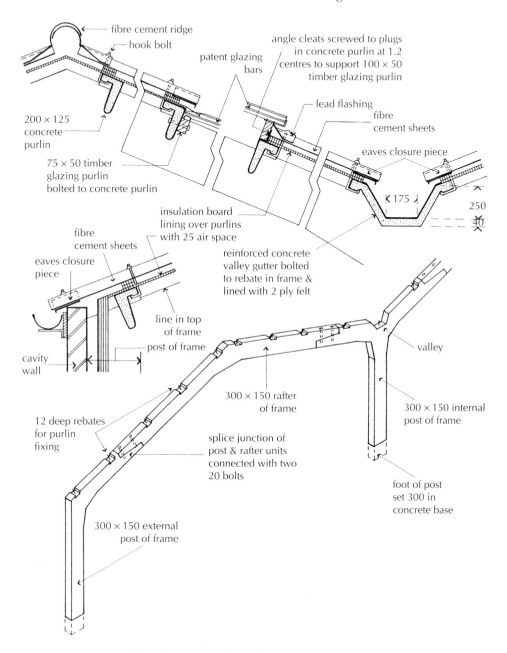

fibre cement ridge

hook bolt

patent glazing bars

angle cleats screwed to plugs in concrete purlin at 1.2 centres to support 100 × 50 timber glazing purlin

lead flashing

fibre cement sheets

200 × 125 concrete purlin

eaves closure piece

75 × 50 timber glazing purlin bolted to concrete purlin

K 175 ⌐

250

40

insulation board lining over purlins with 25 air space

fibre cement sheets

reinforced concrete valley gutter bolted to rebate in frame & lined with 2 ply felt

eaves closure piece

line in top of frame

post of frame

valley

cavity wall

300 × 150 rafter of frame

300 × 150 internal post of frame

12 deep rebates for purlin fixing

splice junction of post & rafter units connected with two 20 bolts

foot of post set 300 in concrete base

300 × 150 external post of frame

Figure 4.33 **Two-bay symmetrical pitch reinforced concrete portal frame.**

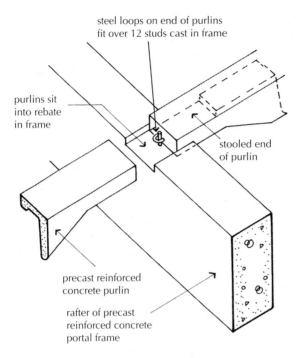

steel loops on end of purlins
fit over 12 studs cast in frame

purlins sit
into rebate
in frame

stooled end
of purlin

precast reinforced
concrete purlin

rafter of precast
reinforced concrete
portal frame

Figure 4.34   Connection of concrete purlins to concrete portal frame.

Timber portal frames

In the middle of the 20th century the technique of gluing timber laminae improved dramatically with the development of powerful, waterproof, synthetic resin adhesives. Later improvements in the technique of selecting wood of uniform properties and gluing laminations together under stringent quality control led to the development of factories in North America and Europe capable of producing laminated timber sections suitable for use in buildings in lieu of steel and reinforced concrete for all but the more heavily loaded structural elements.

Glulam

Glulam is the name that has been adopted for the product of a system of making members such as beams and roof frames from laminae of natural wood glued together to form longer lengths and shapes than is possible with natural wood by itself. Glulam is defined as a structural member made from four or more separate laminations of timber arranged with the grain parallel to the longitudinal axis of the member: the individual pieces being assembled with their grain approximately parallel and glued together to form a member which functions as a single structural unit. The advantage of glulam is that both straight and curved sections can be built up from short, thin sections of timber glued together in long sections, up to 50 m, without appreciable loss of the advantageous properties of the natural wood from which they were cut.

A range of standard glulam straight roof and floor beams are produced in a variety of sizes up to 20 m long and 4.94 m deep. These beams can be cut, holed and notched in the same way as the timber from which they were made. A wide range of purpose-made portal frames, flat pitched and cambered roof beams and arched glulam structures is practical where the curved forms and natural colour and grain can be displayed and where medium to wide clear spans are required.

Glulam structural members
Because of the labour costs involved in the fabrication of glulam members, glulam cannot compete with any of the basic steel frames in initial cost. However, glulam comes into its own in one-off, purpose-designed, medium-span buildings where the durability of glulam and the appearance of natural wood are an intrinsic part of the building design, for example in sports halls, assembly halls and swimming pools (Photograph 4.4). The advantages of timber in this form as a structural material are its low self-weight, minimal maintenance requirements to preserve and maintain its strength, and that it does not suffer from corrosion. Such properties are particularly important where there are levels of high humidity as in swimming pools.

Timber laminae are mostly cut from European white wood, imported from Scandinavia. The knots in this wood are comparatively small, it is widely available in suitable strength grades, has excellent gluing properties and a clear, bright, light creamy colour. The stress (strength) grades are LA, LB and LC, with LC being the weakest of the three. Glulam members are usually composed of LB and LC grades or a combination of LB outer and LC inner laminates. The wood is cut into laminae up to 45 mm finished thickness for straight members and as thin as 13 mm for curved members. Laminates are kiln dried to a moisture content of 12%. Individual lengths of timber are finger jointed at the butt end. The ends of the laminae are cut or stamped to form interlocking protruding fingers that are 50 mm long. The lengths of the end jointed laminates are planed to the required thickness and a waterproof adhesive is applied to the faces to be joined. The adhesive used is, like the wood it is used to bond, resistant to chemical attack in polluted atmospheres and chemical solutions.

The adhesive coated laminates are assembled in sets to suit the straight or curved section member they will form. Before the adhesive hardens, the laminates for curved members are pulled around steel jigs to form the shape required. Both straight and curved sets are hydraulically cramped up until the adhesive is hardened. After assembly the glulam members are cured in controlled conditions of temperature and humidity to the required moisture content. The surfaces of the straight members are then planed to remove adhesive that has been squeezed out and to reduce the section to its required dimensions and surface finish. Curved members are made oversize. The staggered ends of laminae are then cut to the required outside and inside curvature and the faces are then planed in the same manner as that for the straight

Photograph 4.4   Glue-laminated timber structure.

members. The planed natural finish of the wood is usually left untreated to expose the natural colour and grain of the wood.

Timber decking can be used to serve as a natural wood finish to ceilings between glulam frames and rafters and as solid deck to support the roof covering and thermal insulation. The decking is laid across and screwed or nailed to roof beams and portal frames.

Symmetrical-pitch glulam timber portal frame

These portal frames are usually fabricated in two sections for ease of transportation to the site. They are erected and bolted together at the ridge as illustrated in Figure 4.35. The portals are spaced fairly widely apart to support timber or steel purlins, which can be covered with sheet cladding materials, slates or tiles. For the sake of appearance timber decking is used to provide a soffit of natural timber. For buildings that require heating the thermal insulation is placed above the timber soffit. The laminations of the timber from which the portal is made are arranged to taper so that the depth is greatest at the knee, where the frame tends to spread under load and where the depth is most needed. The portal is more slender at the apex and at the base of the post where the least

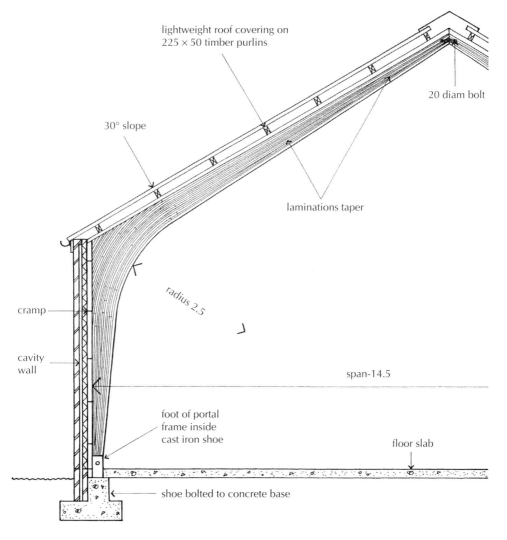

Figure 4.35   Glued laminated timber portal frame.

section is required for strength and rigidity. The maximum radius of curve for shaped members is governed by the thickness of the laminates. A maximum radius of 5625 mm is recommended for 45 mm and 2500 mm for 20 mm thick laminae. Because of the labour involved in the assembly of curved members, they are appreciably more expensive than straight members.

Flat glulam timber portal frame

The flat portal frame illustrated in Figure 4.36 is designed for the most economic use of timber and consists of a web of small section timbers glued

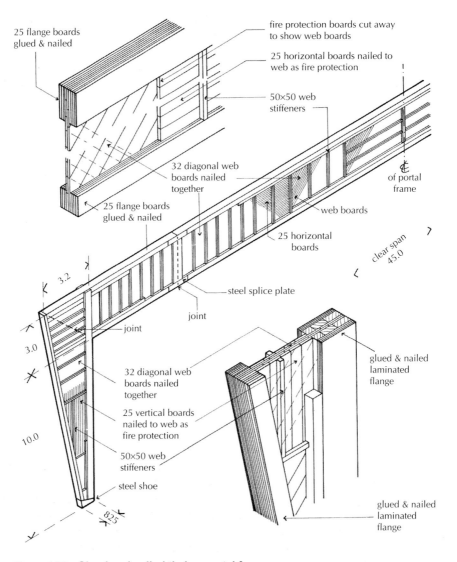

Figure 4.36   Glued and nailed timber portal frame.

together with the top and bottom booms of glued laminate with web stiffeners. The portal frames are used to support metal decking on the roof and profiled sheeting on the walls. This long-span structure is lightweight and free from maintenance.

Flat roof frame construction

Medium- and long-span flat roof structures are less efficient structurally and therefore less economic than truss, lattice or portal frames. The main reason for this is the need to prevent too large a deflection of the flat roof structure under load, thus leading to ponding of water on the surface of the roof. The advantage of a flat roof is that there is little unused roof space to be heated, compared with pitched roofs. Solid web I-section steel beams supported by steel columns may be used for industrial applications where the main beams are used to support lifting gear, but the most common form of framed flat roof construction is with lattice beam or with space frames.

Lattice beam flat roof construction

The terms beam and girder are used in a general sense to describe lattice construction. The term 'beam' is used to describe small depths associated with most roof construction and 'girder' for deeper depths associated with, for example, bridge construction. For flat and low pitch roofs it is convenient to fabricate the top boom to provide a fall for the roof decking or sheeting. Lattice beams are either hot dip galvanised, stove enamel primed or spray primed after manufacture.

Short-span beams that support relatively light loads may be constructed from cold-formed steel strip top and bottom booms with a lattice of steel rods welded between them, as illustrated in Figure 4.37. The top and bottom booms are formed as 'top hat' sections designed to take timber inserts for fixing roof decking and ceiling finishes.

The majority of lattice beams used for flat and low pitch roofs are fabricated from hollow round and rectangular steel sections. For most low pitch roofs to be covered with profiled sheeting a slope of 6° is provided, as illustrated in Figure 4.38.

Space grid flat roof construction

Where there is a requirement for a large unobstructed floor area, such as exhibition areas and sports halls, a space deck roof can be used (Figures 4.39 and 4.40). A two-layer space deck comprises a grid of standard prefabricated units, each in the form of an inverted pyramid as illustrated in Figure 4.41. The units are bolted together and connected with tie bars to form the roof structure. The tie bars can be adjusted to create an upward camber to the top deck to allow for deflection under load and also to provide a fall to the roof to encourage

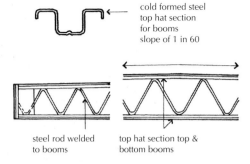

cold formed steel
top hat section
for booms
slope of 1 in 60

steel rod welded
to booms

top hat section top &
bottom booms

Shallow pitch lattice beam

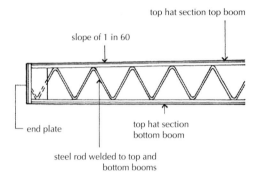

top hat section top boom

slope of 1 in 60

end plate

top hat section
bottom boom

steel rod welded to top and
bottom booms

Figure 4.37   Tapered lattice beam.

Six degree dual pitch lattice beam

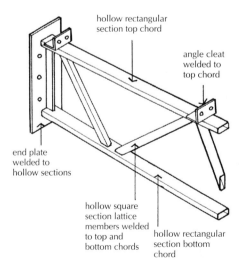

hollow rectangular
section top chord

angle cleat
welded to
top chord

end plate
welded to
hollow sections

hollow square
section lattice
members welded
to top and
bottom chords

hollow rectangular
section bottom
chord

Figure 4.38   Lattice beam.

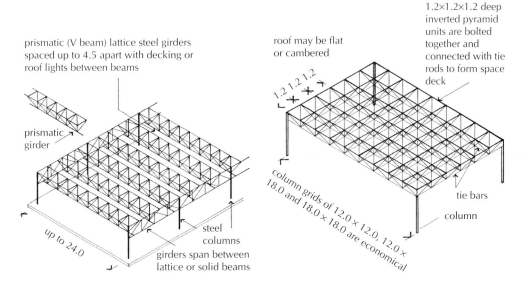

prismatic (V beam) lattice steel girders spaced up to 4.5 apart with decking or roof lights between beams

prismatic girder

up to 24.0

steel columns

girders span between lattice or solid beams

Figure 4.39 Prismatic (V beam) lattice steel roof on steel columns.

roof may be flat or cambered

1.2 1.2 1.2

1.2×1.2×1.2 deep inverted pyramid units are bolted together and connected with tie rods to form space deck

tie bars

column

column grids of 18.0 and 18.0 × 18.0 are economical 12.0 × 12.0, 12.0 × 18.0

Figure 4.40   Steel space deck root.

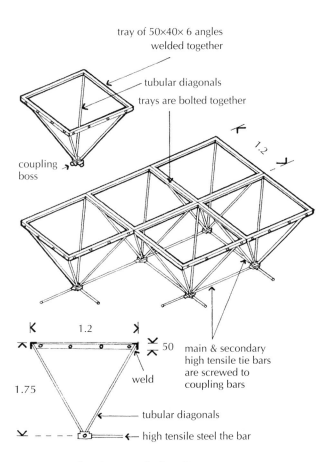

tray of 50×40× 6 angles welded together

tubular diagonals

trays are bolted together

1.2

coupling boss

1.2

50   main & secondary high tensile tie bars are screwed to coupling bars

weld

1.75

tubular diagonals

high tensile steel the bar

Figure 4.41   Steel space deck units.

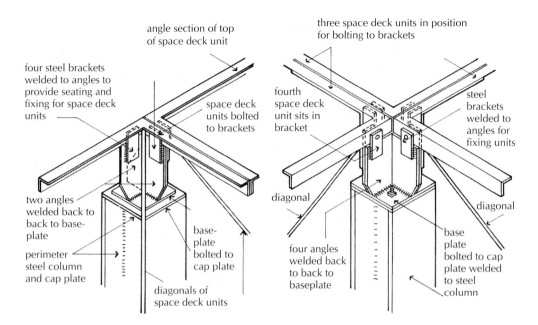

Figure 4.42 Support and fixing of space deck units to perimeter steel columns.

Figure 4.43 Connection of space deck units to an internal column.

rainwater to discharge to gutters and thus avoid ponding. The roof of the structural deck may be covered with thermal insulation and steel decking or sheeting. Rooflights can be easily accommodated within the standard units and the roof can be cantilevered beyond supporting perimeter columns to provide an overhang.

Space deck roofs may be designed as a two-way spanning structure with a square grid, or as a one-way spanning structure with a rectangular grid. Economic grid sizes are 12 × 12 m, 18 × 18 m and 12 × 18 m. The main advantage of the space deck roof is the wide spacing of the supporting columns and the economy of the structure in the use of standard units and the speed of erection. One disadvantage is that the members tend to attract dust and will require regular cleaning. Regular maintenance is also required to prevent rust.

Units are usually connected to the supporting steel columns at the junction of the trays of the units. Figures 4.42 and 4.43 illustrate the fixing of a space deck to perimeter and internal columns respectively. At perimeter columns a steel cap plate is welded to the cap of the column to which a seating is bolted. This seating of steel angles has brackets welded to it into which the flanges of the trays fit and to which the trays are bolted. Similarly, a seating is bolted to a cap plate of internal columns with brackets into which the flanges of the angles of four trays fit.

Composite frame construction

A composite frame construction comprises prefabricated concrete and steel components, usually offered by one supplier as part of a design, manufacture and erection service for both single- and multi-storey framed buildings. Precast reinforced concrete structural beams and columns are used to support lattice steel roof beams. The columns and beams are precast under carefully controlled factory conditions, with frame joints and base fixings etc. cast in as necessary. The advantage of the composite frame construction is that the reinforced concrete columns and beams provide good fire resistance to the main structure and the lattice steel roof provides a lightweight covering. Economy of initial build cost can be made in the extensive use of prefabricated units.

Figure 4.44 is an illustration of a typical two-bay, single-storey composite frame structure. The precast reinforced concrete columns, which have fixed bases, serve as vertical cantilevers to take the major part of the loads from wind pressure. Steel brackets, cast into the column head, support the concrete and lattice steel roof beams. Concrete or lattice steel spine beams are used under the roof valley to provide intermediate support for every other roof beam. The

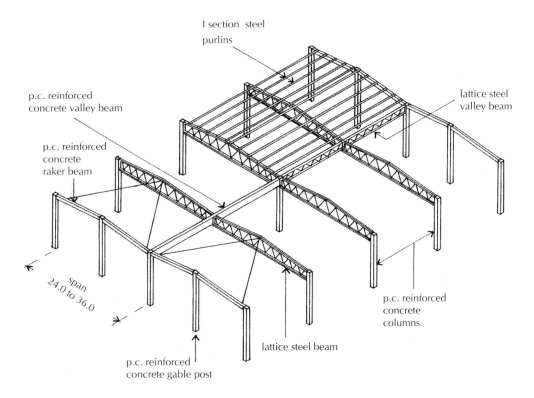

Figure 4.44 Two-bay single-storey composite frame.

top of the lattice steel roof beams, which are pitched at 6° to the horizontal, support the low pitch profiled steel roof sheeting. Fixing slots or brackets cast into the columns provide a fixing and support for sheeting rails, which in turn support the profiled steel cladding to the walls.

4.2 Roof and wall cladding, and decking

Today plastic coated profiled steel sheeting is the principal sheet material used to provide weather protection to single-storey framed buildings. Laminated panels that incorporate thermal insulation are also available (see Chapter 7). Corrugated asbestos cement sheet was also used until health and safety concerns led to its replacement with fibre cement sheet.

Functional requirements

The functional requirements of roofs and walls have already been set out in *Barry's Introduction to Construction of Buildings*. In relation to wall and roof cladding the following functional requirements need to be addressed:

❏ Strength and stability
❏ Resistance to weather
❏ Durability and freedom from maintenance
❏ Safe access during maintenance
❏ Fire safety
❏ Resistance to passage of heat
❏ Resistance to passage of sound
❏ Security
❏ Aesthetics

Strength and stability

The strength of roof and wall cladding and roof decking depends on the properties of the materials used and their ability to support the self-weight of the cladding and the imposed loads of wind and snow between the supporting purlins, rails, bearers and beams. The stability of the cladding and decking depends on the:

❏ Depth and spacing of the profiles of sheeting and decking
❏ Composition of the materials and thickness of the boards and slabs used for decking
❏ Ability of the materials to resist distortion due to wind pressure, wind uplift, snow loads and the weight of personnel engaged in fixing and maintaining the roofs

The strength and stability of the thin sheets of steel or aluminium derive principally from the depth and spacing of the profiles: shallow depth of profile for small spans to deep trapezoidal profiles and standing seams for medium to large spans between supports. Longitudinal and transverse ribs provide additional rigidity against buckling to deep profile sheeting. The comparatively thick corrugated and profiled fibre cement cladding sheets have adequate strength in depth of the profiles for anticipated loads and rigidity in the material to resist distortion and loss of stability over moderate spans between supports. Steel roof cladding sheets fixed across a structural frame act as a diaphragm, which contributes to the stability of the frames in resisting the racking effect of lateral wind forces that act on the sides and roofs of buildings. The extent of the contribution to the stability of the frames depends on the thickness of the sheets and the strength of the fasteners used to fix the sheets, as well as the ability of the sheets to resist the tearing effect of the fasteners fixed through it. Manufacturers provide guidance on the size and thickness of their sheets, minimum end lap, maximum purlin and rail centres, and maximum unsupported overhang of the sheets, as well guidance on the type and spacing of fixings to match the exposure of the site.

Resistance to weather
Sheet steel and aluminium cladding resists the penetration of rainwater through the material's impermeability to water and the ability of the side and end laps to keep water out. The lowest allowable pitch of the roof is dictated by the end lap of the sheets. Thermal and structural movement is accommodated by the profiles, the end lap and designed tolerances at the fixings. Where long sheets are used, the secret fixing of the standing seam will allow for movement. Profiled metal sheets are usually fixed with screws, driven through the sheets into steel purlins and rails. Integral steel and Neoprene washers on the screw head effectively seal the perforation of the sheet against water penetration. Fixing is through the troughs of the profiles (where the rainwater runs) or (preferably) at the ridge of the trough, which takes a little more care and skill in the fixing of the sheets. Top fixing is preferred to bottom fixing because the perforation of the sheet is less exposed to water. Profiled cladding for walling is usually fixed through the troughs of the profile for ease of fixing and where the screw heads will be least visible. Standing seams to the edges of long sheets provide a deep upstand as protection against rain penetration, particularly with very low pitch roofs.

Fibre cement sheets will resist water penetration through the density of the material, the slope of the roof and the end laps. The sheets will absorb some rainwater and should be laid at a pitch of 10° or more to avoid the possibility of frost damage. The sheets will accommodate moisture, thermal and structural movement through the end and side laps, as well as through the relatively large fixing holes for screws or hook bolts.

Flat roof membranes which resist the penetration of rainwater through the impermeability of the two-, three- or single-ply membranes and the sealed joints will, in time, harden and no longer retain sufficient elasticity or tensile strength to resist the thermal movements common to flat roof coverings laid over insulation materials.

Durability and freedom from maintenance

Coated profiled steel sheeting is easily damaged and so its durability depends to a certain extent on the care in handling and fixing on the building site. Damage to protective coatings can lead to corrosion of exposed steel, especially around fixing holes, and fixings driven home too tightly can easily distort the thin metal. Durability also depends on the climate and the colour of the coating material. Sheeting on buildings close to marine environments and in polluted industrial areas will deteriorate more rapidly than those in more sheltered, less polluted areas. Sheeting with light coloured coatings tends to be more durable than dark coatings due to the effect of ultraviolet light on dark hues and the increased heat released from solar radiation on the more absorptive dark coatings. Organic coated sheeting is a relatively short-lived material with a service life of around 25 years in favourable conditions and as low as 10 years in more aggressive climates.

Fibre cement sheeting does not corrode or deteriorate for many years provided it is laid at a sufficiently steep pitch to shed water. The material is, however, relatively brittle and is liable to damage from knocks and undue pressure from people accidentally walking over its surface. Reinforced fibre cement sheets are available that have a higher impact strength. These sheets tend to attract dirt because of the coarse texture of the surface, which is not easily washed away and thus the sheets can become unsightly quite quickly.

Flat roof membranes, laid directly over thermal insulation material, will suffer considerable temperature variations between day and night. In consequence there is considerable expansion and contraction of the membrane, which in time may cause the membrane to tear. Solar radiation also causes oxidation and brittle hardening of bitumen saturated or coated materials, which in time will no longer be impermeable to water. The durability of a roofing membrane in an inverted roof (upside down roof) is much improved by the layer of thermal insulation laid over the membrane, which helps protect it from the destructive effects of solar radiation and less extreme variations in temperature. The useful life of bitumen impregnated felt membranes is from 10 years, for organic fibre felts up to 20 years and for high performance felts up to 25 years: this can be extended by using an inverted roof construction. Mastic asphalt will oxidise and suffer brittle hardening over time, which, combined with thermal movements, will give the material a useful life of around 20 years.

Safe maintenance

The Construction (Design and Management) Regulations (CDM Regulations) 1994 require that buildings should be designed so that they can be constructed, maintained and demolished safely. One in five construction-related accidents are caused by falls from, or through, roofs (HSE, 1998). Accidents occur during both construction and maintenance operations. Roofs pose a real hazard and care should be taken when designing structures to ensure that falling through sheeting materials and from the roof is recognised as a hazard and the risks of such occurrence are reduced. Provision should be made to prevent falls, including adequate access for plant and equipment. Safety rails should be used to prevent falls over the edges of roof structures. Harnesses, fall arrest systems and safety nets do not prevent falls but do reduce the risks of injury in the event of a fall. Such equipment should be used as a secondary mechanism to reduce the risks of injury; the primary mechanism is to prevent the fall. Inclement weather poses a significant risk to those working in exposed positions and at heights. Work at heights should not continue during high winds or conditions that make the risks unacceptable. Debris netting (as well as safety netting) or birdcage scaffolds may be used to offer protection from falling objects and allow work to continue in the zone below the roof area. Debris shoots should also be used to ensure that waste, which presents a hazard if it falls, is quickly removed from the roof. Consideration must be given to maintenance operations once the roof structure is complete. Guarded walkways, access platforms, safety rails etc. will be needed to ensure safe access.

Fire safety

Particular attention should be given to the internal and external fire spread characteristics of sheet materials in relation to the overall design of the building. A further cause for concern in framed buildings is concealed spaces, such as voids above suspended ceilings, roof and wall cavities. Cavity barriers and smoke stops should be fitted in accordance with current regulations.

Resistance to passage of heat and ventilation

Resistance to the passage of heat is provided by thermal insulation materials, either separate from the sheeting material or as an integral part of the sheet in composite panels. Consideration must be given to thermal bridging in steel framed buildings, especially at junctions, and care is required to avoid condensation. The principles of condensation, or rather the manner in which it can be avoided within the roof and wall structures, was discussed in *Barry's Introduction to Construction of Buildings*. Sheet metal may, in time, suffer corrosion from heavy condensation on the underside of the sheet. Ventilation to the space between the sheeting and the insulation, combined with a vapour check to the lining sheets, is the most effective way of minimising the risk of condensation. Fibre cement sheet is permeable to water vapour and thus provides less of a risk from condensation.

Resistance to passage of sound

The thin metal skin of profiled metal sheeting affords no appreciable resistance to sound penetration, thus insulation must be provided via the thermal insulation materials and effective seals around the opening parts of doors and windows. Mineral based insulation materials reduce the passage of sound. If sound insulation is a primary performance requirement it may be advantageous to adopt a denser form of enclosure, such as brick or concrete to help provide the necessary sound reduction. Fibre cement sheets provide much better airborne and impact sound insulation.

Security

Many single-storey framed buildings are only occupied during working hours and are vulnerable to damage by vandalism and forced entry, unless adequately protected through passive and active security measures. Apart from the obvious risk of forced entry through doors, windows and rooflights there is a risk of entry by prising thin profiled sheeting from its fixing and so making an opening large enough to enter. Given that many buildings clad with steel sheeting are for warehousing purposes, this presents a serious challenge to the owners. Where the cost of the goods contained within is high and the likelihood of theft also high, it is wise to use a more solid form of wall construction, such as brick. Roofs are more difficult to protect, and some form of secondary protection is often used, such as a secondary steel cage under the roof (this is outside the scope of this book). Local police crime prevention officers can provide advice to the design team.

Aesthetics

Choice of an appropriate cladding for the building frame will also be determined by the appearance of the sheeting used and its ability to stand up to weathering. Sheet profile and colour will be primary concerns. Curved sheeting is available from most manufacturers and can help to add some interest to otherwise mundane structures. The minimum radius of the internal curve of the sheet will vary according to the profile selected.

Profiled steel sheeting

The advantages of steel as a material for roof and wall sheeting are that its favourable strength-to-weight ratio and ductility make it both practical and economic to use comparatively thin, lightweight sheets that can be cold roll formed to profiles with adequate strength and stiffness (Figures 4.45 and 4.46). The disadvantage of steel as a sheeting material is that it suffers rapid and progressive corrosion unless protected. The corrosive process is a complex electrochemical action that depends on the characteristics of the metal, atmosphere and temperature and is most destructive in conditions of persistent moisture, atmospheric pollution and where different metals are in contact. Typically steel is protected with a zinc coating by the hot-dip galvanising process, which is both effective and economical.

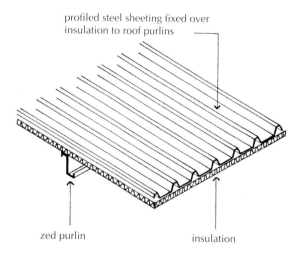

profiled steel sheeting fixed over
insulation to roof purlins

zed purlin   insulation

Figure 4.45   Cladding.

Organic (plastic) coated profiled steel sheets

The majority of profiled sheets used today are coated with an organic plastic coating to provide a protective coating and to provide an attractive finish. The plastic coating is applied to the galvanised zinc-coated steel sheets to serve as a barrier to atmospheric corrosion of the zinc, the erosive effect of wind and rain, and some degree of protection to damage during handling, fixing and in use. Colour is applied to the coated steel sheets by the addition of pigment to the coating material, and a range of colours is available. There will be loss of colour, which tends not to be uniform over the whole sheet, especially on south-facing slopes over time. This spoils the appearance of the building, and cladding sheets may need to be replaced long before there is any danger of corrosion of the steel sheet. Light colours tend to exhibit better colour retention than darker colours. Four organic coatings are available, as described below.

Polyvinyl chloride coatings – uPVC

This is the cheapest and most used of the organic plastic coatings (known as 'plastisol'). The comparatively thick (200 microns) coating that is applied over the zinc coating provides good resistance to normal weathering agents. The material is ultraviolet stabilised to retard the degradation by ultraviolet light and the inevitable loss of colour. The durability of the coating is good as a protection for the zinc coating below but the life expectancy of colour retention is between 10 and 20 years. Polyvinyl chloride is an economic, tough, durable, scratch-resistant coating but has poor colour retention.

Acrylic-polymethyl methacrylate – PMMA

This organic plastic is applied with heat under pressure as a laminate to galvanised zinc steel strip to a thickness of 75 microns. It forms a tough finish

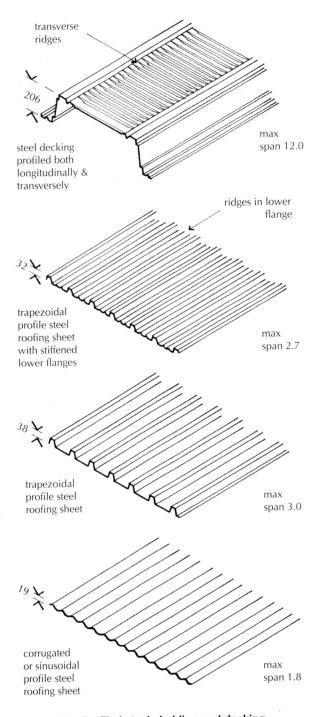

transverse ridges

206

steel decking profiled both longitudinally & transversely

max span 12.0

ridges in lower flange

32

trapezoidal profile steel roofing sheet with stiffened lower flanges

max span 2.7

38

trapezoidal profile steel roofing sheet

max span 3.0

19

corrugated or sinusoidal profile steel roofing sheet

max span 1.8

Figure 4.46   Profiled steel cladding and decking.

with high strength, good impact resistance and good resistance to damage in handling, fixing and in use. It has excellent chemical resistance and its good resistance to ultraviolet radiation gives a life expectancy of colour retention of up to 20 years. The hard smooth finish of this coating is particularly free from dirt staining. It costs about twice as much as uPVC.

Polyvinylidene fluoride – PVF

A comparatively expensive organic plastic coating for profiled steel sheets, which is used as a thin (25 micron) coating for its excellent resistance to weathering, excellent chemical resistance, durability and resistance to all high-energy radiation. Because the coating is thin, careless handling and fixing may damage it. Durability is good and colour retention can be from 15 to 30 years.

Silicone polyester

This is the cheapest of the organic coatings used for galvanised steel sheet. It has a short life, of between 5 and 7 years in a temperate, non-aggressive climate. Galvanised sheets are primed and coated with stoved silicone polyester to a thickness of 25 microns. The coating provides reasonable protection against damage in handling and fixing.

Profiled steel cladding systems for roofs and walls

The term cladding is a general description for materials, such as steel sheets, used to clothe or clad the external faces of framed buildings to provide weather protection. Thermal insulation is fixed under or behind the cladding sheets to provide the required thermal insulation to roofs and walls respectively. A wide range of profiles are available, some of which are illustrated in Figure 4.47.

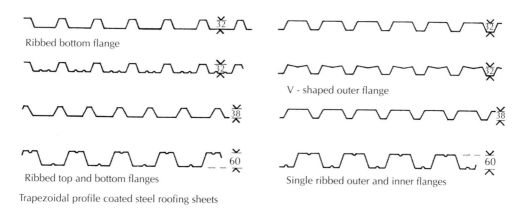

Ribbed bottom flange

V - shaped outer flange

Ribbed top and bottom flanges

Single ribbed outer and inner flanges

Trapezoidal profile coated steel roofing sheets

Figure 4.47   Trapezoidal profile coated steel wall cladding sheets.

Single skin cladding
The simplest system of cladding consists of a single skin of profiled steel sheeting fixed directly to purlins and sheeting rails without thermal insulation. This cheap form of construction is only used for buildings that do not need to be heated, such as warehouses and stores.

Over purlin insulation
The most straightforward and economic system of supporting insulation under cladding is to use semi-rigid or rigid insulation boards laid across roof purlins and sheeting rails as shown in Figure 4.48. Timber spacers are used to provide an air space for passive ventilation between the cladding sheets and the insulation. This system of cladding is suitable for buildings with low to medium levels of humidity and where the appearance of the insulation board is an acceptable finish.

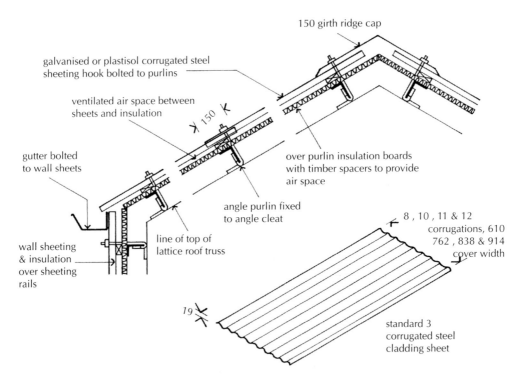

Figure 4.48   Corrugated steel cladding sheets.

Over purlin insulation with inner lining
Where mineral fibre mat insulation is used and where more rigid forms of insulation will not be self-supporting between widely spaced purlins, it is

necessary to use profiled inner lining sheets (or trays) to provide support for the insulation. The lining sheets also help to provide a more attractive finish to the interior.

Linings are cold, rolled formed, steel strips with shallow depth profiles adequate to support the weight of the insulation. The sheets are hot dip galvanised and coated with a protective and decorative organic plastic coating. To prevent compression of the loose mat or quilt, its thickness is maintained by Zed section spacers fixed between cladding and lining panels as illustrated in Figures 4.49 and 4.50. The space between the top of the sheeting and the insulation is passively ventilated to minimise condensation, and a breather paper is usually spread over the top of the insulation. The breather paper protects the insulation from any rain or water condensate yet allows moisture vapour to penetrate it. Some manufacturers also manufacture 'structural' trays, which provide a stronger internal lining and thus help to improve security to the roof.

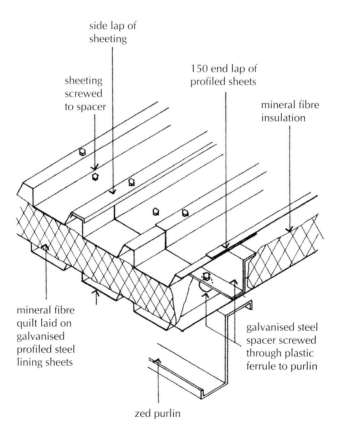

side lap of
sheeting

sheeting
screwed
to spacer

150 end lap of
profiled sheets

mineral fibre
insulation

mineral fibre
quilt laid on
galvanised
profiled steel
lining sheets

galvanised steel
spacer screwed
through plastic
ferrule to purlin

zed purlin

Figure 4.49   Over purlin insulation with inner lining sheets.

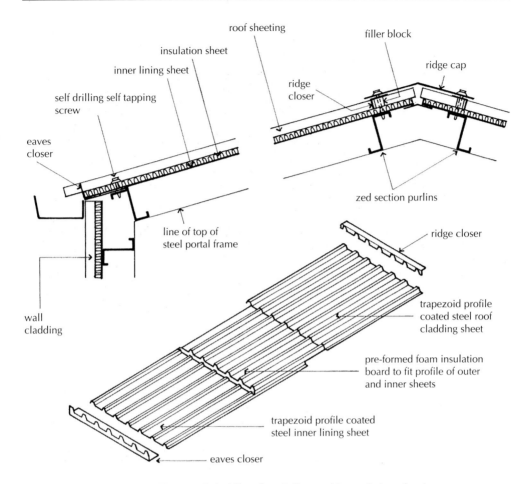

Figure 4.50   Profiled steel cladding, insulation and inner lining sheets.

Over purlin composite (site assembled)

This system comprises a core of rigid preformed lightweight insulation (or mineral wool and spacer), shaped to match the profile of the sheet and the inner lining tray. The separate components are assembled on site and fixed directly to purlins and lining sheets with self-tapping screws (Photograph 4.5). Side and end laps are sealed against the penetration of moisture vapour. Factory formed composite panels have largely replaced this system.

Over purlin composite (factory formed)

Factory formed composite panels consist of a foamed insulation core enclosed and sealed by profiled sheeting and inner lining tray. The two panels and their insulating core act together structurally to improve loadbearing characteristics.

Photograph 4.5   Site assembled over purlin roof cladding (courtesy of G. Throup).

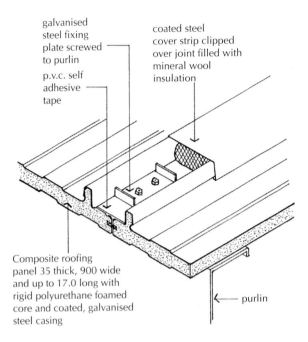

galvanised
steel fixing
plate screwed
to purlin

p.v.c. self
adhesive
tape

coated steel
cover strip clipped
over joint filled with
mineral wool
insulation

Composite roofing
panel 35 thick, 900 wide
and up to 17.0 long with
rigid polyurethane foamed
core and coated, galvanised
steel casing

← purlin

Figure 4.51   Composite roofing panel.

Panels have secret fixings to improve their visual appearance. Figure 4.51 is an illustration of factory formed panels.

Standing seams

Standing seams are principally used for low and very low pitched roofs to provide a deep upstand as weathering to the side joints of sheeting and to allow space for secret fixings. Sheets usually run from ridge to eaves to avoid the complication of detailing at the end laps with standing seams. The standing seam allows some tolerances for thermal movement of the long sheets and also provides some stiffness to the sheets, thus allowing a shallower profile to be used. Figure 4.52 illustrates a standing seam.

Fasteners

Steel cladding, lining sheets and spacers are usually fixed with coated steel or stainless steel self-tapping screws, illustrated in Figure 4.53. The screws are mechanically driven through the sheets into purlins or spacers. These primary fasteners for roof and wall sheeting may have coloured heads to match the colour of the sheeting. Secondary fasteners, which have a shorter tail, are used for fixing sheet to sheet and also flashing to sheet.

Gutters

Gutters are usually made from cold formed organic coated steel, and are laid at a slight fall to rainwater pipes. Gutters are supported on steel brackets screwed

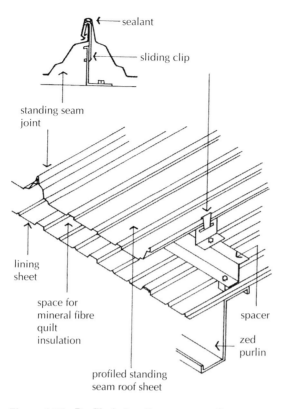

Figure 4.52   Profiled standing seam roofing.

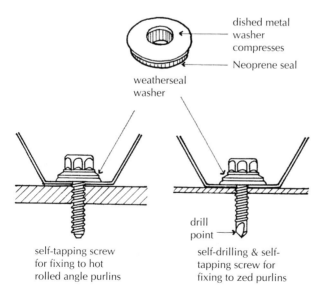

Figure 4.53   Fasteners for profiled steel sheeting and decking.

to eaves purlins. Valley gutters and parapet wall gutters usually have the inside of the gutter painted with bitumen as additional protection against corrosion.

Ridges

Ridges are covered with a cold formed steel strip that is coated to provide the same finish as the roof sheeting. The ridge may be profiled to match the roof profile, or flat with a shaped filler piece to seal the space between sheet and ridge.

Wall cladding

Profiled steel sheeting is usually fixed to walls with the profile vertical, for convenience of fixing to horizontal sheeting rails fixed across the columns. Horizontal fixed sheeting can also be used for a different appearance, although some additional steel support may be required for widely spaced columns. The wall cladding is usually the same profile as that used for the roof. Figures 4.54 and 4.55 illustrate a typical section through a steel framed building with steel sheeting above a lower wall of masonry. A drip flashing helps to keep the top of the wall dry by shedding the rainwater as it runs down the sheets. To provide a flush soffit to the roof cladding, the inner lining and insulation can be fitted under the purlins between the roof frames as illustrated in Figure 4.56.

Profiled aluminium roof and wall cladding

On exposure to the atmosphere, aluminium corrodes to form a thin coating of oxide on its surface. This oxide coating, which is integral with the aluminium, adheres strongly and, being insoluble, protects the metal below from further corrosion so that the useful life of aluminium is 40 years or more. Aluminium is a lightweight, malleable metal with poor mechanical strength, which can be cold formed without damage. Aluminium alloy strip is cold rolled as corrugated and trapezoidal profile sheets for roof and wall cladding. The sheets are supplied as metal mill finish, metal stucco embossed finish, pre-painted or organically coated.

Mill finish is the natural untreated surface of the metal from the rolling mill. It has a smooth, highly reflective metallic silver grey finish, which dulls and darkens with time. Variations in the flat surfaces of the mill finish sheet will be emphasised by the reflective surface. A stucco embossed finish to sheets is produced by embossing the sheets with rollers to form a shallow, irregular raised patterned finish that reduces direct reflection and sun glare and so masks variations in the level of the surface of the sheets. A painted finish is provided by coating the surface of the sheet with a passivity primer and a semi-gloss acrylic or alkyd-amino coating in a wide range of colours. A two-coat PVF acrylic finish to the sheet is applied by roller to produce a low-gloss coating in a wide range of colours.

Aluminium sheeting is more expensive than steel sheeting and is used for its greater durability, particularly where humid internal atmospheres might cause

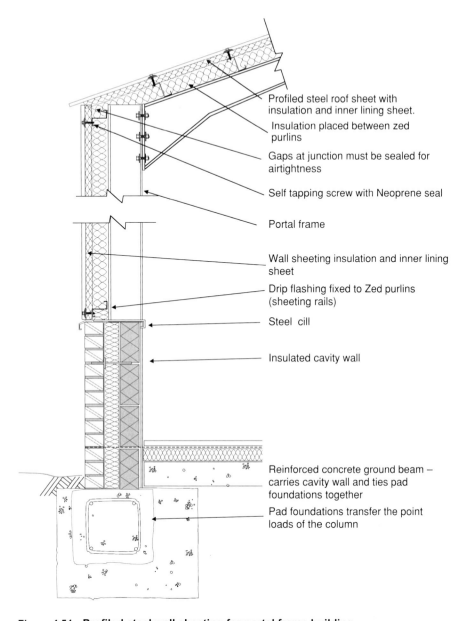

Profiled steel roof sheet with insulation and inner lining sheet.

Insulation placed between zed purlins

Gaps at junction must be sealed for airtightness

Self tapping screw with Neoprene seal

Portal frame

Wall sheeting insulation and inner lining sheet

Drip flashing fixed to Zed purlins (sheeting rails)

Steel cill

Insulated cavity wall

Reinforced concrete ground beam – carries cavity wall and ties pad foundations together

Pad foundations transfer the point loads of the column

Figure 4.54 Profiled steel wall sheeting for portal frame building.

early deterioration of coated steel sheeting. The material also offers some more interesting architectural features and has been used instead of steel sheet for its attractive natural mill finish. Figure 4.57 is an illustration of profiled aluminium roof and wall sheeting, fixed over rigid insulation boards bonded to steel lining trays, to a portal steel frame.

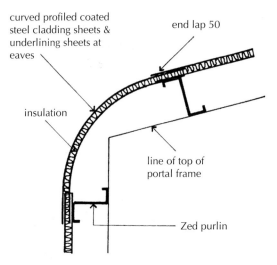

curved profiled coated
steel cladding sheets &
underlining sheets at
eaves

end lap 50

insulation

line of top of
portal frame

Zed purlin

Figure 4.55   Curved profiled steel cladding.

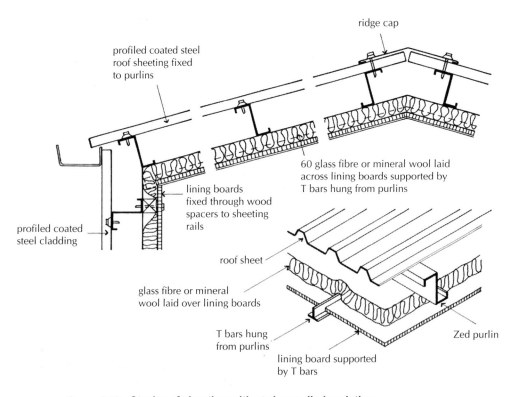

ridge cap

profiled coated steel
roof sheeting fixed
to purlins

60 glass fibre or mineral wool laid
across lining boards supported by
T bars hung from purlins

lining boards
fixed through wood
spacers to sheeting
rails

profiled coated
steel cladding

roof sheet

glass fibre or mineral
wool laid over lining boards

T bars hung
from purlins

lining board supported
by T bars

Zed purlin

Figure 4.56   Steel roof sheeting with under purlin insulation.

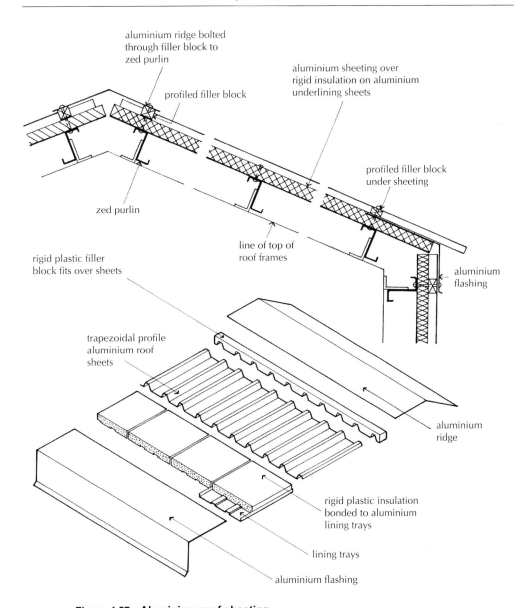

aluminium ridge bolted
through filler block to
zed purlin

aluminium sheeting over
rigid insulation on aluminium
underlining sheets

profiled filler block

profiled filler block
under sheeting

zed purlin

line of top of
roof frames

rigid plastic filler
block fits over sheets

aluminium
flashing

trapezoidal profile
aluminium roof
sheets

aluminium
ridge

rigid plastic insulation
bonded to aluminium
lining trays

lining trays

aluminium flashing

Figure 4.57   Aluminium roof sheeting.

Fibre cement profiled cladding

Original fibre cement corrugated sheets were manufactured using asbestos fibres mixed with water and cement. These sheets were used extensively in the UK from around 1910 until the late 1960s and early 1970s because of their low initial cost, durability and freedom from maintenance. Health and safety legislation led to the replacement of asbestos sheets with fibre reinforced cement sheets, both for use in replacement of asbestos sheets and for new work. These fibre cement sheets are manufactured from cellulose and polymeric fibres, cement and water, and pressed into a range of profiles. High strength fibre cement sheets are made with polypropylene reinforcement strips inserted along precisely engineered locations along the length of the sheet, which provides greater impact strength without affecting the durability of the product.

Sheets are usually finished in a natural grey colour, especially when used for industrial and agricultural buildings, although a range of natural colours and painted finishes are also available from some manufacturers. Fibre cement sheets are vapour permeable, which greatly reduces the risk of condensation. The sheets are a Class 0 material, provide excellent acoustic insulation, have a high level of corrosion resistance, are easy to fix and are maintenance free. Manufacturers provide guarantees for up to 30 years. The reinforced sheets should comply with the requirements for roof safety as set out by the Health and Safety Executive.

Fibre cement sheets are heavier than steel sheets and so require closer centres of support from purlins and sheeting rails. Corrugated fibre cement sheet may be pitched as low as 5° to the horizontal in sheltered locations, although upwards of 10° is more common. The detail shown in Figure 4.58 is typical of the type of construction used in unheated outbuildings such as garages and tool sheds clad with fibre sheets. Typical fixings for fibre cement sheets are illustrated in Figures 4.58 and 4.59. Figure 4.60 is a typical section through a steel structure with profiled fibre cement sheets, insulation and underlining sheets.

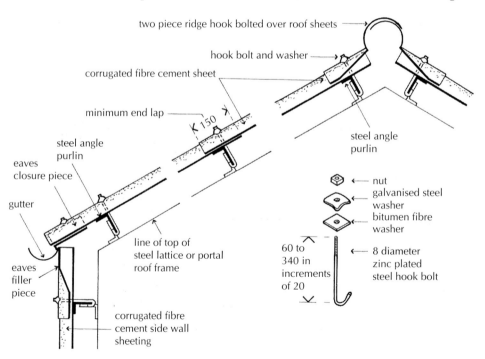

Figure 4.58   Corrugated fibre cement sheet covering to steel framed roof.

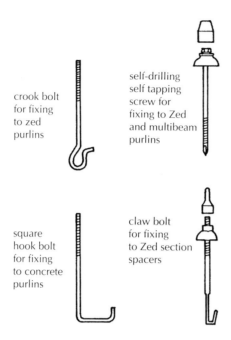

crook bolt for fixing to zed purlins

self-drilling self tapping screw for fixing to Zed and multibeam purlins

square hook bolt for fixing to concrete purlins

claw bolt for fixing to Zed section spacers

Figure 4.59   Fixings for fibre cement sheets.

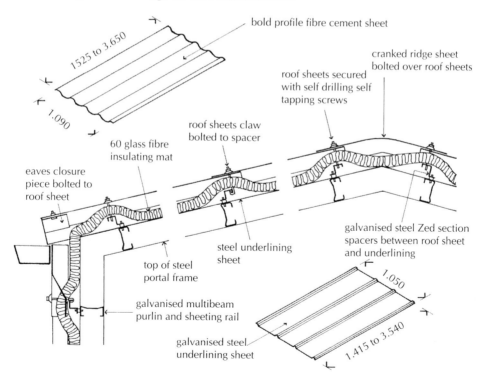

1525 to 3.650

1.090

bold profile fibre cement sheet

cranked ridge sheet bolted over roof sheets

roof sheets secured with self drilling self tapping screws

60 glass fibre insulating mat

roof sheets claw bolted to spacer

eaves closure piece bolted to roof sheet

galvanised steel Zed section spacers between roof sheet and underlining

steel underlining sheet

top of steel portal frame

galvanised multibeam purlin and sheeting rail

galvanised steel underlining sheet

1.050

1.415 to 3.540

Figure 4.60   Fibre cement sheeting with insulation and steel sheet underlining.

Manufacturers of fibre cement sheets offer bespoke systems that combine profiled fibre cement weathering sheets with thermal insulation and an underlining sheet of fibre cement or coated steel. These are offered with a proprietary support bar system, which both supports the roof cladding sheets and helps to maintain a clear cavity into which the insulation blanket is placed. The system is built up on site in accordance with the manufacturer's guidance to provide a highly durable roofing system with tested performance characteristics, giving very good acoustic and thermal insulation as well as high resistance to condensation. Recommended pitch ranges from between 5° and 30°. A typical system is illustrated in Figure 4.61.

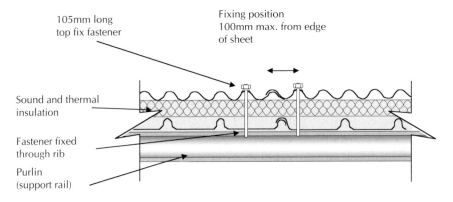

Figure 4.61   Fibre cement roofing sheets.

Roof ventilation to agricultural buildings

Fibre cement sheeting is used quite extensively in agricultural buildings, many of which have very specific ventilation requirements, for example cattle sheds or pig pens. A number of profiled prefabricated ridge fittings, including open ridges, are available that provide high levels of ventilation to the covered area below (Figure 4.62).

Ridge ventilation is usually used in combination with a spaced roof or a breathing roof. A breathing roof is constructed using Tanalised 50 × 25 mm timber battens or strips of nylon mesh to form a spacer between the courses of profiled sheets, thus providing a simple, cheap and effective means of ventilation. A spaced roof is used for buildings that house high unit intensive rearing, which require high levels of natural ventilation. In this roof the profiled sheets are positioned to create a gap of between 15 and 25 mm between the sheets; this provides excellent ventilation but also allows some rain and snow penetration.

Decking

Decking is the general term used for the material or materials used and fixed across roofs to serve as a flat surface on to which one of the flat roof weathering membranes is laid. The decking is also used to support the thermal insulation,

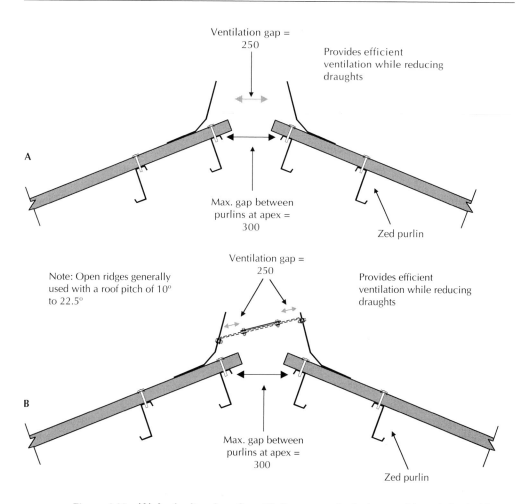

A

Ventilation gap =
250

Provides efficient
ventilation while reducing
draughts

Max. gap between
purlins at apex =
300

Zed purlin

B

Note: Open ridges generally
used with a roof pitch of 10°
to 22.5°

Ventilation gap =
250

Provides efficient
ventilation while reducing
draughts

Max. gap between
purlins at apex =
300

Zed purlin

Figure 4.62   (A) Agricultural roof ventilation: unprotected open ridges (adapted from www.eternit.co.uk). (B) Agricultural roof ventilation: protected open ridges (adapted from www.eternit.co.uk).

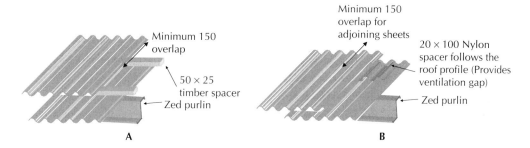

Minimum 150
overlap

50 × 25
timber spacer

Zed purlin

A

Minimum 150
overlap for
adjoining sheets

20 × 100 Nylon
spacer follows the
roof profile (Provides
ventilation gap)

Zed purlin

B

Figure 4.63   (A) Fibre-based agricultural roof: breathing roof with timber spacer. (B) Fibre-based agricultural roof: breathing roof with nylon mesh.

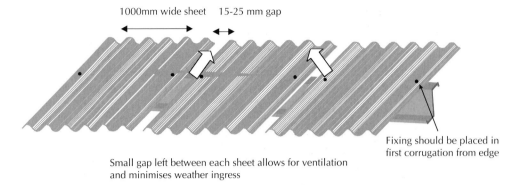

1000mm wide sheet   15-25 mm gap

Fixing should be placed in
first corrugation from edge

Small gap left between each sheet allows for ventilation
and minimises weather ingress

Figure 4.64   Fibre-based agricultural roof: spaced breathing roof.

thus creating a warm roof construction. The decking is designed to support the
weight of the materials of the roof and imposed loads of wind and snow, and
is laid to a shallow fall to encourage rainwater run off. Decking is sometimes
applied to low pitch lattice beam and portal frames. The most common form
of decking is constructed from profiled steel sheeting. Decking can also be
made of timber (for timber structures) or lightweight concrete slabs (for steel
or concrete frames).

Profiled steel decking

The most commonly used form of decking is constructed from galvanised
profiled steel sheeting, which is fixed with screws across beams or purlins. The
underside of the decking may be primed ready for painting or be manufactured
with a coated finish. Typical spans between structural frames or beams are up
to 12 m for 200 mm deep trapezoidal profiles. The decking provides support
for rigid insulation board, which is laid on a vapour check. The weathering
membrane is then bonded to the insulation boards as illustrated in Figure 4.65.
Manufacturers produce a range or proprietary composite steel decking systems
for long spans that provide high thermal insulation values.

Flat roof weathering

There is no economic or practical advantage in the use of a flat roof structure
unless the roof is to be used, e.g. for leisure. A flat roof structure is less efficient
structurally than a pitched roof, and there is little saving on unused roof space
compared with the profiled metal sheeting, which can be laid to pitches as
low as 2.5° to the horizontal. The roof surface must be constructed to create
falls to rainwater outlets to avoid ponding of water on the roof surface, so it is
not entirely 'flat'. In the UK climate flat roofs have not performed particularly
well; however, improvements in flat roof weathering membranes and careful
detailing may help to make flat roofs a viable alternative to profiled sheet metal.
See Chapter 6 of *Barry's Introduction to Building* for further details of materials,
insulation and ventilation for flat roofs.

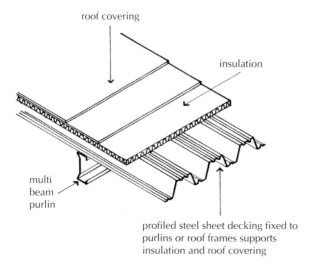

Figure 4.65   Roof decking.

Drainage and falls

Given the importance of removing water from flat roofs it is important to consider how and in which direction the water will fall to eaves, valley and/or central outlets, as illustrated in Figure 4.66. A one direction fall is the simplest to construct, for example from a lattice beam with sloping top boom or with firring pieces of wood or tapered insulation boards laid over the structure to

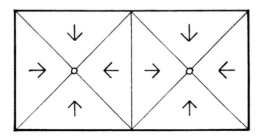

Cross falls to central outlets

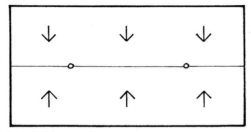

Straight falls to outlets

Figure 4.66   Falls and drainage of flat roofs.

provide the necessary falls. A two directional fall is more complicated and hence more time-consuming to construct because of the need to mitre the ends of the tapered materials. A wet screed of concrete can be laid and finished with cross falls without difficulty.

Flat roof coverings are laid so that they fall directly to rainwater outlets, usually at a fall of 1 in 40. A typical straight-fall rainwater gutter is illustrated in Figure 4.67, where the roof falls to a central valley and rainwater pipes are positioned to run down against the web of structural columns.

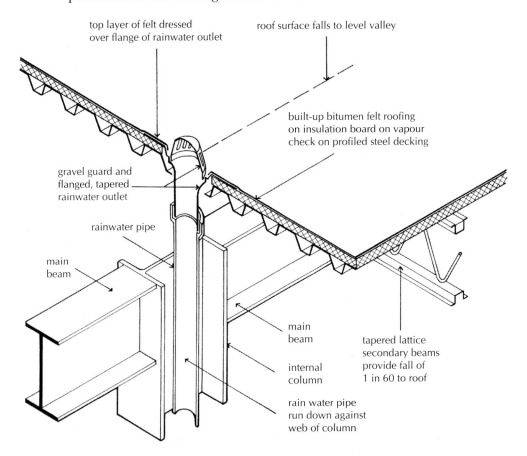

Figure 4.67   Rain water outlet in built-up bitumen felt roof.

Built-up roof coverings to roof decks

The first layer of built-up roof sheeting has to be attached to the surface of the roof deck to resist wind uplift. The manner in which this is done will depend on the nature of the roof deck. Full and partial bond methods were described in *Barry's Introduction to Construction of Buildings*. Particular attention should be given to the detailing and quality of the work to vulnerable areas such as eaves and verges, skirtings and upstands and joints. At control (expansion) joints in the structure it is necessary to make some form of upstand in the roof each side of the joint (Figure 4.68). The roofing is dressed up each side of the joint as a skirting to the upstands. A plastic coated metal capping is then secured with secret fixings to form a weather capping to the joint.

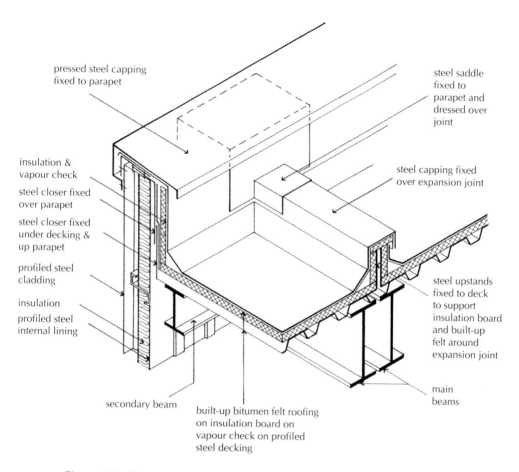

pressed steel capping fixed to parapet

steel saddle fixed to parapet and dressed over joint

insulation & vapour check

steel closer fixed over parapet

steel closer fixed under decking & up parapet

profiled steel cladding

insulation

profiled steel internal lining

steel capping fixed over expansion joint

steel upstands fixed to deck to support insulation board and built-up felt around expansion joint

main beams

secondary beam

built-up bitumen felt roofing on insulation board on vapour check on profiled steel decking

Figure 4.68   Parapet and expansion joint to profiled steel decking covered with built-up bitumen felt roofing.

Single-ply roofing

Single-ply roofing materials provide a tough, flexible, durable lightweight weathering membrane, which is able to accommodate thermal movements without fatigue. To take the maximum advantage of the flexibility and elasticity of the membrane the material should be loose laid over roofs so that it is free to expand and contract independently of the roof deck. To resist wind uplift the membrane is held down either by loose ballast, a system of mechanical fasteners or adhesives. The materials used in the manufacture of single-ply membranes are grouped as thermoplastic, plastic elastic and elastomeric:

❑ Thermoplastic materials include polyvinylchloride (PVC), chlorinated polyethylene (CPE), chlorosulphonated polyethylene (CSM), and vinyl ethylene terpolymer (VET). The materials are tough with good flexibility. All of these materials can be solvent or heat welded.
❑ Plastic elastic materials include polyisobutylene (PIB) and butyl rubber (IIR). PIB can be solvent or heat welded, IIR is joined with adhesive.
❑ Elastomeric, ethylene propylene diene monomer (EPDM). Materials are flexible and elastic with good resistance to oxidation, ozone and ultraviolet degradation. The materials are joined with adhesives.

These single-ply materials are impermeable to water, moderately permeable to moisture vapour, flexible and maintain their useful characteristics over a wider range of temperatures than the materials used for built-up roofing. To enhance tear resistance and strength, these materials may be reinforced with polyester or glass fibre fabric. Manufacturers provide detailed guidance on fixing, exposure and durability, together with conformity to relevant standards and product guarantees.

4.3 Rooflights

The traditional means of providing daylight penetration to the working surfaces of large single-storey buildings is through rooflights, either fixed in the slope of roofs or as upstand lights in flat roofs. Usual practice was to cover the middle third of each north-facing slope with rooflights to provide natural light on working surfaces. With the increase in automated manufacturing and artificial illumination, combined with concerns over poor thermal and sound insulation, unwanted glare and solar heat gain, and concerns over security, the use of rooflights has become much less common.

Functional requirements

The primary function of a rooflight is to allow the admission of daylight. As a component part of the roof, the rooflight also has to satisfy the functional requirements of the roof, being: strength and stability; resistance to weather; durability and freedom from maintenance; fire safety; resistance to the passage of heat; resistance to the passage of sound; and security.

Daylight

Rooflights should be of sufficient area to provide satisfactory daylight, and be spaced to give reasonable uniformity of lighting on the working surface without an excessive direct view of the sky, to minimise glare or penetration of direct sunlight and to avoid excessive solar gain. The area chosen is a compromise between the provision of adequate daylight and the need to limit heat loss through the lights. In pitched roofs, rooflights are usually formed in the slope of the roof to give an area of up to one sixth of the floor area and spaced as indicated in Figure 4.69 to give good uniformity and distribution of light. Rooflights in flat roofs are constructed with upstand curbs to provide a means of finishing and hence weathering the roof covering, and should be designed and positioned to provide an area of up to one sixth of the floor area. North rooflights are used to minimise solar heat gain and solar glare; the area of the rooflight may be up to one third of the floor area as shown in Figure 4.69. Monitor rooflights is a term used to describe vertical or sloping slides to a rooflight, as illustrated in Figure 4.69, and these should have an area of up to one third of the floor area.

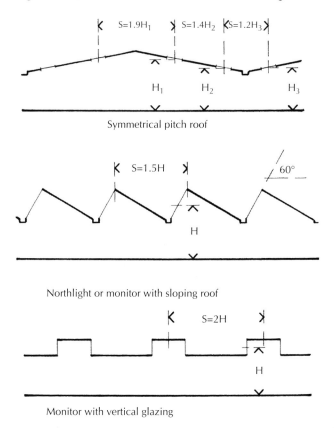

Figure 4.69   **Spacing of rooflights.**

Strength and stability

The materials used for rooflights tend to be used in the form of thin sheets to obtain the maximum transmission of light and also for economy. Glass will require support at relatively close centres to provide adequate strength and stiffness as part of the roof covering. Plastic profiled sheets tend to have less strength than the metal profiled sheets and so as a general rule require support at closer centres. Plastic sheets extruded in the form of double and triple skin cellular flat sheets have good strength and stiffness. Attention must be paid to the safety of rooflights so as to prevent the possibility of anyone falling through the covering.

Resistance to weather

Metal glazing bars, used to provide support for glass, are made with non-ferrous flashings or plastic cappings and gaskets that fit over the glass to exclude wind and rain. A minimum pitch of 15° to the horizontal is recommended. Profiled plastic sheets are designed to provide an adequate side lap and sufficient end lap to give the same resistance to the penetration of wind and rain as the profiled metal cladding in which they are fixed. A minimum pitch of not less than 10° to the horizontal is recommended. For lower pitches it is necessary to seal both side and end laps to profiled metal sheeting with a silicone sealant to exclude wind and rain. Cellular flat plastic sheets are fitted with metal or plastic gaskets to weather the joints between the sheets fixed down the slope and with non-ferrous metal flashings at overlaps at the top and bottom of sheets. Rooflights in flat and low pitch roofs are fixed on a curb (upstand) to which the roof covering is dressed to exclude weather.

Durability and freedom from maintenance

Glass is easily cleaned to maintain a bright and clear finish and is the most durable of materials; however, regular washing is required to maintain adequate daylight penetration to the working surface below. Plastic materials will discolour over time and, depending on the profile of the plastic sheets, may also trap dirt. Regular cleaning is also required to maintain adequate daylight penetration and a regular replacement strategy will be required to replace the discoloured sheets. Manufacturers provide guidance as to the expected life of the translucent sheets.

Fire safety

Fire safety in relation to rooflights is concerned with limiting the internal spread of flame and also limiting the external spread of flame. To limit the spread of fire over the surface of materials it is necessary to limit the use of thermoplastic materials in rooflights. The Building Regulations limit the number, position and use of thermoplastic rooflights (ODPM, 2002, *B4*, section 15). Thermoplastic roof lights must not be used in a protected shaft. Materials for rooflights should be chosen with care and with reference to their spread of flame characteristics. To reduce the risk of a rooflight allowing fire to pass from one building to another there are limitations on the minimum distance within which a rooflight can be placed in relation to the boundary. The distance of the rooflight from the boundary is dependent on the type of rooflight and the type of roof covering used (ODPM, 2002, B4, section 15 and Appendix A).

Resistance to passage of heat
Limiting the number and size of rooflight can mitigate heat transfer through rooflights. Sealed double (or triple) glazed units will go some way in helping to improve the thermal resistance of the roof.

Resistance to passage of sound
The thin sheets of plastic or glass used in rooflights offer very little resistance to the transfer of sound. Although some reduction in sound transfer can be achieved with double and triple glazed units it will be necessary to limit the size and number of rooflights for buildings that house noisy activities. In buildings where sound reduction is a critical requirement, only specifically designed acoustic rooflights should used; normally these are triple glazed units with 90–150 mm cavity between the internal and external sheets of glass.

Security
Single-storey buildings clad with lightweight metal cladding to roofs and walls are vulnerable to forced entry through windows, doors, walls and roof cladding, and through glass and plastic rooflights. Security against forced entry and vandalism is best achieved via secure perimeter fencing and effective 24-hour surveillance. As a general guide, rooflights should be designed and constructed so as not to compromise the security of the roof structure.

Safety – fragile roofs and rooflights
Rooflights and fragile roofs are a potential source of danger when carrying out maintenance on the roof, to trespassers and those constructing the building. Falls through fragile material give rise to more fatal accidents in the construction industry than any other single cause (HSE, 1998). Adequate measures including rooflights, must be taken to prevent people from falling through fragile roofs. Safe access to and over the roof surface must be provided. Platforms and staging may be provided to allow people to access, maintain and inspect the roof. Guarding should be provided to prevent persons who are on the roof from entering into the vicinity of the fragile surface and falling through it. When carrying out refurbishment or maintenance staging, safety nets, bird-cage scaffolds, harnesses and line system, as well as other safe means of access

may need to be provided to sufficiently reduce the risk of falling through the roof. Precautions must also be taken to prevent unauthorised access to fragile roofs. While signage may help reduce the risks of adults entering the roof, this is not sufficient for children. Extra consideration is needed when children may put themselves at risk. In such circumstances access to the fragile roof should be securely blocked. Relevant legislation includes:

❑ The Health and Safety at Work etc. Act 1974
❑ The Management of Health and Safety at Work Regulations 1999
❑ The Construction (Health, Safety and Welfare) Regulations 1996
❑ The Construction (Design and Management) Regulations 1994
❑ The Lifting Operations and Lifting Equipment Regulations 1998

Materials used for rooflights

The traditional material for rooflights was glass laid in continuous bays across the slopes of roofs and lapped under and over slate or tile roofing. The majority of rooflights constructed today are of translucent sheets of plastic, usually formed to the same profile as the roof sheeting.

Glass

The types of glass used for rooflights are float glass, solar control glass, patterned glass, and wired glass, which is used to minimise the danger from broken glass during fires. Glass has poor mechanical strength and must be supported with metal or timber glazing bars, at relatively close centres of about 600 mm for patent glazing. The principal advantage of glass is that it provides a clear view and, with regular washing, maintains a bright surface appearance.

Profiled, cellular and flat plastic sheets

Transparent or translucent plastic sheet material is used as a cheaper alternative to glass. The materials used for profiled sheeting are:

❑ uPVC – polyvinyl chloride – rigid PVC. This is one of the cheapest materials and has a light transmittance of 77%, reasonable impact resistance and good resistance to damage. The material will discolour when exposed to solar radiation.
❑ GRP – glass reinforced polyester. GRP is usually inflammable, has good impact resistance, rigidity and dimensional stability. The material is translucent and has a moderate light transmittance of between 50 and 70%. Translucent GRP sheets comprise thermosetting polyester resins, curing agents, light stabilisers, flame retardants and reinforcing glass fibres. They are roll formed in a range of profiles to match most profiled sheet roofing materials. Three grades of GRP sheet are produced to satisfy the conditions for external fire exposure and surface spread of flame.

The materials used for flat sheet rooflights, laylights and domelights are:

❏ PC – polycarbonate. This material has good light transmittance, up to 88%, good resistance to weathering, reasonable durability and very good impact resistance. Polycarbonate is the most expensive of the materials and is used principally for its high impact resistance.
❏ PMMA – polymethyl methacrylate. This plastic is used for shaped rooflights, having good impact resistance and resistance to ultraviolet radiation, but softens and burns readily when subject to the heat generated by fires.

Rooflights

The most straightforward way of constructing rooflights in pitched roofs covered with profiled sheeting is by the use of GRP or uPVC, which is formed to match the profile of the roof sheeting. The translucent sheets are laid so that they cover the lower sheet and adjacent sheet to form an end and side lap respectively. All side laps should be sealed with self-adhesive closed cell PVC sealing tape to make a weathertight joint. End laps between translucent sheets and between translucent sheet and roof sheets to roofs pitched below 20° should be sealed with extruded mastic sealant. Fixing of sheets is critical to resist wind uplift, in common with all lightweight sheeting materials used for roofing, and the fixing usually follows that used for the main roofing material.

Double skin rooflights are constructed with two sheets of GRP, as illustrated in Figure 4.70, which have the same profile as the sheet roof covering. Profiled, high density foam spacers, bedded top and bottom in silicone mastic, are fitted between the sheets to maintain the airspace and also to seal the cavity. Double-sided adhesive tape is fixed to all side laps of both top and bottom sheets as a seal. The double skin rooflight is secured with fasteners driven through the sheets and foam spacers to the purlins. Stitching screws are then driven through the crown of profiles at side and end laps. Factory-formed sealed double skin GRP rooflight units are made from a profiled top sheet and a flat underside with a spacer and sealer.

Translucent polyvinyl chloride (uPVC) sheets are produced in a range of profiles to match most metal and fibre cement sheeting. For roof pitches of 15° or less, the side and end laps should be sealed with sealing strips and all laps between uPVC sheets should be sealed. Fixing holes should be 3 mm larger in diameter than the fixing to allow for thermal expansion of the material. Fasteners similar to those used for fixing roofing sheets are used. Double skin rooflights are formed in a similar manner to that shown in Figure 4.70.

Flat cellular sheets of polycarbonate are supported by aluminium glazing bars fixed to purlins as illustrated in Figure 4.71 to form a rooflight to a north-facing roof slope. The capping of the glazing bars compresses a Neoprene gasket to the sheets to make a watertight seal.

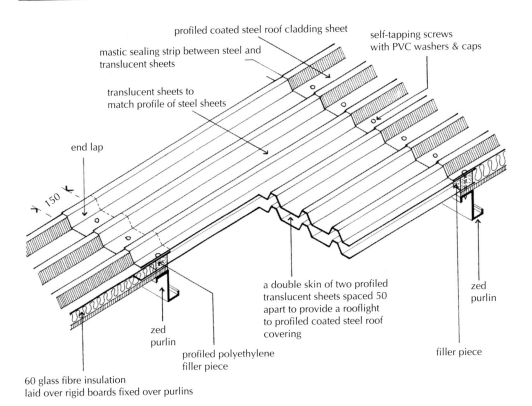

profiled coated steel roof cladding sheet

self-tapping screws
with PVC washers & caps

mastic sealing strip between steel and
translucent sheets

translucent sheets to
match profile of steel sheets

end lap

150

a double skin of two profiled
translucent sheets spaced 50
apart to provide a rooflight
to profiled coated steel roof
covering

zed
purlin

zed
purlin

profiled polyethylene
filler piece

filler piece

60 glass fibre insulation
laid over rigid boards fixed over purlins

Figure 4.70   Rooflights: translucent sheets in profiled steel-covered pitched roof.

Patent glazing

The traditional method of fixing glass in the slopes of roofs to create a rooflight is by means of wood or metal glazing bars that provide support for the glass and form weather flashings, or cappings, to exclude water. The word 'patent' refers to the patents taken out by the original makers of glazing bars. Timber, iron and steel glazing bars have largely been replaced by aluminium and lead or plastic coated steel bars. Likewise, single glazing has been replaced by double glazed units and wired glass. Patent glazing is relatively labour intensive due to the provision and fixing of the glazing bars at relatively close centres; however, the result can be an attractive, durable rooflight with good light transmission.

The most commonly used glazing bars are of extruded aluminium with seatings for glass, condensation channels and a deep web top flange for strength and stiffness in supporting the weight of the glass. The glass is secured with clips, beads or cappings. Figure 4.72 illustrates aluminium glazing bars supporting single wired glass in the slope of a pitched roof. The glazing bars are

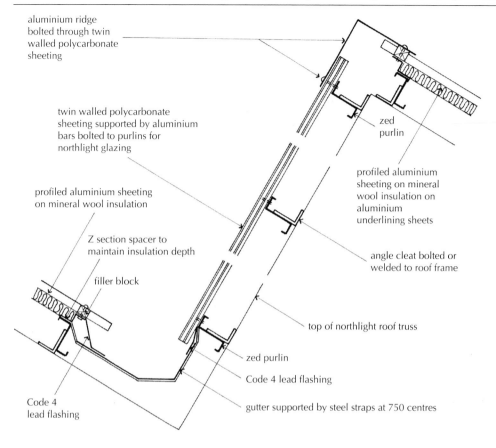

aluminium ridge
bolted through twin
walled polycarbonate
sheeting

twin walled polycarbonate
sheeting supported by aluminium
bars bolted to purlins for
northlight glazing

zed
purlin

profiled aluminium sheeting
on mineral wool insulation

profiled aluminium
sheeting on mineral
wool insulation on
aluminium
underlining sheets

Z section spacer to
maintain insulation depth

angle cleat bolted or
welded to roof frame

filler block

top of northlight roof truss

Code 4
lead flashing

zed purlin

Code 4 lead flashing

gutter supported by steel straps at 750 centres

Figure 4.71   North light roof glazing.

secured in fixing shoes screwed or bolted to angles fixed to purlins and fitted
with aluminium stops to prevent glass from slipping down the slope of the
roof. Aluminium spring clips, fitted to grooves in the bars, keep the glass in
place and serve as weathering between the glass and the bar. Also illustrated is
a system of steel battens and angles, an angle and a purlin to provide fixing for
the glass and sheeting at their overlap. Lead flashings are fixed as weathering
at the overlap of the glass and sheeting.

Figure 4.73 shows six different types of glazing bar. Aluminium glazing bar
for sealed double glazing (Figure 4.73A) and single glazing (Figure 4.73B) are
secured with aluminium beads bolted to the bar and weathered with butyl
strips. Aluminium glazing bars with bolted aluminium capping and snap-on
aluminium cappings to the bars are illustrated in Figures 4.73C and 4.73D.
Cappings are used to secure glass in position on steep slopes and for vertical
glazing as they afford a more secure fixing than spring clips; visually they give
greater emphasis to the bars. Steel bars covered with lead and PVC sheathing

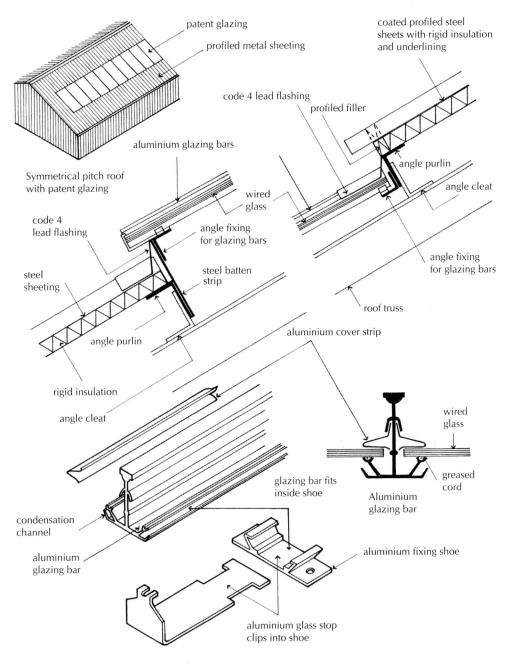

patent glazing

profiled metal sheeting

coated profiled steel
sheets with rigid insulation
and underlining

code 4 lead flashing

profiled filler

aluminium glazing bars

angle purlin

Symmetrical pitch roof
with patent glazing

angle cleat

code 4
lead flashing

wired
glass

angle fixing
for glazing bars

steel
sheeting

steel batten
strip

angle fixing
for glazing bars

angle purlin

roof truss

aluminium cover strip

rigid insulation

angle cleat

wired
glass

glazing bar fits
inside shoe

greased
cord

condensation
channel

Aluminium
glazing bar

aluminium
glazing bar

aluminium fixing shoe

aluminium glass stop
clips into shoe

Figure 4.72   Patent glazing.

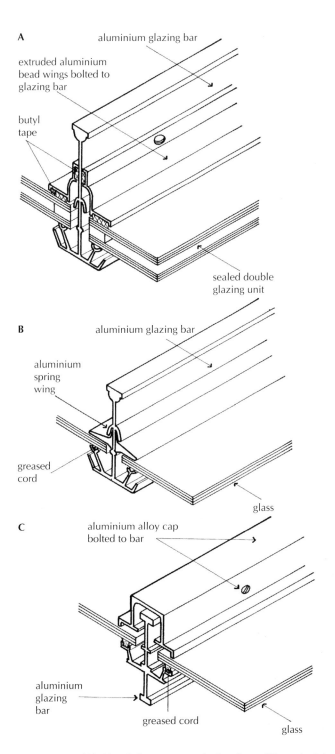

A

aluminium glazing bar

extruded aluminium
bead wings bolted to
glazing bar

butyl
tape

sealed double
glazing unit

B

aluminium glazing bar

aluminium
spring
wing

greased
cord

glass

C

aluminium alloy cap
bolted to bar

aluminium
glazing
bar

greased cord

glass

Figure 4.73 **(A) Aluminium patent glazing bar with sealed double glazing. (B) Aluminium patent glazing bar for single glazing. (C) Aluminium glazing bar with aluminium cap. (D) Aluminium glazing bar with snap-on capping. (E) Lead clothed steel core patent glazing. (F) PVC sheathed steel core glazing bar.**

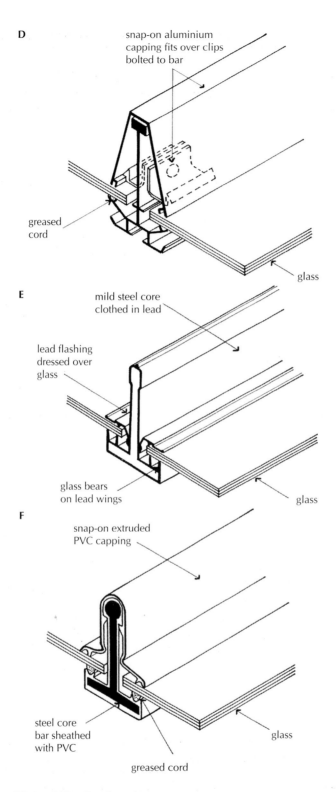

D

snap-on aluminium capping fits over clips bolted to bar

greased cord

glass

E

mild steel core clothed in lead

lead flashing dressed over glass

glass bears on lead wings

glass

F

snap-on extruded PVC capping

steel core bar sheathed with PVC

greased cord

glass

Figure 4.73 *Continued*

as protection against corrosion are shown in Figures 4.73E and 4.73F. Steel bars are used for mechanical strength of the material and the advantage of more widely spaced supports than is possible with aluminium bars of similar depth.

Rooflights in flat and low pitch roofs

Before the introduction of plastic and fibre glass as material for rooflights, the majority of rooflights to flat roofs were constructed as lantern lights or deck lights, which were framed in timber or steel and covered with glass.

A lantern light is constructed with glazed vertical sides and a hipped or gable-ended glazed roof. The vertical sides of the lantern light are used as opening lights for ventilation. Lantern lights were often used to cover considerable areas, the light being framed with substantial timbers of iron or steel, to provide top light to large stairwells and internal rooms. Ventilation from the opening upstand sides is controlled by cord or winding gear from below to suit the requirements of the occupants of the space below. The light requires relatively frequent maintenance if it is to remain sound and watertight, and many have been replaced by domelights for economy of installation and maintenance. Figure 4.74 is an illustration of an aluminium lantern light constructed with standard aluminium window frame and sash sections, aluminium corner posts, aluminium patent glazing to the pitched roof and an aluminium ridge section. The lantern light is bolted to an upstand curb (in common with all rooflights fixed in a flat roof) to resist wind uplift, and the roof covering is dressed to a height of at least 150 mm above the surface of the flat roof.

Deck lights are constructed as a hipped or gable-ended glazed roof with no upstand sides, thus they provide daylight to the space below but no ventilation, as shown in Figures 4.75 and 4.76. This deck light is constructed with lead sheathed steel glazing bars pitched and fixed to a ridge and bolted to a steel tee fixed to the upstand curb.

A variety of shapes are produced to serve as rooflights for flat roofs, as illustrated in Figure 4.77. The advantage of the square and rectangular shapes over the circular and ovoid ones is that they require straightforward trimming of the roof structure around the openings and upstands. Plastic rooflights are made as either single skin lights or as sealed double skin lights, which improves their resistance to the transfer of heat. Plastic rooflights are bolted or screwed to upstand curbs to resist wind uplift, and the roof covering dressed against the upstand as illustrated in Figure 4.78.

Lens light

Lens light consists of square or round glass blocks or lenses that are cast into reinforced concrete ribs, as illustrated in Figure 4.79, to provide diffused daylight through concrete roofs. The lens light can be prefabricated and bedded in place on site, or it can be cast insitu in the concrete roof. Although light

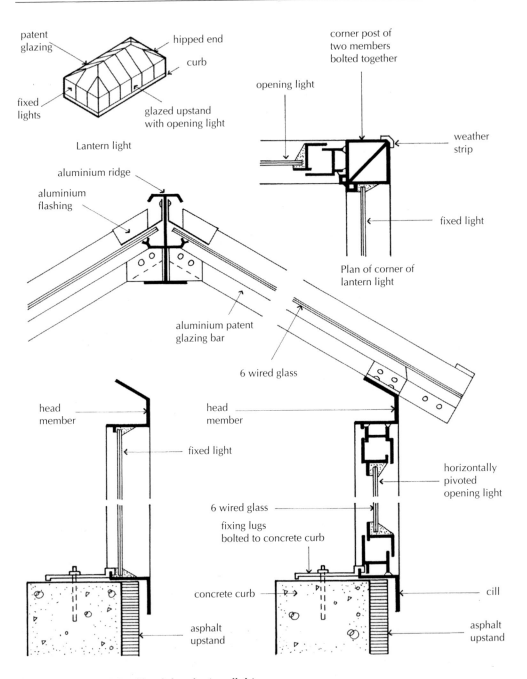

Figure 4.74 Aluminium lantern light.

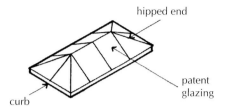

Figure 4.75   Deck light.

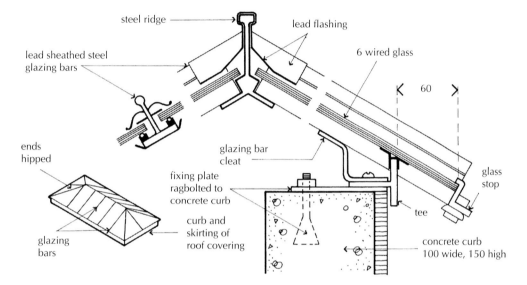

Figure 4.76   Deck light.

transmission is poor, these rooflights are used primarily to provide resistance to fire, to improve security and to reduce sound transmission through the roof.

4.4 Diaphragm, fin wall and tilt-up construction

The majority of tall, single-storey buildings that enclose large open areas such as sports halls, warehouses, supermarkets and factories with walls more than 5 m high are constructed with a frame of lattice steel or a portal frame covered with lightweight steel cladding and infill brick walls at lower level. An alternative approach is to use diaphragm walls and fin walls constructed of brickwork or blockwork. Brickwork is preferred to blockwork because the smaller unit of the brick facilitates bonding and avoids cutting of blocks. Some of the advantages

Single or double skin domelight in polycarbonate, acrylic
or uPVC.

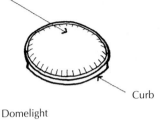

Curb

Domelight

rectangular base single or double
skin domelight in polycarbonate,
acrylic or uPVC.

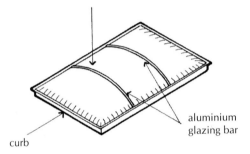

aluminium
glazing bar

curb

Rectangular base domelight

Single or double skin pyramid
roof light in polycarbonate acrylic
or uPVC.

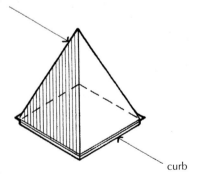

curb

Figure 4.77   Pyramid roof light.

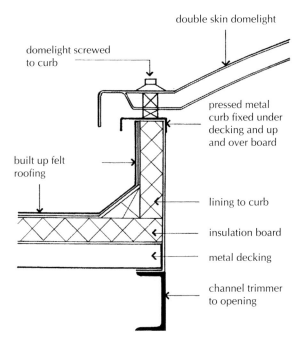

double skin domelight

domelight screwed
to curb

built up felt
roofing

pressed metal
curb fixed under
decking and up
and over board

lining to curb

insulation board

metal decking

channel trimmer
to opening

Figure 4.78   Upstand to domelight.

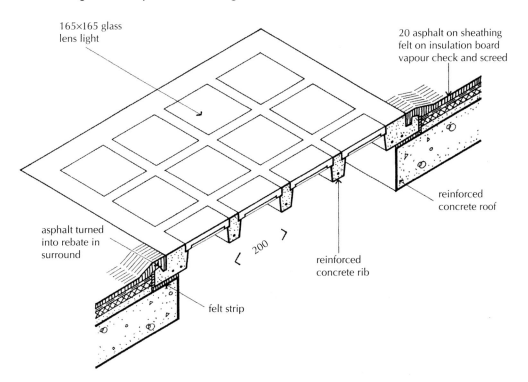

165×165 glass
lens light

20 asphalt on sheathing
felt on insulation board
vapour check and screed

asphalt turned
into rebate in
surround

200

reinforced
concrete rib

reinforced
concrete roof

felt strip

Figure 4.79   Reinforced concrete and glass rooflights.

of diaphragm and fin wall construction include durability, security, thermal insulation, sound insulation and resistance to fire. Visual appearance of the wall can be enhanced with the use of special bricks and creative design of fin walls.

Brick diaphragm walls

A diaphragm wall is built with two leaves of brickwork bonded to brick cross ribs (diaphragms) inside a wide cavity between the leaves, thus forming a series of stiff box or I-sections structurally, as illustrated in Figure 4.80. The compressive strength of the bricks and mortar is considerable in relation to the comparatively small dead load of the wall, roof and imposed loads. Stability is provided by the width of the cavity and the spacing of the cross ribs, together with the roof, which is tied to the top of the wall to act as a horizontal plate to resist lateral forces.

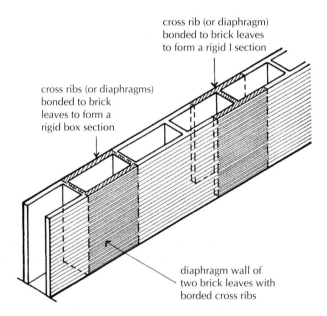

cross rib (or diaphragm)
bonded to brick leaves
to form a rigid I section

cross ribs (or diaphragms)
bonded to brick
leaves to form a
rigid box section

diaphragm wall of
two brick leaves with
borded cross ribs

Figure 4.80   Brick diaphragm wall.

Construction

The width of the cavity and the spacing of the cross ribs is determined by the size of the box section required for stability and the need for economy in the use of materials by using whole bricks. Cross ribs are usually placed four or five whole brick lengths (with mortar joints) apart and the cavity one-and-a-half or two-and-a-half whole bricks (with mortar joints) apart so that the cross ribs can be bonded in alternate courses to the outer and inner leaves, as illustrated in

Figure 4.81. Loads on the foundations are relatively slight, thus a simple strip foundation can be used in good ground conditions.

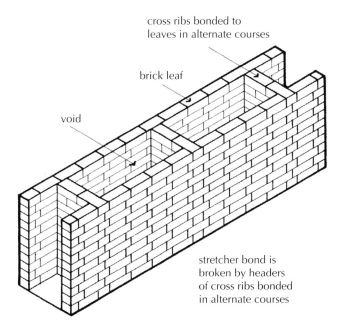

cross ribs bonded to
leaves in alternate courses

brick leaf

void

stretcher bond is
broken by headers
of cross ribs bonded
in alternate courses

Figure 4.81   Bonding of diaphragm wall.

The roof is tied to the top of the diaphragm wall to act as a prop in resisting the overturning action of lateral wind pressure, by transferring the horizontal forces on the long walls to the end walls of the building that act as shear walls. The roof structure is tied to a reinforced concrete capping beam by bolts, as illustrated in Figure 4.82. Care is required at this junction to ensure that thermal bridging does not occur across the capping beam. Roof beams are braced by horizontal lattice steel wind girders, which are connected to roof beams, as illustrated in Figure 4.83.

Door and window openings should be designed to fit between the cross ribs so that the ribs can form the jambs of the opening. Large door and window openings will cause large local loadings, thus double ribs (or thicker ribs) are built to take the additional load, as illustrated in Figure 4.84. Vertical movement joints are formed by the construction of double ribs at the necessary centres to accommodate thermal movement (Figure 4.85).

Diaphragm walls built in positions of severe exposure will resist moisture penetration, although the cavity should be ventilated to assist with the drying out of the brickwork. Given the problem of thermal bridging inherent in the brick diaphragms, the most convenient method of insulation is to fix insulation to the inside face of the wall. A long, high diaphragm wall with flat panels

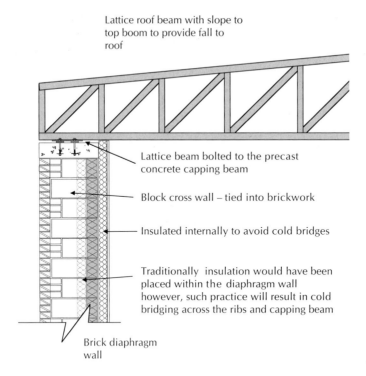

Lattice roof beam with slope to top boom to provide fall to roof

Lattice beam bolted to the precast concrete capping beam

Block cross wall – tied into brickwork

Insulated internally to avoid cold bridges

Traditionally insulation would have been placed within the diaphragm wall however, such practice will result in cold bridging across the ribs and capping beam

Brick diaphragm wall

Figure 4.82   Connection of roof beams to diaphragm wall.

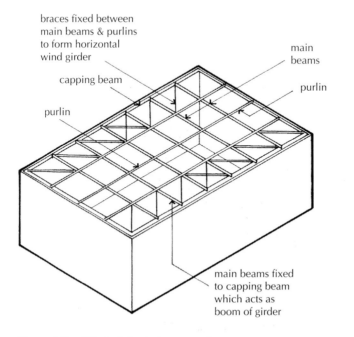

braces fixed between main beams & purlins to form horizontal wind girder

capping beam

purlin

main beams

purlin

main beams fixed to capping beam which acts as boom of girder

Figure 4.83   Wind girder to beam roof.

of brickwork may have a rather uninspiring appearance. Variations in the depth of the cavity wall, the use of projecting brick fins and polychromatic brick-work may go some way to alleviate the monotony, although there will be cost implications.

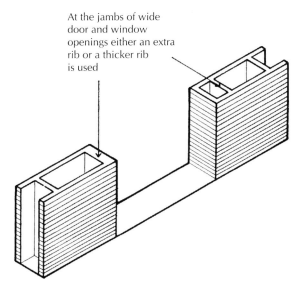

At the jambs of wide door and window openings either an extra rib or a thicker rib is used

Figure 4.84   Openings in diaphragm walls.

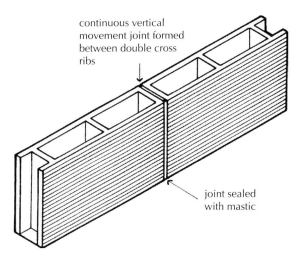

continuous vertical movement joint formed between double cross ribs

joint sealed with mastic

Figure 4.85   Movement joint in diaphragm wall.

Brick fin walls

A fin wall is built as a cavity wall buttressed with piers (fins), which are bonded to the external leaf of the cavity wall to buttress and hence stiffen the wall against overturning. A fin wall acts structurally as a series of T-sections, as illustrated in Figure 4.86. The compressive strength of the bricks and mortar is considerable in relation to the comparatively small dead load of the wall, roof and imposed loads. Stability against lateral forces from wind pressure is provided by the T-sections of the fins and the prop effect of the roof, which is usually tied to the top of the wall to act as a horizontal plate to transfer forces to the end walls. The minimum dimensions and spacing of the fins are determined by the cross-sectional area of the T-section of the wall required to resist the tensile stress from lateral pressure and by considerations for the appearance of the building. Spacing and dimensions of the fins can be varied to suit a chosen external appearance. Some typical profiles for brick fins are illustrated in Figure 4.87, with brick specials use for maximum effect.

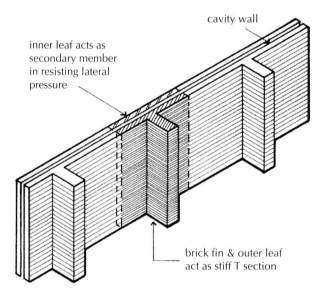

inner leaf acts as
secondary member
in resisting lateral
pressure

cavity wall

brick fin & outer leaf
act as stiff T section

Figure 4.86   Fin wall.

Construction

The wall is constructed as a cavity wall, with inner and outer leaves of brick tied with wall ties and thermal insulation positioned within the cavity. The fins are bonded to the outer leaf in alternate courses. Thickness of the fin will typically be one brick thick with a projection of four or more brick lengths,

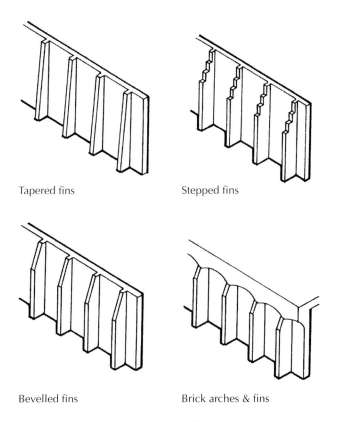

Tapered fins   Stepped fins

Bevelled fins   Brick arches & fins

Figure 4.87   Typical profiles for brick fins.

with the size of the fin varying to suit structural and aesthetic requirements. The fins should be spaced to suit whole brick sizes, thus minimising the cutting of bricks, and at regular centres necessary for stability and for appearance. The loads on the foundation of a fin wall are relatively slight, and a continuous concrete strip foundation should provide adequate support and stability on good bearing ground. The foundation will extend under the fin, as illustrated in Figure 4.88.

Roof beams are usually positioned to coincide with the centres of the fins and tied to a continuous reinforced capping beam that is cast or bedded on the top of the wall, or to concrete padstones cast or bedded on top of the fins as illustrated in Figure 4.89. To resist wind uplift on lightweight roofs, the beams are anchored to the brick fins through bolts built into the fins, cast or threaded through the padstones and bolted to the beams. Horizontal bracing to the roof beams is provided by lattice wind girders fixed to the beams to act as a plate in propping the top of the wall.

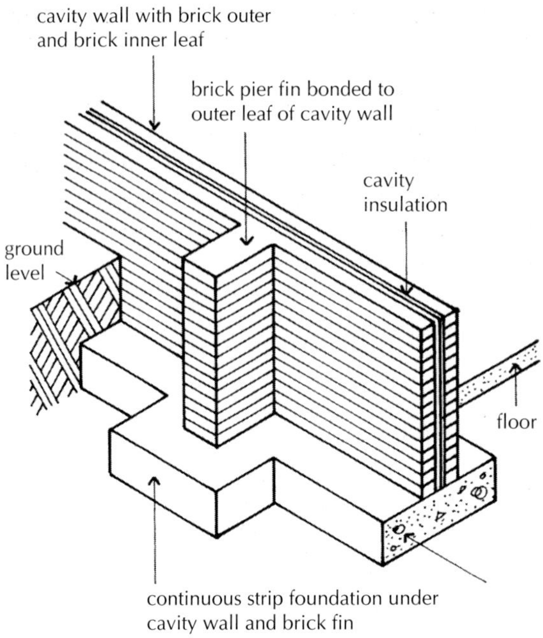

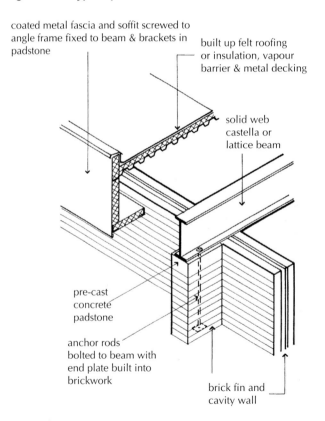

cavity wall with brick outer
and brick inner leaf

brick pier fin bonded to
outer leaf of cavity wall

cavity
insulation

ground
level

floor

continuous strip foundation under
cavity wall and brick fin

Figure 4.88   Typical profiles for brick fins.

coated metal fascia and soffit screwed to
angle frame fixed to beam & brackets in
padstone

built up felt roofing
or insulation, vapour
barrier & metal decking

solid web
castella or
lattice beam

pre-cast
concrete
padstone

anchor rods
bolted to beam with
end plate built into
brickwork

brick fin and
cavity wall

Figure 4.89   Fin wall, beams and roofing.

Door and window openings should be the same width as the distance between the fins for simplicity and economy of construction. To allow sufficient cross-section of brickwork at the jambs of wide openings, a thicker fin or a double fin is built, as illustrated in Figure 4.90. Movement joints are usually formed between double brick fins as illustrated in Figure 4.91. In addition to the usual resistance to weather provided by brickwork, the projecting fins may provide some additional shelter to the wall from driving rain.

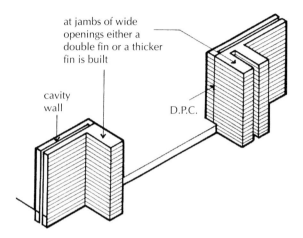

Figure 4.90   Openings in fin walls.

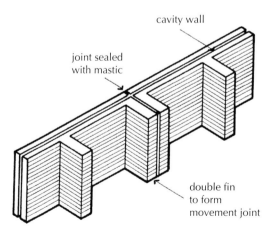

Figure 4.91   Movement joint in fin wall.

Concrete tilt-up construction

Tilt-up construction is a technique of precasting large, slender reinforced concrete wall panels on site (on a temporary casting bed or on the concrete floor slab) which, when cured, are tilted by crane into position. This technique has been used principally for the construction of single-storey commercial and industrial buildings on open sites where there is room for casting and the necessary lifting equipment. Tilt-up construction has been used quite extensively in the USA, where it originated, and many other countries such as Australia and New Zealand, for the speed of casting and speed of erection of the panels. The technique is most economical when there is a high degree of repetitiveness in the structure and the walls are used in a loadbearing capacity. Typical applications include low rise warehouses, offices and factories. There are few examples of this type of construction in the UK.

Tilt up concrete panels vary in size, shape and thickness, but typically will be around 7 × 5 m, 160 mm thick, and weigh between 20 and 30 tonnes. Panel size is limited by the strength of the reinforced concrete panel necessary to accommodate the stresses induced in the panel as it is tilted from the horizontal to the vertical and also by the lifting capacity of the cranes. Wall panels may vary in design from plain, flat slabs to frames with wide openings for glazing provided that there is adequate reinforced concrete to carry the anticipated loads. A variety of shapes and features are made possible by repetitive use of the formwork in the casting bed, and a variety of external finishes can be produced, ranging from smooth to textured finishes.

Sequence of assembly

The sequence of operations is shown in Figure 4.92. The site slab of concrete is cast over the completed foundations, drainage and service pipe work and accurately levelled to provide a level surface on which the wall panels can be cast. A bond breaker/cure-coat is then applied to the concrete slab and the panels cast around reinforcement inside steel (or timber) edge shuttering, which is placed as near as possible to the final position of the wall panel. Lifting lugs and other fittings are usually cast into the upper face of the panels (which will be covered by insulation and internal finishes). Wall panels may be cast individually or as a continuous strip. If the panels are cast as a continuous strip they are cut to size once the concrete has gone off but during the early stages of the concrete's maturity (one or two days). Panels may also be cast as a stack, one on top of the other, separated by a bond breaker. Once cured, the hardened panels are then gently lifted or tilted into position and propped or braced ready to receive the roof deck. The panels are tilted up and positioned on the levelled foundations against a rebate in the concrete, or up to timber runners or on to a sheathing angle and then set level on steel levelling shims. A mechanical connection between the foot of the slabs to the foundation and/or

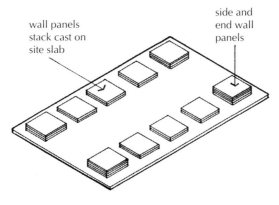

wall panels
stack cast on
site slab

side and
end wall
panels

Casting wall panels

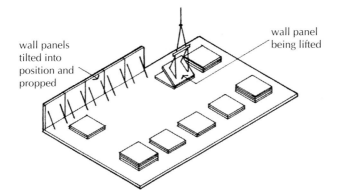

wall panels
tilted into
position and
propped

wall panel
being lifted

Wall panels tilted into position

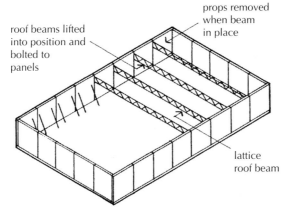

roof beams lifted
into position and
bolted to
panels

props removed
when beam
in place

lattice
roof beam

Figure 4.92   Tilt-up construction.

floor slab is usually employed. Cast in metal, dowels projecting from the foot of the panels are set into slots or holes in the foundations and grouted in position. Alternatively a plate welded to studs or bar anchors, cast into the foot of the panel, provides a means of welded connection to rods cast into the site slab as illustrated in Figure 4.93. The roof deck serves as a diaphragm to give support to the top of the wall panels and to transmit lateral wind forces back to the foundation. Lattice beam roof decks are welded to seat angles, welded to a plate and cast in studs as shown in Figure 4.93. A continuous chord angle is

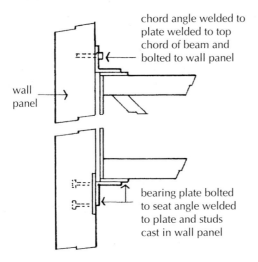

chord angle welded to
plate welded to top
chord of beam and
bolted to wall panel

wall
panel

bearing plate bolted
to seat angle welded
to plate and studs
cast in wall panel

Connection of roof beams to
wall panel

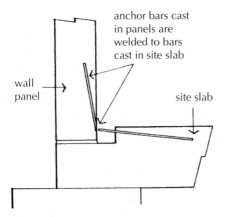

anchor bars cast
in panels are
welded to bars
cast in site slab

wall
panel

site slab

Figure 4.93   Connection of wall panels to site slab.

welded to the top of the lattice beams and to bolts cast or fixed in the panel. The chord angle serves as a transverse tie across the panels and is secured to them with bolts set into slots in the angle to allow for shrinkage movements of the panels.

Functional properties

The concrete panels provide good resistance to the penetration of rain and also provide good durability and freedom from maintenance. Panels also provide good fire resistance, resistance to the passage of sound and relatively good security against forced entry. The reinforced concrete panels do not provide adequate thermal insulation for heated buildings. Thermal insulation is usually applied to the internal face of the panels with a moisture vapour check between wall and insulation. Insulation boards are used to provide both insulation and an internal finish to the building, fixed to timber battens which are shot fired to the panel.

4.5 Shell structures

A shell structure is a thin, curved membrane or slab, usually of reinforced concrete, that functions both as a structure and covering, the structure deriving its strength and rigidity from the curved shell form. The term 'shell' is used to describe these structures by reference to the considerable strength and rigidity of thin, natural, curved forms such as the shell of an egg. The material most suited to the construction of a shell structure is concrete, which is a highly plastic material when wet and which can take up any shape inside formwork (also known as centring). Small section reinforcing bars can readily be bent to follow the curvature of shells. Wet concrete is spread over the centring and around the reinforcement, and compacted to the required thickness with the stiffness of the concrete mix and the reinforcement preventing the concrete from running down the slope of the curvature of the shell while the concrete is wet. Once the concrete has hardened, the reinforced concrete membrane or slab acts as a strong, rigid shell, which serves as both structure and covering to the building. The strength and rigidity of curved shell structures makes it possible to construct single curved barrel vaults 60 mm thick and double curved hyperbolic paraboloids 40 mm thick in reinforced concrete for clear spans up to 30 m.

The attraction of shell structures lies in the elegant simplicity of the curved shell form that utilises the natural strength and stiffness of shell forms with great economy in the use of material. The main disadvantages relate to their cost and poor thermal insulation properties. A shell structure is more expensive than, for example, a portal framed structure covering the same floor area because of the considerable labour required to construct the centring on which the shell is cast. Shell structures cast in concrete are also difficult to insulate

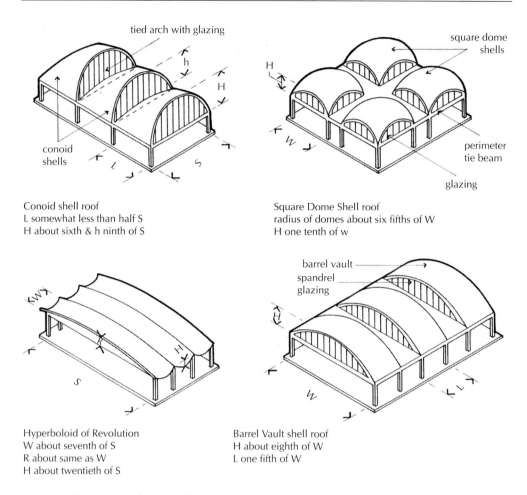

tied arch with glazing

square dome
shells

conoid
shells

perimeter
tie beam

glazing

Conoid shell roof
L somewhat less than half S
H about sixth & h ninth of S

Square Dome Shell roof
radius of domes about six fifths of W
H one tenth of w

barrel vault
spandrel
glazing

Hyperboloid of Revolution
W about seventh of S
R about same as W
H about twentieth of S

Barrel Vault shell roof
H about eighth of W
L one fifth of W

Figure 4.94   Some typical shell roof forms.

economically because of their geometry and so are mainly suited to unheated spaces.

Shell structures tend to be described as single or double curvature shells. Single curvature shell structures are curved on one linear axis and form part of a cylinder in the form of a barrel vault or conoid shell; double curvature shells are either part of a sphere as a dome or a hyperboloid of revolution (see Figure 4.94). The terms are used to differentiate the comparative rigidity of the two forms and the complexity of the formwork (centring) necessary to construct the shell form. Double curvature of a shell adds considerably to its stiffness, resistance to deformation under load and reduction in the need for restraint against deformation.

Centring (or formwork) is the term used to describe the necessary temporary support on which a curved reinforced concrete shell structure is cast. The centring for a single curvature barrel vault is less complex than that for a dome, which is curved from a centre point. Advances in computer software have made the design of shell structures and the setting out of formwork much easier; however, there is still a considerable demand on labour to make and erect the centring, and the more complex the shape the greater the amount of cutting and potential waste of material. The simplest, and hence most economic, of all shell structures is the barrel vault, constructed in concrete or timber.

Reinforced concrete barrel vaults

These consist of a thin membrane of reinforced concrete positively curved in one direction so that the vault acts as both structure and roof surface. The most common form of barrel vault is the long span vault, illustrated in Figure 4.95, where the strength and stiffness of the shell lie at right angles to the curvature.

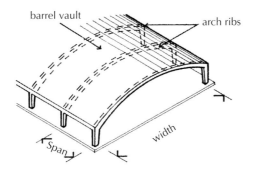

Short Span Barrel Vault

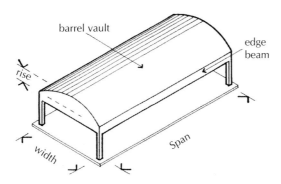

Long Span Barrel Vault

Figure 4.95   Reinforced concrete barrel vaults.

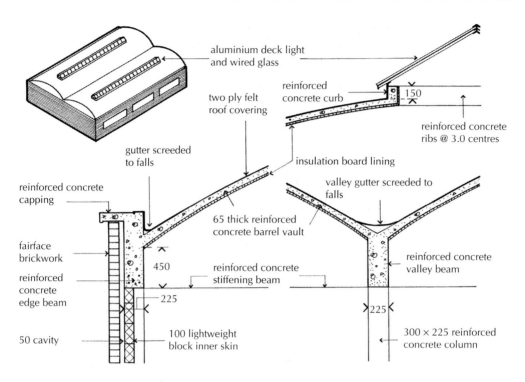

Figure 4.96   Reinforced concrete barrel vault.

Typical spans range from 12 to 30 m, with the width being about half the span and the rise about one fifth of the width. To cover large areas, multi-span, multi-bay barrel vault roofs can be used (see Figure 4.96). The concrete shell may be from 57 to 75 mm thick for spans of 12 and 30 m respectively. The thickness of the concrete provides sufficient cover of concrete to protect the reinforcement against damage by fire and corrosion.

Stiffening beams and arches

Under local loads the thin shell of the barrel vault will tend to distort and lose shape and, if this distortion were of sufficient magnitude, the resultant increase in local stress would cause the shell to progressively collapse. To strengthen the shell against this possibility, stiffening beams or arches are cast integrally with the shell. Figure 4.97 illustrates the four types of stiffening members generally used, with common practice being to provide a stiffening member between the columns supporting the shell. The downstand reinforced concrete beam, which is usually 150 or 225 mm thick, is the most efficient of the four because of its depth. To avoid the interruption of the line of the soffit of the vaults caused by a downstand beam, an upstand beam is sometimes used. The disadvantage of an upstand beam is that it breaks up the line of the roof and also needs

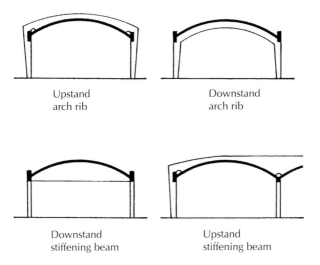

Upstand
arch rib

Downstand
arch rib

Downstand
stiffening beam

Upstand
stiffening beam

Figure 4.97   Stiffening beams and arches for reinforced concrete barrel vaults.

protection against the weather. Arch ribs are sometimes used because they follow the curve of the shell and therefore do not interrupt the line of the vault; however, these are less efficient structurally because they have less depth than beams.

Edge and valley beams

Reinforced concrete edge beams are cast between columns as an integral part of the shell to resist the tendency of the thin shell to spread and its curvature to flatten out due to self-weight and imposed loads. The edge beams may be cast as dropped beams, upstand beams, or partly upstand or partly dropped beams, as illustrated in Figure 4.98. Between multi-bay vaults the loads on the vaults are largely transmitted to adjacent shells and then to the edge beams, thus allowing the use of comparatively slender feather edge beams.

Rooflights

Natural light through the shell structure can be provided by decklights formed in the crown of the vault, as illustrated in Figure 4.96, or by domelights. Rooflights are fixed to an upstand curb cast integrally with the shell, as illustrated in Figure 4.96. Care is required to avoid overheating and glare. One way of providing natural light and avoiding glare and overheating is to use a system of northlight barrel vaults, as illustrated in Figures 4.99 and 4.100. The roof consists of a thin reinforced concrete shell on the south-facing side of the roof, with a reinforced concrete framed north-facing slope, and pitched at between 60 and 80°. This construction is less efficient structurally than a barrel vault because the rigidity of the shell is interrupted by the northlights.

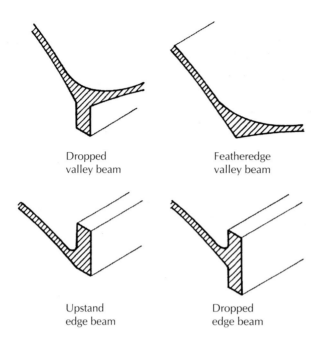

Dropped
valley beam

Featheredge
valley beam

Upstand
edge beam

Dropped
edge beam

Figure 4.98   Edge and valley beams for reinforced concrete barrel vaults.

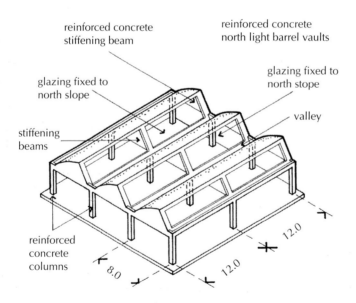

reinforced concrete
stiffening beam

reinforced concrete
north light barrel vaults

glazing fixed to
north slope

glazing fixed to
north stope

stiffening
beams

valley

reinforced
concrete
columns

8.0

12.0

12.0

12.0

Figure 4.99   Three-bay reinforced concrete north light barrel vault.

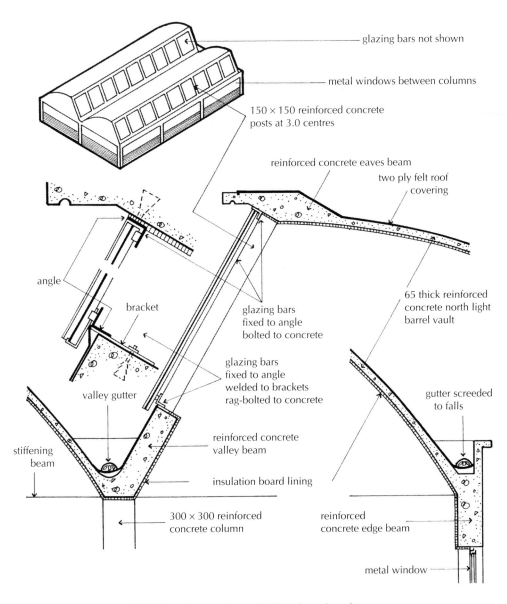

glazing bars not shown

metal windows between columns

150 × 150 reinforced concrete
posts at 3.0 centres

reinforced concrete eaves beam

two ply felt roof
covering

angle

bracket

glazing bars
fixed to angle
bolted to concrete

65 thick reinforced
concrete north light
barrel vault

glazing bars
fixed to angle
welded to brackets
rag-bolted to concrete

valley gutter

gutter screeded
to falls

stiffening
beam

reinforced concrete
valley beam

insulation board lining

300 × 300 reinforced
concrete column

reinforced
concrete edge beam

metal window

Figure 4.100   Reinforced concrete north light barrel vault.

Thermal insulation

The thin concrete shell offers poor resistance to the transfer of heat, and some form of insulating soffit lining is necessary to meet the requirements of the Building Regulations. This is difficult to achieve without causing thermal bridges and also avoiding interstitial condensation between the insulation and the concrete structure, which adds considerably to the cost of the shell.

Combined this makes concrete shells largely unsuitable for buildings which are to be heated.

Expansion joints

To limit expansion and contraction caused by changes in temperature, continuous expansion joints are formed at intervals of approximately 30 m along the span and across the width of multi-bay, multi-span barrel roofs. The expansion joints are formed by erecting separate shell structures, each with its own supports and with a flexible joint material between neighbouring elements (see Figure 4.101). Vertical expansion joints are made so as to form a continuous joint to the ground with double columns either side of the joint. Longitudinal expansion joints are formed in a valley with upstands weathered with non-ferrous cappings over the joint.

Roof covering

A variety of materials may be used to cover concrete shells, the choice depending on the use of the building and to a certain extent the position of the thermal insulation. Lightweight materials such as thin non-ferrous sheet metal, bitumen felt and plastic membranes may be used.

Walls

The walls of shell structures between the columns are non-loadbearing, their purpose being to provide shelter, security and privacy as well as thermal and sound insulation. Thus a variety of partition wall constructions may be used, from brick and blockwork to timber and steel studwork with facing panels.

Conoid and hyperboloid shell roofs

Reinforced concrete conoid shell

In this shell form the curvature and rise of the shell increases from a shallow curve to a steeply curved end in which the north light glazing is fixed, as illustrated in Figure 4.94. The glazed end of each shell consists of a reinforced concrete or steel lattice, which serves as a stiffening beam to resist deformation of the shell. Edge beams resist spreading of the shell as previously described.

Hyperbolic paraboloid shells

The hyperbolic paraboloid shells provide dramatic shapes and structural possibilities of doubly curved shells. The name hyperbolic paraboloid comes from the geometry of the shape: the horizontal sections through the surface are hyperbolas and the vertical sections parabolas, as illustrated in Figures 4.102 and 4.103. The structural significance of this shape is that at every point on the

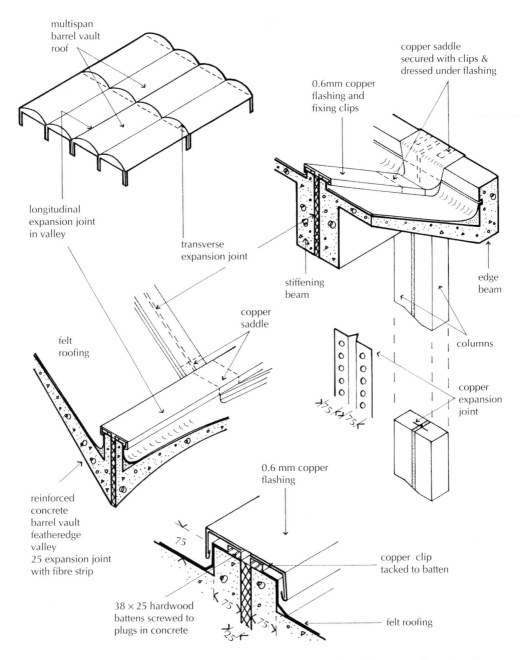

multispan barrel vault roof

copper saddle secured with clips & dressed under flashing

0.6mm copper flashing and fixing clips

longitudinal expansion joint in valley

transverse expansion joint

stiffening beam

edge beam

copper saddle

felt roofing

columns

copper expansion joint

0.6 mm copper flashing

reinforced concrete barrel vault featheredge valley 25 expansion joint with fibre strip

copper clip tacked to batten

75

38 × 25 hardwood battens screwed to plugs in concrete

75

75

25

felt roofing

Figure 4.101   Expansion joints and flashings in reinforced concrete barrel vaults. Expansion joints at intervals of not more than 30 m.

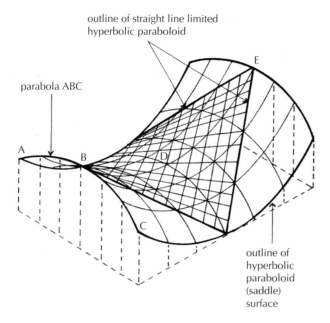

outline of straight line limited
hyperbolic paraboloid

parabola ABC

E

A

B

D

C

outline of
hyperbolic
paraboloid
(saddle)
surface

Figure 4.102   Hyperbolic paraboloid (saddle) surface.

surface, straight lines, which lie in the surface, intersect so that in effect the surface is made up of a network of intersecting straight lines. Thus the centring (formwork) can consist of thin straight sections of timber, which are simple to fix and support.

Reinforced concrete hyperbolic paraboloid shell

Figure 4.104 illustrates an umbrella roof formed from four hyperbolic paraboloid surfaces supported on one column. The small section reinforcing mesh in the surface of the shell resists tensile and compressive stress and the heavier reinforcement around the edges and between the four hyperbolic paraboloid surfaces resists shear forces developed by the tensile and compressive stress in the shell. A series of these roofs can be combined, with glazing between them, to provide shelter to the area below.

Timber shell structures

Timber barrel vaults

Single- and multi-bay barrel vaults can be constructed from small section timber with spans and widths similar to reinforced concrete barrel vaults (Figure 4.105). The vault is formed from layers of boards glued and mechanically fixed together and stiffened with ribs at close centres. The timber ribs serve to both

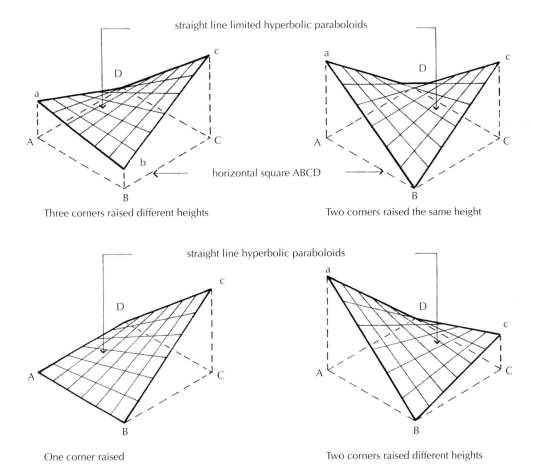

Figure 4.103 **Setting out straight line limited hyperbolic paraboloid surfaces on a square base.**

stiffen the shell and also to maintain the boards' curvature over the vault. Glued laminated edge and valley beams are formed to resist spreading of the vault. Timber barrel vaults have some advantage over concrete in that the material performs better in terms of providing some thermal insulation. Indeed, it is easier to include thermal insulation within the construction while maintaining the visual integrity of the shell.

Timber hyperbolic paraboloid shell

Timber can also be used to form hyperbolic paraboloid shell structures (Figure 4.106). Laminated boards and edge beams are used. Low points of the shell are usually anchored to concrete abutments/ground beams to prevent the shell from spreading under load.

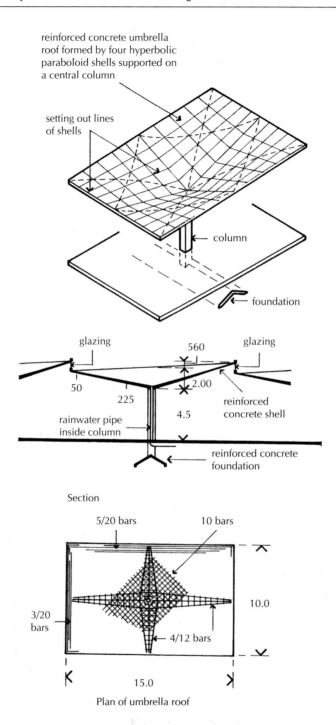

reinforced concrete umbrella
roof formed by four hyperbolic
paraboloid shells supported on
a central column

setting out lines
of shells

column

foundation

glazing   560   glazing

50   2.00

225   reinforced
concrete shell

rainwater pipe   4.5
inside column

reinforced concrete
foundation

Section

5/20 bars   10 bars

3/20
bars   10.0

4/12 bars

15.0

Plan of umbrella roof

Figure 4.104   Reinforced concrete hyperbolic paraboloid.

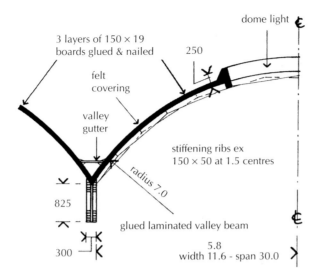

3 layers of 150 × 19
boards glued & nailed

250

dome light

felt
covering

valley
gutter

stiffening ribs ex
150 × 50 at 1.5 centres

radius 7.0

825

300

glued laminated valley beam

5.8
width 11.6 - span 30.0

Figure 4.105   Timber barrel vault.

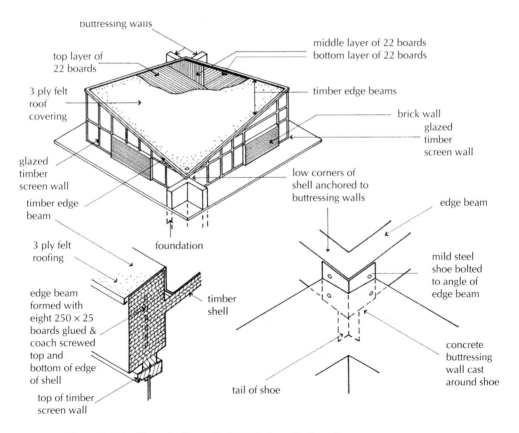

buttressing walls

middle layer of 22 boards
bottom layer of 22 boards

top layer of
22 boards

3 ply felt
roof
covering

timber edge beams

brick wall
glazed
timber
screen wall

glazed
timber
screen wall

timber edge
beam

low corners of
shell anchored to
buttressing walls

edge beam

3 ply felt
roofing

foundation

mild steel
shoe bolted
to angle of
edge beam

edge beam
formed with
eight 250 × 25
boards glued &
coach screwed
top and
bottom of edge
of shell

timber
shell

top of timber
screen wall

tail of shoe

concrete
buttressing
wall cast
around shoe

Figure 4.106   Hyperbolic parboloid timber shell roof.

5 Structural Steel Frames

Following the inventions in 1856 by Bessemer and in 1865 by Siemens and Martin of processes of converting iron into steel, mild steel began to compete with wrought iron and cast iron as a structural material. Up to the beginning of the Second World War (1939) the majority of tall buildings in this country were constructed with structural steel frames, generally clad with brick or masonry to give the appearance of large brick or masonry loadbearing structures. The shortage of steel that followed the Second World War encouraged the use of reinforced concrete frames for buildings up to about 1980. Since 1980, due to overproduction of steel and resulting competition, structural steel frames may be cheaper than comparable reinforced concrete frames. The growing economies and markets around the world have an effect on steel prices. For example, China's recent expansion has placed significant demands on construction materials.

The advantages of the structural steel frame are the speed of erection of the ready prepared steel members and the accuracy of setting out and connections that is a tradition in engineering works. Accurate placing of steel members, with small tolerances, facilitates the fixing of cladding materials. With the use of sprayed on or dry lining materials to encase steel members to provide protection against damage by fire, a structural steel frame may be cheaper than a reinforced concrete structural frame because of speed of erection and economy in material and construction labour costs.

5.1 Functional requirements

The functional requirements of a structural frame are:

❑ Strength and stability
❑ Durability and freedom from maintenance
❑ Fire safety

Strength and stability

The requirements from the Building Regulations are that buildings be constructed so that the loadbearing elements, foundations, walls, floors and roofs have adequate strength and stability to support the dead loads of the construction and anticipated imposed loads on roofs, floors and walls without such undue deflection or deformation as might adversely affect the strength and stability of parts or the whole of the building. The strength of the loadbearing elements of the structure is assumed either from knowledge of the behaviour of similar traditional elements, such as walls and floors under load, or by calculations of the behaviour of parts or the whole of a structure under load, based on data from experimental tests, with various factors of safety to make allowance for unforeseen construction or design errors. The strength of individual elements of a structure may be reasonably accurately assessed taking account of tests on materials and making allowance for variations of strength in both natural and manmade materials.

The strength of combinations of elements such as columns and beams depends on the rigidity of the connection and the consequent interaction of the elements. Simple calculations, based on test results, of the likely behaviour of the joined elements, or a more complex calculation of the behaviour of the parts of the whole of the structure can be made. Various factors of safety are included in calculations to allow for unforeseen circumstances. Calculations of structural strength and stability provide a mathematical justification for an assumption of a minimum strength and stability of structures in use.

Imposed loads are those loads that it is assumed the building or structure is designed to support taking account of the expected occupation or use of the building or structure. Assumptions are made of the likely maximum loads that the floors of a category of building may be expected to support. The load of the occupants and their furniture on the floors of residential buildings will generally be less than that of goods stored on a warehouse floor.

The loads imposed on roofs by snow are determined by taking account of expected snow loads in the geographical location of the building. Loads imposed on walls and roofs by wind (wind loads) are determined by reference to the situation of the building on a map of the United Kingdom on which basic wind speeds have been plotted. These basic wind speeds are the maximum gust speeds averaged over 3-second periods, which are likely to be exceeded on average only once in 50 years. In the calculation of the wind pressure on buildings, a correction factor is used to take account of the shelter from wind afforded by obstructions and ground roughness.

The stability of a building depends initially on a reasonably firm, stable foundation. The stability of a structure depends on the strength of the materials of the loadbearing elements in supporting, without undue deflection or deformation, both concentric and eccentric loads on vertical elements and the ability of the structure to resist lateral pressure of wind on walls and roofs.

The very considerable dead weight of walls of traditional masonry or brick construction is generally sufficient, by itself, to support concentric and eccentric loads and the lateral pressure of wind. Generally, the dead weight of skeleton framed multi-storey buildings is not, by itself, capable of resisting lateral wind pressure without undue deflection and deformation. Some form of bracing is required to enhance the stability of skeleton framed buildings. Unlike the joints in a reinforced concrete structural frame, the normal joints between vertical and horizontal members of a structural steel frame do not provide much stiffness in resisting lateral wind pressure.

Disproportionate collapse

A requirement from the Building Regulations is that a building shall be constructed so that, in the event of an accident, the building will not suffer collapse to an extent disproportionate to the cause. This requirement applies only to a building having five or more storeys (each basement level being counted as one storey), excluding a storey within the roof space where the slope of the roof does not exceed 70° to the horizontal.

Durability and freedom from maintenance

The members of a structural steel frame are usually inside the wall fabric of buildings so that in usual circumstances the steel is in a comparatively dry atmosphere, which is unlikely to cause progressive, destructive corrosion of steel. Structural steel will, therefore, provide reasonable durability for the expected life of the majority of buildings and require no maintenance. Where the structural steel frame is partially or wholly built into the enclosing masonry or brick walls, the external wall thickness is generally adequate to prevent such penetration of moisture as is likely to cause corrosion of steel. Where there is some likelihood of penetration of moisture to the structural steel, it is practice to provide protection by the application of paint or bitumen coatings or the application of a damp-proof layer. Where it is anticipated that moisture may cause corrosion of the steel, either externally or from a moisture-laden interior, weathering steels, that are much less subject to corrosion, are used.

Fire safety

The requirements from the Building Regulations are concerned to ensure a reasonable standard of safety in case of fire. The application of the Regulations, as set out in the practical guidance given in Approved Document B, is directed to the safe escape of people from buildings in case of fire rather than the protection of the building and its contents. Insurance companies that provide cover against the risks of damage to the building and contents by fire may require additional fire protection such as sprinklers.

Internal fire spread (structures)

The requirement from the Regulations relevant to structure is to limit internal fire spread (structure). As a measure of ability to withstand the effects of fire, the elements of a structure are given notional fire resistance times, in minutes, based on tests. Elements are tested for their ability to withstand the effects of fire in relation to:

❏ Resistance to collapse (loadbearing capacity), which applies to loadbearing elements
❏ Resistance to fire penetration (integrity), which applies to fire separating elements
❏ Resistance to the transfer of excessive heat (insulation), which applies to fire separating elements

The notional fire resisting times, which depend on the size, height, number of basements and use of buildings, are chosen as being sufficient for the escape of occupants in the event of fire. The requirements for the fire resistance of elements of a structure do not apply to:

❏ A structure that supports only a roof unless:
   (a) the roof acts as a floor, e.g. car parking, or as a means of escape; and
   (b) the structure is essential for the stability of an external wall, which needs to have fire resistance.
❏ The lowest floor of the building.

5.2 Methods of design

There are a number of established approaches to the method of design of structural steel frames.

Permissible stress design method

With the introduction of steel as a structural material in the late 19th and early years of the 20th century, the permissible stress method of design was accepted as a basis for the calculation of the sizes of structural members. Having established and agreed a yield stress for mild steel, the permissible tensile stress was taken as the yield stress divided by a factor of safety to allow for unforeseen overloading, defective workmanship and variations in steel. The yield stress in steel is that stress at which the steel no longer behaves elastically and suffers irrecoverable elongation, as shown in Figure 5.1, which is a typical stress/strain curve for mild steel.

The loads to be carried by a structural steel frame are dead, imposed and wind loads. Dead loads comprise the weight of the structure including walls, floors, roof and all permanent fixtures. Imposed loads include all moveable items that are stored on or usually supported by floors, such as goods, people, furniture and moveable equipment. Wind loads are those applied by wind pressure or suction on the building. Dead loads can be accurately calculated. Imposed loads are assumed from the usual use of the building to give reasonable

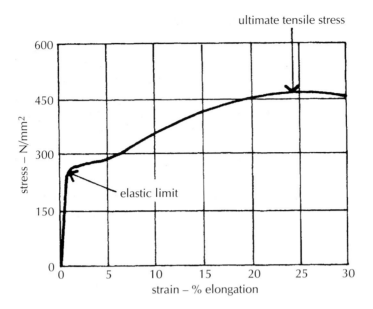

Figure 5.1 Stress/strain curve for mild steel.

maximum loads that are likely to occur. Wind loads are derived from the maximum wind speeds.

Having determined the combination of loads that are likely to cause the worst working conditions the structure is to support, the forces acting on the structural members are calculated by the elastic method of analysis to predict the maximum elastic working stresses in the members of the structural frame. Beam sections are then selected so that the maximum predicted stress does not exceed the permissible stress. In this calculation a factor of safety is applied to the stress in the material of the structural frame. The permissible compressive stress depends on whether a column fails due to buckling or yielding and is determined from the slenderness ratio of the column, Young's modulus and the yield stress divided by a factor of safety. The permissible stress method of design provides a safe and reasonably economic method of design for simply connected frames and is the most commonly used method of design for structural steel frames.

A simply connected frame is a frame in which the beams are assumed to be simply supported by columns to the extent that while the columns support beam ends, the beam is not fixed to the column and in consequence when the beam bends (deflects) under load, bending is not restrained by the column. Where a beam bears on a shelf angle fixed to a column and the top of the beam is fixed to the column by means of a small top cleat designed to maintain the beam in a vertical position, it is reasonable to assume that the beam is simply supported and will largely behave as if it had a pin jointed connection to the column.

Collapse or load factor method of design

Where beams are rigidly fixed to columns and where the horizontal or near horizontal members of a frame, such as the portal frame, are rigidly fixed to posts or columns then beams do not suffer the same bending under load that they would if simply supported by columns or posts. The effect of the rigid connection of beam ends to columns is to restrain simple bending, as illustrated in Figure 5.2. The fixed end beam bends in two directions, upwards near fixed ends and downwards at the centre. The upward bending is termed negative bending and the downward positive bending. It will be seen that bending at the ends of the beam is prevented by the rigid connections that take some of the stress due to loading and transfer it to the supporting columns. Just as the rigid connection of beam to column causes negative or upward bending of the beam at the ends, so a comparable, but smaller, deformation of the column will occur.

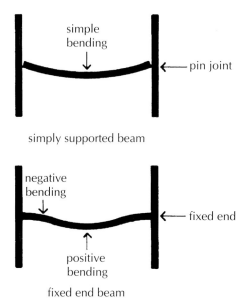

Figure 5.2   Comparison of pin-jointed and fixed end beams.

Using the elastic method of analysis to determine working stress in a fixed end beam, to select a beam section adequate for the permissible stress the design method produces a section greater than is needed to provide a reasonable factor of safety against collapse, because in practice the permissible stress is not reached and in consequence the beam could safely support a greater load.

The collapse or load factor method of design seeks to provide a load factor, that is a safety factor, against collapse applied to particular types of structural frame for economy in the use of materials by using the load factor which is applied to the loads instead of stress in materials. The load factor method was developed principally for use in the design of reinforced concrete and welded connection steel frames with rigid connections as an alternative to the permissible stress method, as a means to economy in the selection of structural sections. In the use of the load factor method of design, plastic analysis is used. In this method of analysis of the forces acting in members it is presumed that extreme fibre stress will reach or exceed yield stress and the fibres behave plastically. This is a valid assumption as in practice the fibres of the whole section play a part in sustaining stress, and under working loads extreme fibre stress would not reach yield point.

Limit state method of design

The purpose of structural analysis is to predict the conditions applicable to a structure that would cause it to become either unserviceable in use or unable to support loads to the extent that members might fail.

In the permissible stress method a limit is set on the predicted working stress in the members of the frame by the use of a factor of safety applied to the predicted yield stress of the materials used. In the load factor method of design a limit is set on the working loads to ensure that they do not exceed a limit determined by the application of a factor of safety to the loads that would cause collapse of the structure. The limit state method of design seeks to determine the limiting states of both materials and loads that would cause a particular structure to become unserviceable in use or unsafe due to excessive load. The limiting conditions that are considered are serviceability during the useful life of the building and the ultimate limit state of strength.

Serviceability limit states set limits to the behaviour of the structure to limit excessive deflection, excessive vibration and irreparable damage due to material fatigue or corrosion that would otherwise make the building unserviceable in use. Ultimate limit states of strength set limits to strength in resisting yielding, rupture, buckling and transformation into a mechanism, and stability against overturning and fracture due to fatigue or low temperature brittleness.

In use the limit state method of design sets characteristic loads and characteristic strengths, which are those loads and strengths that have an acceptable chance of not being exceeded during the life of the building. To take account of the variability of loads and strength of materials in actual use, a number of partial safety factors may be applied to the characteristic loads and strengths, to determine safe working loads and strengths.

The limit state method of design has not been accepted wholeheartedly by structural engineers because, they say, it is academic, highly mathematical, increases design time and does not lead to economic structures. The opponents of limit state design prefer and use permissible stress design for simplicity in execution and the knowledge that the use of the more complex limit state design method may be rewarded with little significant reduction in frame sections. Structural engineers profess to predict the likely behaviour of a structure from an acceptance of working loads and yield stresses in materials so as to design a structure that will be both safe and serviceable during the life of a building. There is often little reward in employing other than the permissible stress method of design for the majority of buildings so that the use of the limit state method is confined in the main to larger and more complex structures where the additional design time is justified by more adventurous and economic design.

5.3 Steel sections

Mild steel is the material generally used for constructional steelwork. It is produced in several basic strength grades of which those designated as 43, 50 and 55 are most commonly used. The strength grades 43, 50 and 55 indicate minimum ultimate tensile strengths of 430, 500 and 550 N/mm², respectively. Each strength grade has several sub-grades indicated by a letter between A and E; the grades that are normally available are 43A, 43B, 43C, 43D, 43E, 50A, 50B, 50C, 50D and 55C. In each strength grade the sub-grades have similar ultimate tensile strengths and as the sub-grades change from A to E the specification becomes more stringent, the chemical composition changes and the notch ductility improves. The improvement in notch ductility (reduction in brittleness), particularly at low temperatures, assists in the design of welded connections and reduces the risk of brittle and fatigue failure, which is of particular concern in structures subject to low temperatures.

Properties of mild steel

Strength
Steel is strong in both tension and compression with permitted working stresses of 165, 230 and 280 N/mm² for grades 43, 50 and 55, respectively. The strength to weight ratio of mild steel is good so that mild steel is able to sustain heavy loads with comparatively small self weight.

Elasticity
Under stress induced by loads a structural material will stretch or contract by elastic deformation and return to its former state once the load is removed. The ratio of stress to strain, which is known as Young's modulus (the modulus of elasticity), gives an indication of the resistance of the material to elastic deformation. If the modulus of elasticity is high the deformation under stress will be low. Steel has a high modulus of elasticity, 200 kN/mm², and is therefore a comparatively stiff material, which will suffer less elastic deformation than aluminium, which has a modulus of elasticity of 69 kN/mm². Under stress induced by loads beams bend or deflect and in practice this deflection under load is limited to avoid cracking of materials fixed to beams. The sectional area of a mild steel beam can be less than that of other structural materials for given load, span and limit of deflection.

Ductility

Mild steel is a ductile material which is not brittle and can suffer strain beyond the elastic limit through what is known as plastic flow, which transfers stress to surrounding material so that at no point will stress failure in the material be reached. Because of the ductility of steel the plastic method of analysis can be used for structures with rigid connections, which makes allowance for transfer of stress by plastic flow and so results in a section less than would be determined by the elastic method of analysis, which does not make allowance for the ductility of steel.

Resistance to corrosion

Corrosion of steel occurs as a chemical reaction between iron, water and oxygen to form hydrated iron oxide, commonly known as rust. Because rust is open grained and porous a continuing reaction will cause progressive corrosion of steel. The chemical reaction that starts the process of corrosion of iron is affected by an electrical process through electrons liberated in the reaction, whereby small currents flow from the area of corrosion to unaffected areas and so spread the process of corrosion. In addition, pollutants in air accelerate corrosion as sulphur dioxides from industrial atmospheres and salt in marine atmospheres increase the electrical conductivity of water and so encourage corrosion. The continuing process of corrosion may eventually, over the course of several years, affect the strength of steel. Mild steel should therefore be given protection against corrosion in atmospheres likely to cause corrosion.

Fire resistance

Although steel is non-combustible and does not contribute to fire it may lose strength when its temperature reaches a critical point in a fire in a building. A temperature of 550°C is generally accepted as the critical temperature for steel, which will generally be reached in the early stages of a fire. To give protection against damage by fire, Building Regulations require fire protection of structural steelwork in certain situations.

Weathering steels

The addition of small quantities of certain elements modifies the structure of the rust layer that forms. The alloys encourage the formation of a dense, fine-grained rust film and also react chemically with sulphur in atmospheres to form insoluble basic sulphate salts which block the pores on the film and so prevent further rusting. The thin, tightly-adherent film that forms on this low alloy steel is of such low permeability that the rate of corrosion is reduced almost to zero. The film forms a patina of a deep brown colour on the surface of steel. The low permeability rust film forms under normal wet/dry cyclical conditions. In conditions approaching constant wetness and in conditions exposed to severe marine or salt spray conditions the rust film may remain porous and not prevent further corrosion. Weathering steels are produced under the brand names

'Cor-Ten' for rolled sections and 'Stalcrest' for hollow sections. Cor-Ten is a particular favourite of architects for its appearance.

Standard rolled steel sections

The steel sections most used in structural steelwork are standard hot rolled steel universal beams and columns together with a range of tees, channels and angles illustrated in Figure 5.3. Universal beams and columns are produced in

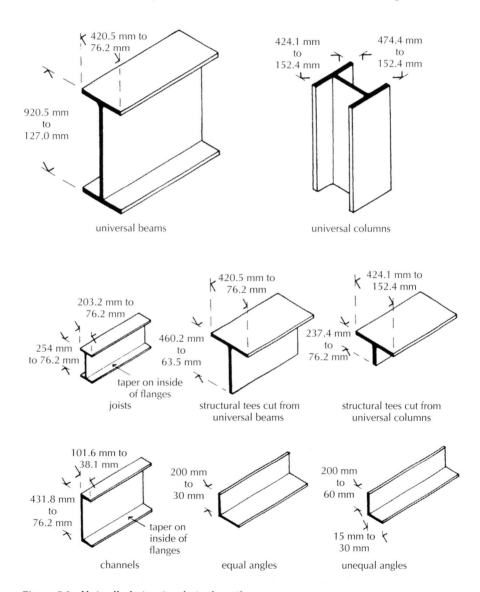

Figure 5.3   Hot rolled structural steel sections.

a range of standard sizes and weights designated by serial sizes. Within each serial size the inside dimensions between flanges and flange edge and web remain constant and the overall dimensions and weights vary, as illustrated in Figure 5.4. This grouping of sections in serial sizes is convenient for production within a range of rolling sizes and for the selection of a suitable size and weight by the designer. The deep web to flange dimensions of beams and the near similar flange to web dimensions of columns are chosen to suit the functions of the structural elements. Because of the close similarity of the width of the flange to the web of column sections, they are sometimes known as 'broad flange sections'.

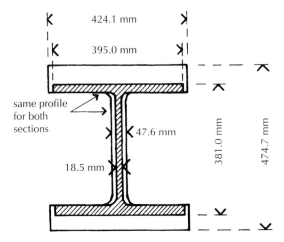

Figure 5.4   Universal columns.

A range of comparatively small section 'joists' is also available, which have shallowly tapered flanges and are produced for use as beams for small to medium spans. The series of structural tees is produced from cuts that are half the web depth of standard universal beams and columns. The range of standard hot rolled structural steel angles and channels have tapered flanges. The standard rolled steel sections are usually supplied in strength grade 43A material with strength grades 50 and 55 available for all sections at an additional cost per tonne. All of the standard sections are available in Cor-Ten B weathering steel.

Castella beam

An open web beam can be fabricated by cutting the web of a standard section mild steel beam along a castellated line illustrated in Figure 5.5. The two cut sections of the beam are then welded together to form an open web, castellated beam, which is one and a half times the depth of the beam from which it was

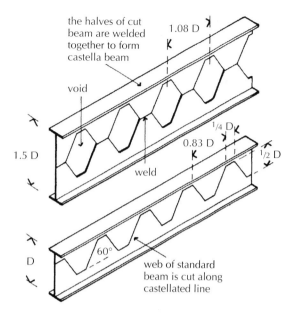

Figure 5.5   Castella beam.

formed. Because of the increase in depth, the castella beam will suffer less deflection (bending) under light loads. The castella beam is no stronger than the beam from which it was cut but will suffer less deflection under load. The increase in cost due to fabrication and the reduction in weight of the beam as compared to a solid web beam of the same depth and section justify the use of these beams for long span, lightly loaded beams particularly for roofs. The voids in the web of these beams are convenient for housing runs of electrical and heating services.

Steel tubes

A range of seamless and welded seam steel tubes is manufactured for use as columns, struts and ties. The use of these tubes as columns is limited by the difficulty of making beam connections to a round section column. These round sections are the most efficient and compact structural sections available and are extensively used in the fabrication of lattice girders, columns, frames, roof decks and trusses for economy, appearance and comparative freedom from dust traps. Connections are generally made by scribing the ends of the tube to fit around the round sections to which they are welded. For long-span members such as roof trusses, bolted plate connections are made at mid span for convenience in transporting and erecting long-span members in sections.

Hollow rectangular and square sections

Hollow rectangular and square steel sections are made from hollow round sections of steel tube, which are heated until they are sufficiently malleable to be deformed. The heated tubes are passed through a series of rollers, which progressively change the shape of the tube to square or rectangular sections with rounded edges, as illustrated in Figure 5.6. To provide different wall thicknesses, the heated tube can be gradually stretched. The advantage of these sections is that they are ideal for use as columns as the material is uniformly

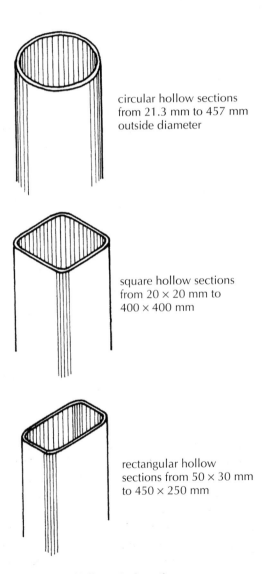

circular hollow sections
from 21.3 mm to 457 mm
outside diameter

square hollow sections
from 20 × 20 mm to
400 × 400 mm

rectangular hollow
sections from 50 × 30 mm
to 450 × 250 mm

Figure 5.6   Hollow steel sections.

disposed around the long axis, and the square or rectangular section facilitates beam connections.

Hollow square and rectangular sections are much used as the members of lattice roof trusses and lightly loaded framed structures with the square sections for columns and the rectangular sections as beams. The economy in material and neat appearance of the sections, which with welded connections have a more elegant appearance than angle connections, recommend their use. These sections are also much used in the fabrication of railings, balustrades, gates and fences with welded connections for the neat, robust appearance of the material. Prefabricated sections can be hot dip galvanised to inhibit rusting prior to painting.

Cold roll-formed steel sections
Cold roll-formed structural steel sections are made from hot rolled steel strip, which is passed through a series of rollers. Each pair of rollers progressively takes part in gradually shaping the strip to the required shape. As the strip is cold formed it has to be passed through a series of rollers to avoid the thin material being torn or sheared in the forming process, which produces sections with slightly rounded angles to this end. There is no theoretical limit to the length of steel strip that can be formed. The thickness of steel strip commonly used is from 0.3 to 0.8 mm and the width of strip up to about 1 m. A very wide range of sections is possible with cold rolled forming, some of which are illustrated in Figure 5.7.

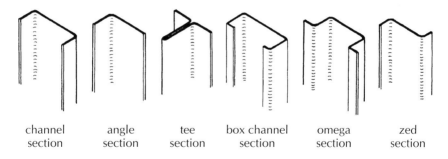

| channel section | angle section | tee section | box channel section | omega section | zed section |

Figure 5.7   Cold roll-formed sections.

The advantage of cold rolled forming is that any shape can be produced to the exact dimensions to suit a particular use or design. Figure 5.8 is an illustration of cold-formed sections, spot welded back to back to form structural steel beam sections, and sections welded together to form box form column sections. Connections of cold-formed sections are made by welding self-tapping screws or bolts to plate cleats welded to one section. Because of the comparatively thin material from which the sections are formed, it is necessary to use some coating that will inhibit corrosion and some form of casing as protection against early damage by fire where regulations so require. Cold formed steel sections

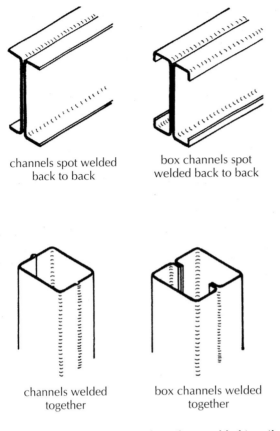

channels spot welded
back to back

box channels spot
welded back to back

channels welded
together

box channels welded
together

Figure 5.8 Cold roll-formed sections welded together.

are extensively used in the manufacture of roof trusses and lattice beams and frames. The fabricated sections are protected with a galvanised coating where practical. Cold-formed, pressed steel sections are much used for floor and roof decking for the floors of framed buildings and also for metal doors, frames and metal trim such as skirtings.

5.4 Structural steel frames

The earliest structural steel frame was erected in Chicago in the year 1883 for the Home Insurance building. At that time there were no limitations to the height of buildings, and the introduction of the passenger lift provided access to multi-storey buildings. Property taxes at that time were levied on site area so that developers were encouraged to obtain the maximum lettable floor area. At the time, the traditional method of building was solid loadbearing walls of stone or brick. To use this system of building for multi-storey buildings would have necessitated walls of such thickness that there would have been an appreciable loss of floor area. The skeleton steel frame was introduced to reduce the thickness of external walls and so gain valuable floor space. A skeleton of steel columns and beams carried the whole of the load of floors and the solid masonry or brick walls, the least thickness of which was dictated by weather resistance rather than loadbearing requirements. Since then the steel frame has been one of the principal methods of constructing multi-storey buildings.

Skeleton frame

The conventional steel frame is constructed with hot rolled section beams and columns in the form of a skeleton designed to support the whole of the imposed and dead loads of floors, external walling or cladding and wind pressure. The arrangement of the columns is determined by the floor plans, horizontal and vertical circulation spaces and the requirements for natural light to penetrate the interior of the building. Figure 5.9 is an illustration of a typical rectangular grid skeleton steel frame. In general, the most economic arrangement of columns is on a regular rectangular grid with columns spaced at 3.0 to 4.0 m apart, parallel to the span of floors which bear on floor beams spanning up to 7.5 m with floors designed to span one way between main beams. This arrangement provides the smallest economic thickness of floor slab and least

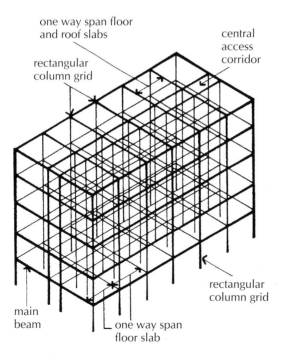

Figure 5.9   Rectangular grid steel frame.

depth of floor beams, and therefore least height of building for a given clear height at each floor level.

Figure 5.10 is an illustration of a typical small skeleton steel frame designed to support one-way span floors on main beams and beams to support solid walls at each floor level on the external faces of the building. This rectangular grid can be extended in both directions to provide the required floor area.

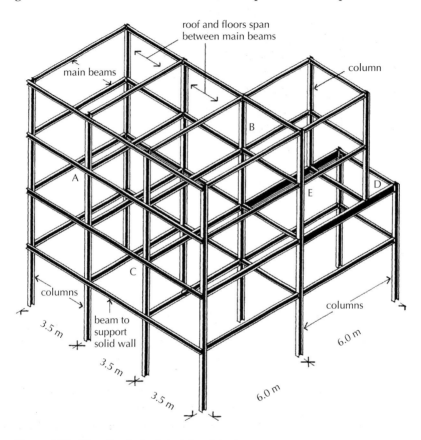

Figure 5.10   Structural steel skeleton frame.

Where comparatively closely-spaced columns may obstruct internal floor space, a larger rectangular or square grid of columns is used. The columns support main beams, which in turn support secondary beams spaced at up to 4.5 m apart to carry one-way span floor slabs, as illustrated in Figure 5.11. This arrangement allows for the least span and thickness of floor slab and the least weight of construction.

A disadvantage of this layout is that the increased span of main beams requires an increase in their depths so that they will project below the underside of the secondary beams. Heating, ventilating and electrical services, which are suspended, and run below the main beams are usually hidden above a suspended ceiling. The consequence is that the requirement for comparatively

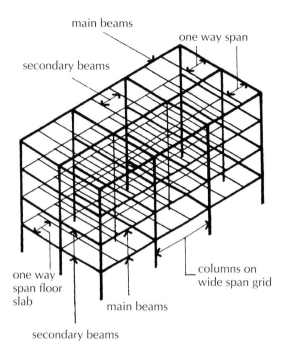

main beams

one way span

secondary beams

one way span floor slab

columns on wide span grid

main beams

secondary beams

Figure 5.11   Wide span column grid.

unobstructed floor space causes an increase in the overall height of a building because of the increase in depth of floor from floor finish to suspended ceiling at each floor level. Where there is a requirement for a large floor area, unobstructed by columns, either a deep long-span solid web beam, a deep lattice girder or Vierendeel girders are used.

The advantage of using deep lattice girders or Vierendeel girders is that they may be designed so that their depth occupies the height of a floor and does not, therefore, increase the overall height of construction. The Vierendeel girder illustrated in Figure 5.12 is fabricated from mild steel plates, angles, channels and beam sections, which are cut and welded together to form an open web beam. The advantage of the open web form is that it can accommodate both windows externally and door openings internally, unlike the diagonals of a lattice girder of the same depth. The solid parts of the web of this girder are located under the columns they are designed to support. The specialist fabrication of this girder together with the cost of transporting and hoisting into position involves considerable cost.

The conventional structural steel frame comprises continuous columns which support short lengths of beam that are supported on shelf angles bolted to the columns. Where there is a requirement, for example, for the structural frame to overhang a pavement, the frame has to be cantilevered out, as illustrated in Figure 5.13. To support the columns on the external face of the

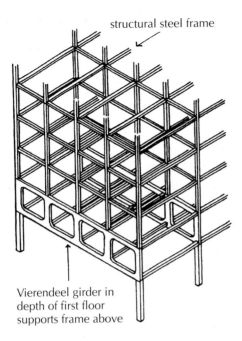

Figure 5.12   **Vierendeel girder.**

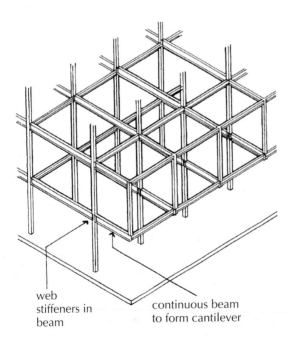

Figure 5.13   **Continuous beam to form cantilever.**

cantilever, it is necessary to use shorter lengths of main external column carried on continuous cantilever beams. The continuous cantilever beam is carried back and connected to an internal column. To support the cantilever a short length of column is connected under the cantilever beam to which web stiffeners are welded to reinforce the beam against web buckling.

Parallel beam structural steel frame

This type of structural steel frame uses double main or spine beams fixed each side of internal columns to support secondary rib beams that support the floor. The principal advantage of this form of structure is improved flexibility for the services, which can be located in both directions within the grid between the spine beams in one direction and the rib beams in the other. The advantage of using two parallel main spine beams is simplicity of connections to columns and the use of continuous long lengths of beam independent of column grid, which reduces fabrication and erection complexities and the overall weight of steel by the use of continuity of beams.

The most economical arrangement of the frame is a rectangular grid with the more lightly loaded rib beams spanning the greater distance between the more heavily-loaded spine or main beams. Where long-span ribs are used, for reasons of convenience in internal layout or for convenience in running services or both, a square grid may be most suitable. The square grid illustrated in Figure 5.14 uses double spine or main beams to internal columns with pairs of rib beams fixed to each side of columns with profiled steel decking and composite construction structural concrete topping fixed across the top of the rib beams. The spine beams are site bolted to end plates welded to short lengths of channel section steel that are shop welded to the columns. At the perimeter of the building a single spine beam is bolted to the end plate of channel sections welded to the column.

The parallel beam structural frame may be used, with standard I section beams and columns or with hollow rectangular section columns and light section rolled steel sections or cold formed strip steel beams and ribs, for smaller buildings supporting moderate floor loads in which there is need for provision for the full range of electric and electronic cables and air conditioning. Although the number of steel sections used for each grid of the framework in this system is greater than that needed for the conventional steel frame, there is generally some appreciable saving in the total weight and, therefore, the cost of the frame, and appreciable saving in the erection time due to the simplicity of connections. The overall depth of the structural floor is greater than that of a similar conventional structural steel frame. As all the services common to modern buildings may be housed within the structural depth, rather than being slung below the structural floor of a conventional frame above a suspended ceiling, there may well be less overall height of building for given clear height between finished floor and ceiling level.

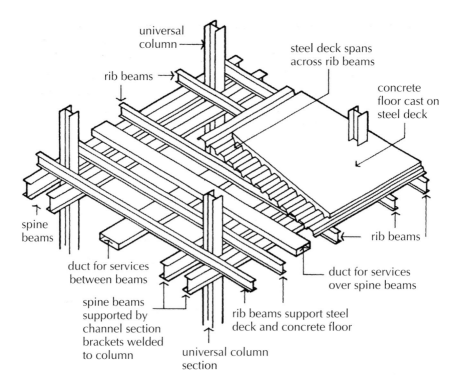

universal
column

rib beams

steel deck spans
across rib beams

concrete
floor cast on
steel deck

spine
beams

rib beams

duct for services
between beams

duct for services
over spine beams

spine beams
supported by
channel section
brackets welded
to column

rib beams support steel
deck and concrete floor

universal column
section

Figure 5.14   Parallel beam structural steel frame.

Pin jointed structural steel frames

The shortage of materials and skilled craftsmen that followed the Second World War encouraged local authorities in the UK to develop systems of building employing standardised components that culminated in the CLASP (Consortium of Local Authorities Special Programme) system of building. The early development was carried out by the Hertfordshire Country Council in 1945 in order to fulfil their school building programme. A system of prefabricated building components based on a square grid was developed, to utilise light engineering prefabrication techniques, aimed at economy by mass production and the reduction of site labour. Some ten years later the Nottinghamshire County Council, faced with a similar problem and in addition the problem of designing a structure to accommodate subsidence due to mining operations, developed a system of building based on a pin jointed steel frame and prefabricated components. The pin jointed frame, with spring loaded diagonal braces, was designed to accommodate earth movements.

In order to gain the benefits of economy in mass production of component parts, the Nottinghamshire County Council joined with other local authorities to form CLASP, which was able to order, well in advance, considerable quantities of standard components at reasonable cost. The CLASP system of building has since been used for schools, offices, housing and industrial buildings of up to four storeys. The system retained the pin jointed frame, originally designed for mining subsidence areas, as being the cheapest light structural steel frame. The CLASP system is remarkable in that it was designed by architects for architects and allows a degree of freedom of design, within standard modules and using a variety of standard components, that no other system of prefabrication has yet to achieve. The CLASP building system is illustrated in Figure 5.15.

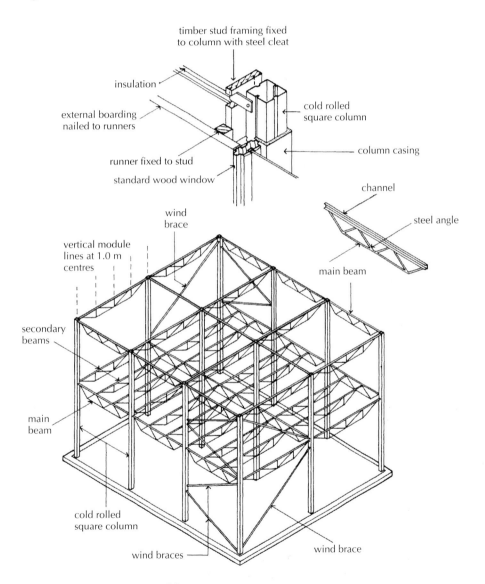

Figure 5.15   Pin jointed steel frame.

Wind bracing

The connections of beams to columns in multi-storey skeleton steel frames do not generally provide a sufficiently rigid connection to resist the considerable lateral wind forces that tend to cause the frame to rack. The word 'rack' is used to describe the tendency of a frame to be distorted by lateral forces that cause right-angled connections to close up against the direction of the force (in the same way that books on a shelf will tend to fall over if not firmly packed in place). To resist racking caused by the very considerable wind forces acting on the faces of a multi-storey building, it is necessary to include some system of cross bracing between the members of the frame to maintain the right-angled connection of members. The system of bracing used will depend on the rigidity of the connections, the exposure, height, shape and construction of the building.

The frame for a 'point block' building, where the access and service core is in the centre of the building and the plan is square or near square, is commonly braced against lateral forces by connecting cross braces in the two adjacent sides of the steel frame around the centre core which are not required for access, as illustrated in Figure 5.16. Wind loads are transferred to the braced centre

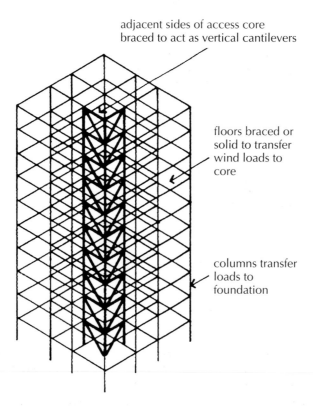

adjacent sides of access core
braced to act as vertical cantilevers

floors braced or
solid to transfer
wind loads to
core

columns transfer
loads to
foundation

Figure 5.16   Wind bracing to central core.

core through solid concrete floors acting as plates or by bracing steel framed floors.

With the access and service core on one face of a structural frame, as illustrated in Figure 5.17, it may be convenient to provide cross bracing to the opposite sides of the service core, leaving the other two sides free for access and natural lighting for toilets, respectively. The bracing to the service core makes it a vertical cantilever anchored to the ground. To transfer wind forces acting on the four faces of the building either one or more of the floors or roof are framed with horizontal cross bracing which is tied to the vertical bracing of the service core. The action of the vertical cross bracing to the service core and the connected horizontal floor cross bracing to a floor or floor and roof will generally provide adequate stiffness against wind forces.

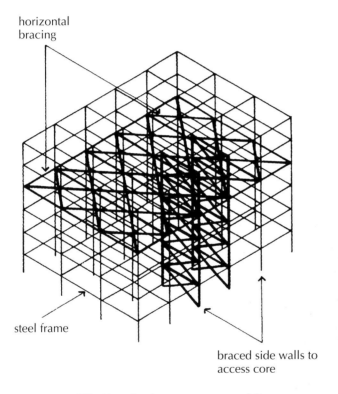

horizontal bracing

steel frame

braced side walls to access core

Figure 5.17   Wind bracing to access core and floors.

A slab block is a building that is rectangular on plan with two main wall faces much wider than the end walls, as illustrated in Figure 5.18. With this design it may be reasonable to accept that the smaller wind forces acting on the end walls will be resisted by the many connections of the two main walls and the horizontal solid plate floors. Here cross bracing to the end walls acting with the

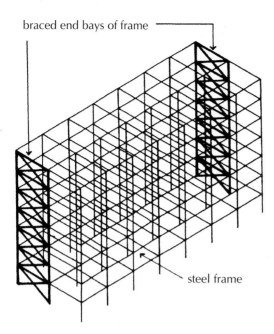

braced end bays of frame

steel frame

Figure 5.18   Wind bracing to end walls.

horizontal plates of the many solid floors may well provide adequate bracing against wind forces. To provide fire protection to means of escape service and access cores to multi-storey buildings, it is common to construct a solid cast insitu reinforced concrete core to contain lifts and escape stairs. A reinforced concrete core by its construction and foundation will act as a very stiff vertical cantilever capable of taking wind forces.

To provide wind bracing to a point block, multi-storey structural steel frame with a central reinforced concrete access core, illustrated in Figure 5.19, it is necessary to transfer wind forces, acting on the walls, to the core. The systems of bracing that are used combine bracing through solid concrete floor plates and cross bracing to structural steel floors. The type of cross bracing illustrated in Figure 5.19 takes the form of braced girders hung from the frame to the four corners of the building and carried back, below floor level, and tied to the core to act as hung, cantilevered cross wind bracing.

Connections and fasteners

Usual practice is to use long lengths of steel column between which shorter lengths of beam are connected to minimise the number of column to column joints and for the convenience of setting beam ends on shelf angles bolted to columns. In making the connections of four beam ends to a column, it is usual

concrete core and steel hangers
support floor beams

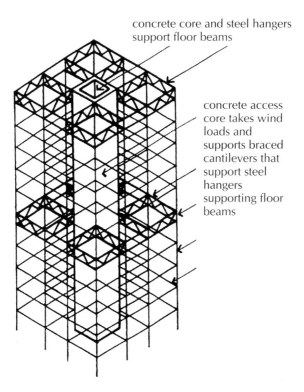

concrete access
core takes wind
loads and
supports braced
cantilevers that
support steel
hangers
supporting floor
beams

Figure 5.19   Reinforced concrete core supporting cantilever beams and steel hangers.

to connect the ends of main beams to the thicker material of column flanges and the secondary, more lightly loaded beam ends to the thinner web material. The ready cut beams are placed on the shelf or seating angles, which have been shop or site bolted to columns, as illustrated in Figure 5.20. The beam ends are bolted to the projecting flanges of the shelf angles. Angle side cleats are bolted to the flange of columns and webs to main beams, and angle top cleats to the web of columns and flanges of secondary beams. The side and top cleats serve the purpose of maintaining beams in their correct position. Where convenient, angle cleats are bolted to columns and beams in the fabricator's shop to reduce site connections to a minimum. These simple cleat connections can be accurately and quickly made to provide support and connections between beams and columns.

An alternative to the simple cleat connection is to use end plates welded to the ends of the beams (Figure 5.21 and Photograph 5.1). The end plate is predrilled with holes that are accurately positioned to line in with predrilled holes in the steel column. The plate can then be bolted to the beam to form a more rigid connection, which will transfer some bending forces (Photograph 5.2).

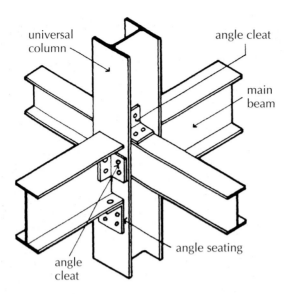

Figure 5.20   Four beam to column connection.

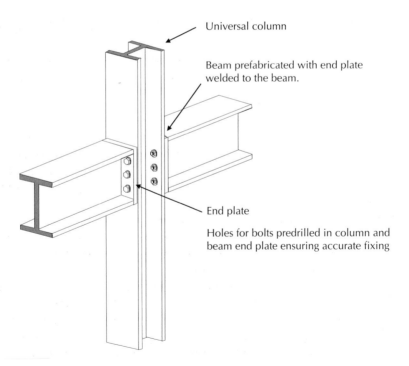

Figure 5.21   Beam connection using end plates welded to the beam.

Photograph 5.1   Column and beam connection using end plates welded to the beam.

Cleat connections of beam to column (Figure 5.20) are generally assumed to provide a simple connection in structural analysis and calculation, as there is little restraint to simple bending by this type of end connection of beams. This simply supported, that is unrestrained, connection is the usual basis of design calculations for structural steel frames as the simple connection provides little restraint to bending, whereas a welded end connection is rigid and affects beam bending.

Figure 5.22 is an illustration of the connection of a main beam to a column on the external face of a building with the external face beam connected across the outside flanges of external columns. The internal beam is supported by a bottom seating angle cleat and top angle cleat. The external beam is fixed continuously across the outer face of columns to provide support for external walling or cladding that is built across the face of the frame. This beam is supported on a beam cutting to which a plate has been welded to provide a level seating for the beam and for bolt fixing to the beam. The beam cutting is bolted to the flange of columns. A top angle cleat is bolted to the beam and column to maintain the beam in its correct upright position.

The connection of long column lengths up the height of a structural steel frame is usually made some little distance above floor beam connections, as

A crane lifts the beam into position. Fitters, elevated on cherry pickers or MEWPs (mobile elevated working platforms), guide the beam into place.

A spiked tool is placed through one beam and column drill holes to align all of the holes.

All of the bolts are securely fastened, checked with a torque wrench and the next beam is lifted into place.

Photograph 5.2   Rigid steel frame assembly — fixing beams to columns.

illustrated in Figure 5.23. The ends of columns are accurately machined flat and level. Cap plates are welded to the ends of columns in the fabricator's shop and drilled ready for site bolted connections. Columns for the top floors of multi-storey buildings will be less heavily loaded than those to the floors below and it may be possible to use a smaller section of column. The connection of these dissimilar section columns is effected through a thick bearing plate welded to the machined end of the lower column and splice plates welded to the outer flange faces. The thick bearing plate will transfer the load from the smaller section column to that of the larger column. The splice plates provide a means of joining the columns. The upper column is hoisted into position on the bearing plate and packing plates are fitted into place to make up the

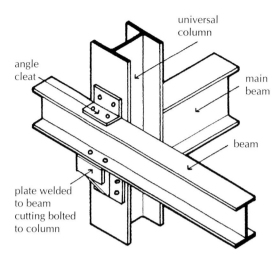

Figure 5.22   External beam to column connection.

difference between the column sections. The connection is then made by bolts through the splice plates, packing pieces and flanges of the upper column, as illustrated in Figure 5.24. It is also common to splice columns together using a bolted connection as shown in Figure 5.25 and Photograph 5.3. Where for design purposes a column is required to take its bearing on a main beam, a simple connection will suffice. A bearing plate is welded to the machined end of the column, ready for bolting to the top flange of the main beam, as illustrated

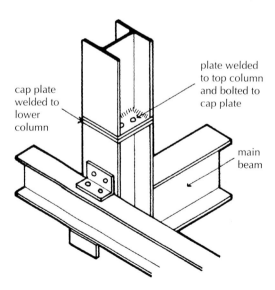

Figure 5.23   Column to column connection.

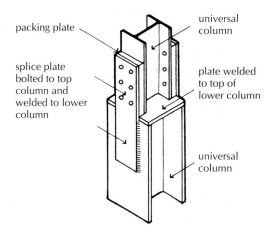

packing plate

universal column

splice plate bolted to top column and welded to lower column

plate welded to top of lower column

universal column

Figure 5.24 Small to larger column connection.

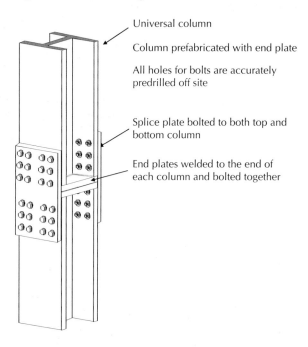

Universal column

Column prefabricated with end plate

All holes for bolts are accurately predrilled off site

Splice plate bolted to both top and bottom column

End plates welded to the end of each column and bolted together

Figure 5.25 Column to column spliced–bolted connection.

in Figure 5.26A. Where a secondary beam is required to take a bearing from a main beam, as for example where a floor is trimmed for a stair well, the end of the secondary beam is notched to fit under and around the top flange of the main beam. The connection is made with angle cleats bolted each side of the web of the secondary beam and to the web of the main beam as illustrated in Figure 5.26B.

Photograph 5.3 **Spliced column connection.**

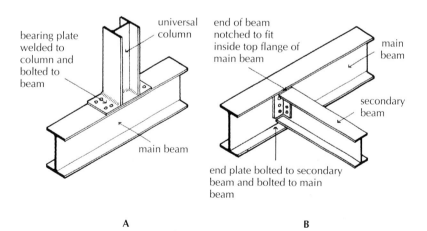

Figure 5.26 **A, Column to beam connection. B, Beam to beam connection.**

Fasteners

In the early days of the use of structural steel for buildings, connections were made with angle cleats riveted to cleats and members. Wrought iron rivets were used as fasteners, the rivets being either dome or countersunk headed. The rivets were heated until red hot, fitted to holes in the connecting metals and closed by hammering the shank end to a dome head. As the rivet cooled it shrank in length and drew the connecting plates together. Rivets were used as both shop and field (site) fasteners for structural steelwork up to the early 1950s. Today, rivet fasteners are rarely used and bolts are used as fasteners for site connections with welding for some shop connections. Site bolting requires less site labour than riveting, requires less skill, is quieter and eliminates fire risk.

Hexagon headed black bolts

For many years hexagon headed, black bolts and nuts, illustrated in Figure 5.27, were used for structural steel connections made on site. These bolts were fitted to holes 2 mm larger in diameter than that of the shank of the bolt for ease of fitting. The nut was tightened by hand and the protruding end of the shank of the bolt was burred over the nut by hammering to prevent the nut working loose. The operation of fitting these bolts, which does not require any great degree of skill, can be quickly completed. The disadvantages of this connection are that the bolts do not make a tight fit into the holes to which they are fitted and there is the possibility of some slight movement in the connection. For this reason, black bolts are presumed to have less strength than fitted bolts and their strength is taken as 80 N/mm^2. These bolts are little used today for structural steel connections.

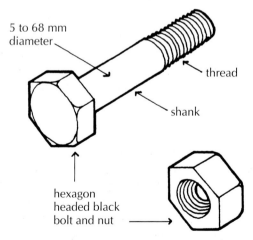

Figure 5.27   **Black hexagon bolt.**

Turned and fitted bolts

To obtain more strength from a bolted connection it may be economical to use steel bolts that have been accurately turned. These bolts are fitted to holes the same diameter as their shank and the bolt is driven home by hammering and then secured with a nut. Because of their tight fit the strength of these bolts is taken as 95 N/mm^2. These bolts are more expensive than black bolts and have largely been superseded by the high strength bolts described below.

High strength friction grip bolts

These bolts are made from high strength steel, which enables them to suffer greater stress due to tightening than ordinary bolts. The combined effect of the greater strength of the bolt itself and the increased friction due to the firm clamping together of the plates being joined makes these bolts capable of taking greater loads than ordinary bolts. These bolts are tightened with a torque wrench which measures the tightness of the bolt by reference to the torque applied, which in turn gives an accurate indication of the strength of the connection, whereas hand tightening would give no measure of strength. Though more expensive than ordinary bolts, these bolts and their associated washers are commonly used.

Strength of bolted connections – single shear, double shear

Bolted connections may fail under load for one of two reasons. First they may fail by the shearing of their shank. Shear is caused by the action of two opposite and equal forces acting on a material. The simplest analogy is the action of the blades of a pair of scissors or shears on a sheet of paper. As the blades close they exert equal and opposite forces which tear through the fibres of the paper, forcing one part up and the other down. In the same way, if the two plates joined by a bolt move with sufficient force in opposite directions then the bolt will fail in single shear, as illustrated in Figure 5.28. The strength of a bolt is determined by its resistance to shear in accordance with the strengths previously noted. Where a bolt joins three plates it is liable to failure by the movement of adjacent plates in opposite directions, as illustrated in Figure 5.28. It will be seen that the failure is caused by the shank failing in shear at two points simultaneously, hence the term double shear. It is presumed that a bolt is twice as strong in double as in single shear.

Bearing strength

A second type of failure that may occur at a connection is caused by the shank of a bolt bearing so heavily on the metal of the member or members it is joining that the metal becomes crushed, as illustrated in Figure 5.29. The strength of the mild steel used in the majority of steel frames and the connections, in resisting crushing, is taken as 200 N/mm^2. The bearing area of a bolt on the mild steel of a connection is the product of the diameter of the bolt and the thickness of the thinnest member of the joint. When selecting the diameter and the number of

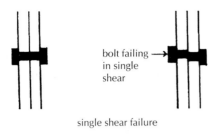

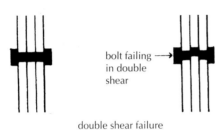

Figure 5.28 Shear failure.

bolts required for a connection, the shear resistance of the bolts and the bearing area of the thinnest plate have to be taken into account.

Bolt pitch (spacing)

If bolts are too closely spaced they may bear so heavily on the section of the members around them that they tear through the metal, with the result that, instead of the load being borne by all, it may be transferred to a few bolts which may then fail in shear. To prevent the possibility of this type of failure it is usual to space bolts at least two and a half times their diameter apart. The distance apart is measured centre to centre. Bolts should be at least one and three quarter times their diameter from the edge of the steel member.

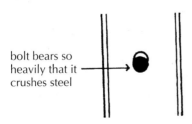

Figure 5.29 Bearing failure.

5.5 Welding

The word 'welding' describes the operation of running molten weld metal into the heated junction of steel plates or members so that, when the weld metal has cooled and solidified, it strongly binds them together. The edges of the members to be joined are cleaned and also shaped for certain types of weld. For a short period the weld metal is molten as it runs into the joint, and for this reason it is obvious that a weld can be formed more readily with the operator working above the joint than in any other position. It will be seen that welding can be carried out more quickly and accurately in a workshop where the members can be manipulated more conveniently for welding than they can be on site. Welding is most used in the prefabrication of built-up beams, trusses and lattice frames. The use of shop welded connections for angle cleats to conventional skeleton frames is less than it was due to the possibility of damage to the protruding cleats during transport, lifting and handling of members.

In the design of welded structures it is practice to prefabricate as far as practical in the workshop and make site connections either by bolting or by designing joints that can readily be welded on site. The advantage of welding as applied to structural steel frames is that members can be built up to give the required strength for minimum weight of steel, whereas standard members do not always provide the most economical section. The labour cost in fabricating welded sections is such that it can only be justified in the main for long span and non-traditional frames. The reduction in weight of steel in welded frames may often justify higher labour costs in large, heavily loaded structures. In buildings where the structural frame is partly or wholly exposed, the neat appearance of the welded joints and connections is an advantage. It is difficult to tell from a visual examination whether a weld has made a secure connection, and X-ray or sonic equipment is the only exact way of testing a weld for adequate bond between weld and parent metal. This equipment is somewhat bulky to use on site and this is one of the reasons why site welding is not favoured.

Surfaces to be welded must be clean and dry if the weld metal is to bond to the parent metal. These conditions are difficult to achieve in the UK's wet climate out on site. The process of welding used in structural steelwork is 'fusion welding', in which the surface of the metal to be joined is raised to a plastic or liquid condition so that the molten weld metal fuses with the plastic or molten parent metal to form a solid weld or join. For fusion welding the requirements are a heat source, usually electrical, to melt the metal, a consumable electrode to provide the weld metal to fill the gap between the members to be joined, and some form of protection against the entry of atmospheric gases which can adversely affect the strength of the weld. The metal of the members to be joined is described as the parent or base metal and the metal deposited from the consumable electrode, the weld metal. The fusion zone is the area of fusion of weld metal to parent metal.

The method of welding most used for structural steelwork is the arc welding process, where an electric current is passed from a consumable electrode to the parent metals and back to the power source. The electric arc from the electrode to the parent metals generates sufficient heat to melt the weld metal and the parent metal to form a fusion weld. The processes of welding most used are:

❑ Manual metal-arc (MMA) welding
❑ Metal inert-gas (MIG) and metal active-gas (MAG) welding
❑ Submerged arc (SA) welding

Manual metal-arc (MMA) welding
This manually operated process is the oldest and the most widely used process of arc welding. The equipment for MMA welding is simple and relatively inexpensive, and the process is fully positional in that welding can be carried out vertically and even overhead due to the force with which the arc propels drops of weld metal on to the parent metal. Because of its adaptability this process is suitable for complex shapes, welds where access is difficult and on-site welding. The equipment consists of a power supply and a hand held, flux covered, consumable electrode, as illustrated in Figure 5.30. As the electrode

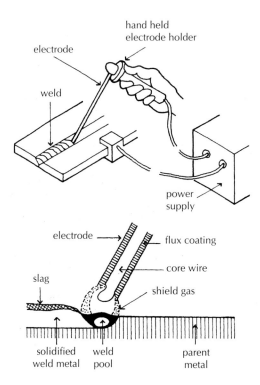

Figure 5.30 Manual metal arc welding.

is held by hand the soundness of the weld depends largely on the skill of the operator in controlling the arc length and speed of movement of the electrode. The purpose of the flux coating to the electrode is to stabilise the arc, provide a gas envelope or shield around the weld to inhibit pick up of atmospheric gases, and produce a slag over the weld metal to protect it from the atmosphere. Because this weld process depends on the skill of the operator there is a high potential for defects.

Metal inert-gas (MIG) and metal active-gas (MAG) welding

These processes use the same equipment, which is more complicated and expensive than that needed for MMA welding. In this process the electrode is continuously fed with a bare wire electrode to provide weld metal, and a cylinder to provide gas through an annulus to the electrode tip to form a gas shield around the weld, as illustrated in Figure 5.31A. The advantage of the continuous electrode wire feed is that there is no break in welding to replace electrodes

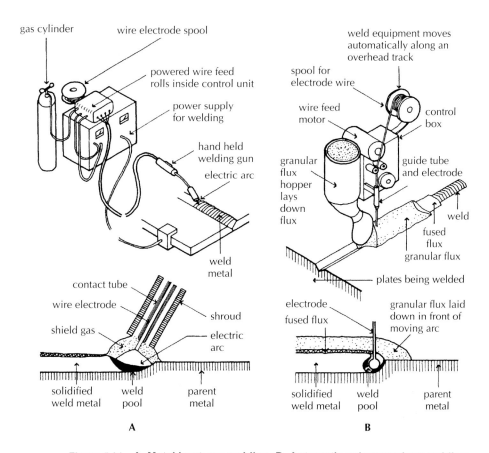

Figure 5.31   A, Metal inert-gas welding. B, Automatic submerged arc welding.

as there is with MMA welding, which can cause weakness in the weld run, and the continuous gas supply ensures a constant gas shield protection against the entry of atmospheric gases which could weaken the weld. The manually operated electrode of this type of welding equipment can be used by less highly trained welders than the MMA electrode. The bulk of the equipment and the need for shelter to protect the gas envelope limit the use of this process to shop welding.

Submerged arc (SA) welding

Submerged arc (SA) welding is a fully automatic bare wire process of welding where the arc is shielded by a blanket of flux that is continuously fed from a hopper around the weld, as illustrated in Figure 5.31B. The equipment is mounted on a gantry that travels over the weld bench to lay down flux over the continuous weld run. The equipment, which is bulky and expensive, is used for long continuous shop weld runs of high quality, requiring less skilled welders.

Types of weld

Two types of weld are used, the fillet weld and the butt weld.

Fillet weld

This weld takes the form of a fillet of weld metal deposited at the junction of two parent metal membranes to be joined at an angle, the angle usually being a right angle in structural steelwork. The surfaces of the members to be joined are cleaned and the members fixed in position. The parent metals to be joined are connected to one electrode of the supply and the filler rod to the other. When the filler rod electrode is brought up to the join, the resulting arc causes the weld metal to run in to form the typical fillet weld illustrated in Figure 5.32.

The strength of a fillet weld is determined by the throat thickness multiplied by the length of the weld to give the cross-sectional area of the weld, the strength of which is taken as $115\,\mathrm{N/mm^2}$. The throat thickness is used to determine the strength of the weld, as it is along a line bisecting the angle of the join that a weld usually fails. The throat thickness does not extend to the convex surface of the weld over the reinforcement weld metal because this reinforcement metal contains the slag of minerals other than iron that form on the surface of the molten weld metal, which are of uncertain strength. The dotted lines in Figure 5.32 represent the depth of penetration of the weld metal into the parent metal and enclose that part of the parent metal that becomes molten during welding and fuses with the molten weld metal.

The leg lengths of fillet weld used in structural steelwork are 3, 4, 5, 6, 8, 10, 12, 15, 18, 20, 22 and 25 mm (Figure 5.32). Throat thickness is the leg length multiplied by 0.7 mm. Fillet welds 5 to 22 mm are those most commonly used in structural steelwork, the larger sizes being used at heavily loaded connections.

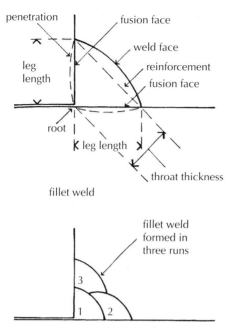

Figure 5.32   Fillet weld.

Fillet welds of up to 10 mm are formed by one run of the filler rod in the arc welding process and the larger welds by two or more runs, as illustrated in Figure 5.32. When filled welds are specified by leg length, the steel fabricator has to calculate the gauge of the filler rod and the current to be used to form the weld. An alternative method is to specify the weld as, for example, a 1–10/225 weld, which signifies that it is a 1 run weld with a 10 gauge filler rod to form 225 mm of weld for each filler rod. As filler rods are of standard length, this specifies the volume of the weld metal used for specified length of weld and therefore determines the size of the weld. Intermittent fillet welds are generally used in structural steelwork, common lengths being 150, 225 and 300 mm.

Butt welds
These welds are used to join plates at their edges. The weld metal fills the gap between them. The section of the butt weld employed depends on the thickness of the plates to be joined and whether welding can be executed from one side only or from both sides. The edges of the plates to be joined are cleaned and shaped as necessary, the plates are fixed in position and the weld metal run in from the filler rod. Thin plates up to 5 mm thick require no shaping of their edges and the weld is formed as illustrated in Figure 5.33. Plates up to 12 mm thick have their edges shaped to form a single V weld as illustrated

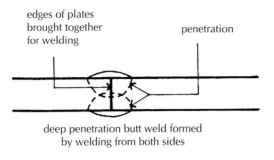

edges of plates
brought together
for welding

penetration

deep penetration butt weld formed
by welding from both sides

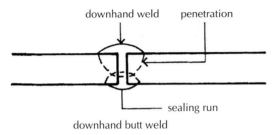

downhand weld   penetration

sealing run

downhand butt weld

Figure 5.33   Butt welds.

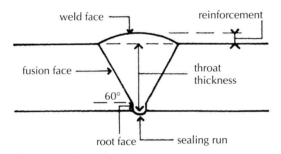

weld face   reinforcement

fusion face   throat
thickness

60°

root face   sealing run

Figure 5.34   Single V butt weld.

in Figure 5.34. The purpose of the V section is to allow the filler rod to be manipulated inside the V to deposit weld metal throughout the depth of the weld without difficulty. Plates up to 24 mm thick are joined together either with a double V weld, where welding can be carried out from both sides, or by a single U where welding can only be carried out from one side. Figure 5.35 is an illustration of a double V and a single U weld. The U-shaped weld section provides room to manipulate the filler rod in the root of the weld but uses less of the expensive weld metal than would a single V weld of similar depth. It is more costly to form the edges of plates to the U-shaped weld than it is to form the V-shaped weld, and the U-shaped weld uses less weld metal than does a

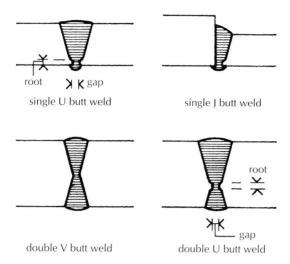

Figure 5.35   Butt welds.

V weld of similar depth. Here the designer has to choose the weld that will be the cheapest. Plates over 24 mm thick are joined with a double U weld, as illustrated in Figure 5.35. Butt welds between plates of dissimilar thickness are illustrated in Figure 5.35.

The throat thickness of a butt weld is equal to the thickness of the thinnest plate joined by the weld, and the strength of the weld is determined by the throat thickness multiplied by the length of the weld to give the cross-sectional area of throat. The size of a butt weld is specified by the throat thickness, that is the thickness of the thinnest plate joined by the weld. The shape of the weld may be described in words as, for example, a double V butt weld or by symbols.

Uses of welding in structural frames

As has been stated, welding can often be used economically in fabricating large span beams, whereas it is generally cheaper to use standard beam sections for medium and small spans. Figure 5.36 is an illustration of a built-up beam section fabricated from mild steel strip and plates, fillet and butt welded together. It will be seen that the material can be disposed to give maximum thickness of flange plates at mid span where it is needed. Figure 5.37 illustrates a welded beam end connection where strength is provided by increasing the size of the plates, which are shaped for welding to the column.

Built-up columns

Columns particularly lend themselves to fabrication by welding where a fabricated column may be preferable to standard rolled steel sections. The advantages of these fabricated hollow section columns are that the sections may be designed to suit the actual loads, connections for beams and roof frames can

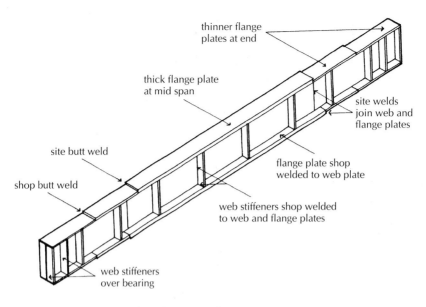

Figure 5.36   Welded built-up long span beam.

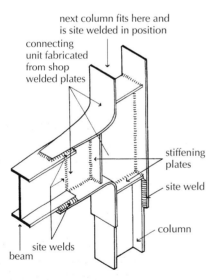

Figure 5.37   Welded beam to column connection.

be simply made to the square faces, and the appearance of the column may be preferred where there is no necessity for fire resistant casing. Hollow section columns are fabricated by welding together angle or channel sections or plates, as illustrated in Figure 5.38. The advantage of columns fabricated by welding two angle or channel sections together is the least weld length necessary. The disadvantage is the limited range of sections available. The benefit of welding

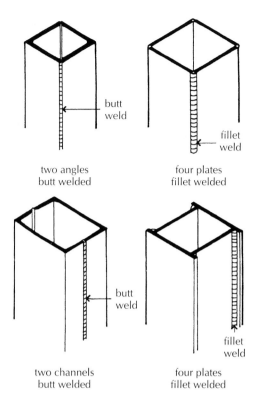

Figure 5.38   Welded built-up columns.

four plates together is the facility of selecting the precise thickness and width of plate necessary structurally, and the disadvantage is the length of weld necessary. The considerable extra cost of fabricating built-up sections limits their use to one-off special structural designs.

Column bases and foundations

Because of the comparatively small section area of a steel column, it is necessary to weld a steel base plate to it to provide a flat base to bear on the foundation and so spread the load, and to provide a means of fixing with holding down bolts. The bases of steel columns are accurately machined so that they bear truly on the steel base plates to which they are welded. The three types of steel base plate that are used are the plate base, the gussetted plate base and the slab or bloom base. For comparatively light loads, it is usual to use a 12 mm thick steel base plate fillet welded to the column. The thin plate is sufficient to spread the light loads over its area without buckling. The plate, illustrated in Figures 5.39 and 5.40, is of sufficient area to provide holes for holding down bolts.

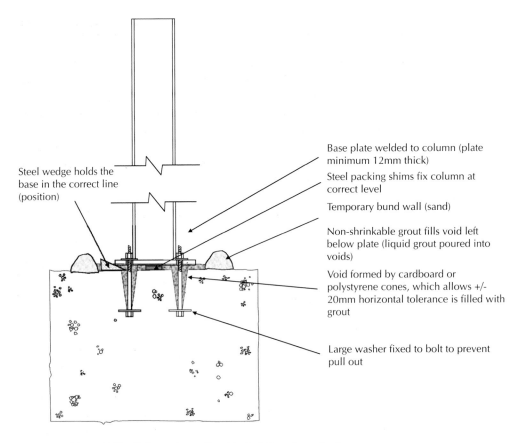

Base plate welded to column (plate minimum 12mm thick)

Steel packing shims fix column at correct level

Temporary bund wall (sand)

Non-shrinkable grout fills void left below plate (liquid grout poured into voids)

Void formed by cardboard or polystyrene cones, which allows +/- 20mm horizontal tolerance is filled with grout

Large washer fixed to bolt to prevent pull out

Steel wedge holds the base in the correct line (position)

Figure 5.39   Column base fixed to holding-down bolts.

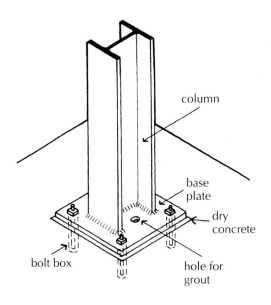

column

base plate

dry concrete

bolt box

hole for grout

Figure 5.40   Column base plate.

Prior to the column being positioned, the foundation base is checked for level, and steel shims are used to ensure that the base of the column sits at the required level. The column is hoisted into position over the concrete base so that it is plumb (vertical). The column is then lowered over the bolts. Because the bolts are cast with a void around them, which allow a small amount of moment (lateral tolerance), wedges are used to move the base into the correct position. The base is checked for line and level before being grouted in. A small bund wall of sand is then positioned around the column base and non-shrinking grout is then poured between the base and the foundation. The grout fills all the voids, including those made by the cones, which had allowed movement. Once the grout has set, the column is securely held in position.

A gussetted base plate may also be used to spread the load of the column over a sufficient area of plate. The machined column base is fillet welded to the base plate and four shaped steel gusset plates are welded to the flanges of the column and the base plate, as illustrated in Figure 5.41. The gusset plates effectively spread the loads from the column over the area of the base plate. The column is hoisted into position so that it is plumb, and steel wedges are driven in between the plate and the concrete base ready for levelling concrete or grout. The word 'slab' or 'bloom' is used to describe comparatively thick, flat sections of steel that are produced by the process of hot rolling steel ingots to shape. A slab or bloom is thicker than a plate. Slab or bloom bases for steel columns are used for the benefit of their ability to spread heavy loads over their area without buckling. The machined ends of columns are fillet welded to slab bases and the columns hoisted into position until plumb.

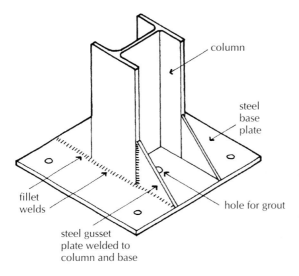

Figure 5.41   Gussetted base plate.

Steel wedges are driven in between the base and the concrete foundation ready for holding-down bolts and grouting. Figure 5.42 is an illustration of a slab base.

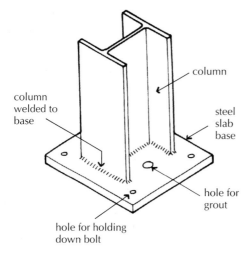

column

column welded to base

steel slab base

hole for grout

hole for holding down bolt

Figure 5.42   Slab base plate.

The purpose of the levelling concrete or grout that is run between the base plate and the concrete is to provide uniform contact between the level underside of the plate and the irregular surface of the concrete foundation. Traditionally, a comparatively dry mix of graded fine aggregate, cement and water was hammered in between the plate and the top of the concrete foundation. The dry mix was rammed in with a batten of wood from all four sides of the base. As concrete dries and hardens, it shrinks. The purpose of using dry concrete was to limit shrinkage. As an alternative to a dry mix, and more commonly nowadays, a self-levelling compound of expanding cement grout is used. A wet (liquid) mix of expanding cement is poured between the column and foundation, and a temporary bund wall is used to ensure that the liquid grout is contained. As the grout dries and hardens, it expands slightly. Steel column bases are secured in position on concrete foundations with two, four or more holding-down bolts. These bolts are termed holding-down because they hold the columns in position and may hold columns down against uplift that may occur due to the effect of wind pressure on the face of tall buildings.

Holding-down bolts may be cast into concrete foundations by themselves or inside collars which make allowance for locating bolts in the correct position. To cast the expanded metal collars in concrete they are supported by timber templates which are supported at the sides of bases. Figure 5.43 is an illustration of a timber template supporting collars ready for casting into a concrete base. The advantage of these collars is that they allow some little adjustment for

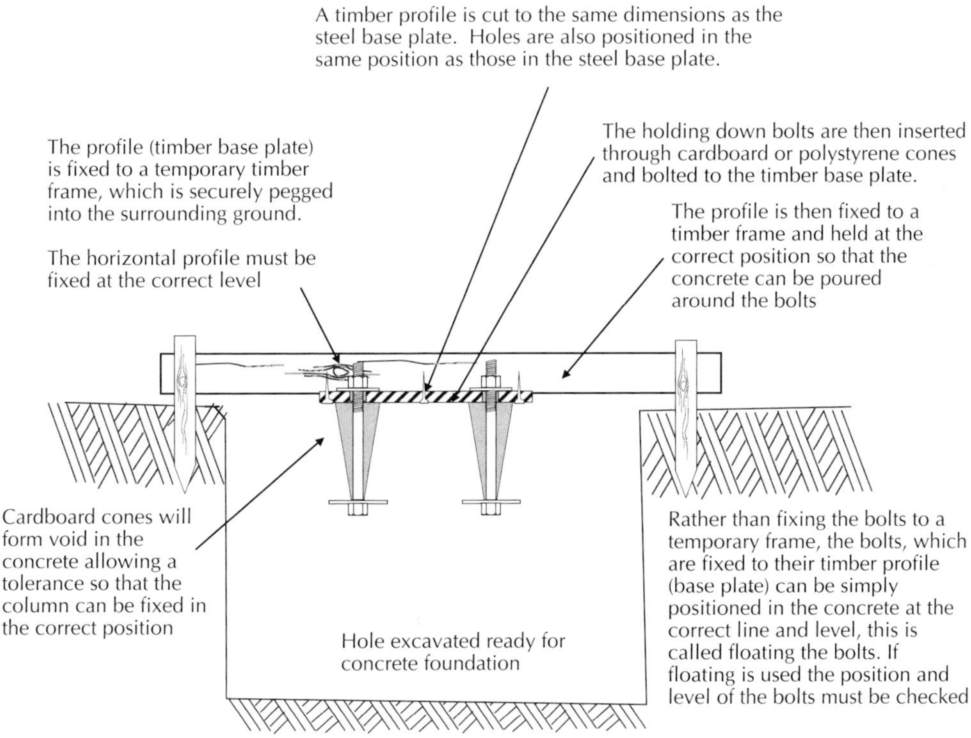

A timber profile is cut to the same dimensions as the
steel base plate. Holes are also positioned in the
same position as those in the steel base plate.

The profile (timber base plate)
is fixed to a temporary timber
frame, which is securely pegged
into the surrounding ground.

The horizontal profile must be
fixed at the correct level

The holding down bolts are then inserted
through cardboard or polystyrene cones
and bolted to the timber base plate.

The profile is then fixed to a
timber frame and held at the
correct position so that the
concrete can be poured
around the bolts

Cardboard cones will
form void in the
concrete allowing a
tolerance so that the
column can be fixed in
the correct position

Hole excavated ready for
concrete foundation

Rather than fixing the bolts to a
temporary frame, the bolts, which
are fixed to their timber profile
(base plate) can be simply
positioned in the concrete at the
correct line and level, this is
called floating the bolts. If
floating is used the position and
level of the bolts must be checked

Figure 5.43   Temporary bolt boxes.

aligning the holes in steel bases with the position for casting in holding-down
bolts. The boss shown in the diagram at the top of collars is a wood plug to
hold the bolts in the correct position. The frame holds the bolts at the correct
line and level. Once the column is in position, grout is run to fill the collars
around the holding-down bolts which are fitted with anchor plates to improve
contact with the concrete (Photograph 5.4). Another method of fixing holding-
down bolts is to drill holes in dry concrete bases and use expanding bolts in
the drilled holes. This necessitates a degree of accuracy in locating the position
for drilling holes to coincide with the holes in base plates.

Mass concrete foundation to columns

The base of columns carrying moderate loads of up to, say, 400 kN bearing on
soils of good bearing capacity can be formed economically of mass concrete.
The size of the base depends on the bearing capacity of the soil and load
on the column base, and the depth of the concrete is equal to the projection
of the concrete beyond the base plate, assuming an angle of dispersion of load
in concrete of 45°.

The bolt boxes are held in position by a temporary timber frame and checked for line and level. Once the concrete is poured the bases are checked for line and level again

Photograph 5.4   Positioning of holding-down bolts – bolt boxes (courtesy of G. Throup).

Reinforced concrete base

The area of the base required to spread the load from heavily loaded columns on subsoils of poor to moderate bearing capacity is such that it is generally more economical to use a reinforced concrete base than a mass concrete one. The steel column base plate is fixed as it is to a mass concrete base. Where column bases are large and closely spaced it is often economical to combine them in a continuous base or raft. When a heavily reinforced concrete base is used it may be possible to tie and position the bolt boxes to the reinforcement cage prior to pouring the concrete, rather than erecting a temporary timber frame.

Steel grillage foundation

Steel grillage foundation is a base in which a grillage of steel beams transmits the column load to the subsoil. The base consists of two layers of steel beams, two or three in the top layer under the foot of the column and a lower cross layer of several beams so that the area covered by the lower layer is sufficient to spread column loads to the requisite area of subsoil. The whole of the steel beam grillage is encased in concrete. This type of base is rarely used today as a reinforced concrete base is much cheaper.

Hollow rectangular sections

Beam to column connections

Bolted connections to closed box section columns may be made with long bolts passing through the section. Long bolts are expensive and difficult to use as they necessitate raising beams on opposite sides of the column at the same time in order to position the bolts. Beam connections to hollow rectangular and square section columns may be made through plates, angles or tees welded to the columns. Standard beam sections are bolted to tee section cleats welded to columns and lattice beams by bolting end plates welded to beams to plates welded to columns, as illustrated in Figure 5.44.

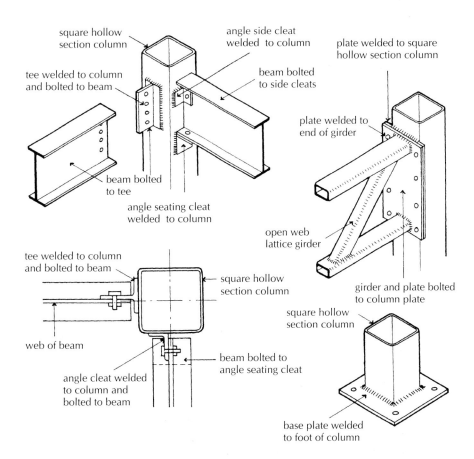

Figure 5.44   Connections to hollow section columns.

Flowdrill jointing

A recent innovation in making joints to hollow rectangular steel sections (HRS) is the use of the flowdrill technique as an alternative to the use of either long bolting through the hollow sections or welding and site bolting. The flowdrill technique depends on the use of a tungsten carbide bit (drill) which can be used in a conventional power operated drill. As the tungsten carbide bit rotates at high speed on the surface of the steel, friction generates sufficient heat to soften the steel. As the bit penetrates the now softened wall of the steel section, it redistributes the metal to form an internal bush, as illustrated in Figure 5.45. Once the metal has cooled, the formed internal bush is threaded with a coldform flowtap bit to make a threaded hole ready for a bolt. The beam connection to the hollow steel column is completed by bolting welded on end plates or bolting to web cleats welded to the column through the ready drilled holes. The execution of this form of connection requires a good deal of skill in setting out centre punched holes accurately in the face of the hollow section to align exactly the holes to be drilled. Flowdrill jointing is the preferred method of making site connections to hollow sections for the benefit of economy in materials and site labour, and the security of the bolted connection.

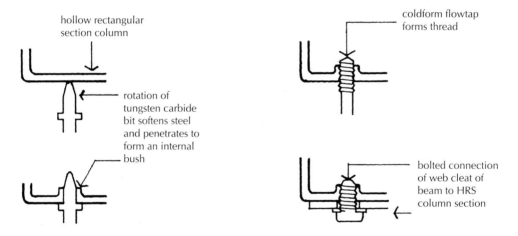

Figure 5.45   Flowdrill jointing.

Cold strip sections

Beam to column connections

These connections are made by means of protruding studs or Ts welded to the columns and bolted to the beams. Studs welded to columns are bolted to small section beams and ties, and larger section beams to tee section cleats welded to columns, as illustrated in Figure 5.46. The T section cleat is required for larger beams to spread the bearing area over a sufficient area of thin column wall to resist buckling.

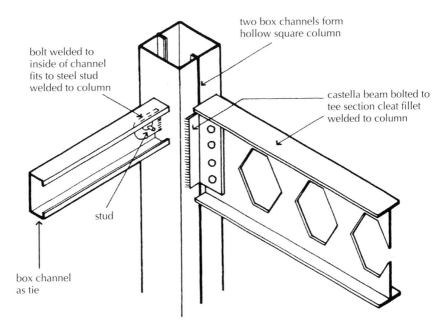

two box channels form
hollow square column

bolt welded to
inside of channel
fits to steel stud
welded to column

castella beam bolted to
tee section cleat fillet
welded to column

stud

box channel
as tie

Figure 5.46   Cold roll-formed sections — connections.

5.6 Fire protection of structural steelwork

Building Regulations are mainly concerned with controlling the spread of fire to ensure the safety of those in the building and their safe escape in a notional period of time that varies from a half-hour to six hours, depending on the use of the building, its construction and size. To limit the growth and spread of fires in buildings the Regulations classify materials in accordance with the tendency of the materials to support spread of flame over their surface, which is also an indication of the combustibility of the materials. Regulations also impose conditions to contain fires inside compartments to limit the spread of flame. To provide safe means of escape, the Regulations set standards for the containment of fires and the associated smoke and fumes from escape routes for notional periods of time deemed adequate for escape from buildings.

One aspect of fire regulations is to specify notional periods of fire resistance for the loadbearing elements of a building so that they will maintain their strength and stability for a stated period during fires in buildings for the safety of those in the building. Steel, which is non-combustible and makes no contribution to fire, loses so much of its strength at a temperature of 550°C that a loaded steel member would begin to deform, twist and sag and no longer support its load. Because a temperature of 550°C may be reached early in the development of fires in buildings, regulations may require a casing to structural steel members to reduce the amount of heat getting to the steel. The larger the section of a structural steel member, the less it will be affected by heat from fires by absorbing heat before it loses strength. The greater the mass and the smaller the perimeter of a steel section, the longer it will be before it reaches a temperature at which it will fail. This is due to the fact that larger sections will absorb more heat than smaller ones before reaching a critical temperature.

The traditional method of protecting structural steelwork from damage by fire is to cast concrete around beams and columns or to build brick or blockwork around columns with concrete casing to beams. These heavy, bulky and comparatively expensive casings have by and large been replaced by lightweight systems of fire protection employing sprays, boards, preformed casing and intumescent coatings. The materials used for fire protection of structural steelwork may be grouped as:

❏ Spray coatings
❏ Board casings
❏ Preformed casings
❏ Plaster and lath
❏ Concrete, brick or block casings

Spray coatings

A wide range of products is available for application by spraying on the surface of structural steel sections to provide fire protection. The materials are sprayed on to the surface of the steel sections so that the finished result is a lightweight coating that takes the profile of the coated steel, as illustrated in Figure 5.47. This is one of the cheapest methods of providing a fire protection coating or casing to steel for protection of up to four hours, depending on the thickness of

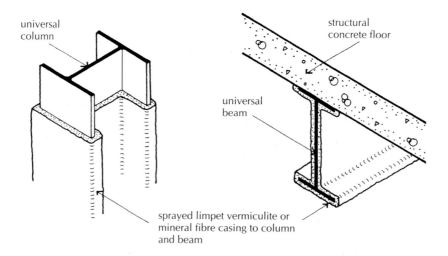

Figure 5.47 Fire protection of structural steelwork by sprayed limpet casing.

the coating. The finished surface of these materials is generally coarse textured and, because of the lightweight nature of the materials, these coatings are easily damaged by knocks and abrasions. They provide some protection against corrosion of steel and, being lightweight, assist in controlling condensation. These sprayed systems of protection are suitable for use where appearance is not a prime consideration and for beams in floors above suspended ceilings. Being lightweight and porous, spray coatings are not generally suited to external use. Spray coatings may be divided into three broad groups as described below.

Mineral fibre coatings

Mineral fibre coatings consist of mineral fibres that are mixed with inorganic binders, the wet mix being sprayed directly on to the clean, dry surface of the steel. The material dries to form a permanent, homogenous insulation that can be applied to any steel profile.

Vermiculite/gypsum/cement coatings

Vermiculite/gypsum/cement coatings consist of mixes of vermiculite or aerated magnesium oxychloride with cement or vermiculite with gypsum plaster. The materials are premixed and water is added on site for spray application directly to the clean, dry surface of steel. The mix dries to a hard, homogenous insulation that can be left rough textured from spraying or trowelled to a smooth finish. These materials are somewhat more robust than mineral spray coatings but will not withstand knocks.

Intumescent coatings

These coatings include mastics and paints which swell when heated to form an insulating protective coat which acts as a heat shield. The materials are applied by spray or trowel to form a thin coating over the profile of the steel section. They provide a hard finish which can be left textured from spraying or trowelled smooth, and provide protection of up to two hours.

Board casings

There is a wide choice of systems based on the use of various preformed boards that are cut to size and fixed around steel sections as a hollow, insulating fire protection. Board casings may be grouped in relation to the materials that are used in the manufacture of the boards that are used as:

❑ Mineral fibre boards or batts
❑ Vermiculite/gypsum boards
❑ Plasterboard

For these board casings to be effective as fire protection they must be securely fixed around the steel sections, and joints between boards must be covered, lapped or filled to provide an effective seal to the joints in the board casing. These board casings, which are only moderately robust, can suffer abrasion but are readily damaged by moderate knocks and are not suitable for external use. Board casings are particularly suitable for use in conjunction with ceiling and wall finishes of the same or like materials.

Mineral fibre boards and batts

Mineral fibre boards and batts are made of mineral fibres bound with calcium silicate or cement. The surface of the boards and batts, which is coarse textured, can be plastered. These comparatively thick boards are screwed to light steel framing around the steel sections. Mineral fibre batts are semi-rigid slabs which are fixed by means of spot-welded pins and lock washers. Mineral fibre boards are moderately robust and are used where appearance is not a prime consideration.

Vermiculite/gypsum boards

Vermiculite/gypsum boards are manufactured from exfoliated vermiculite and gypsum or non-combustible binders. The boards are cut to size and fixed around steelwork, either to timber noggins wedged inside the webs of beams and columns or screwed together and secured to steel angles or strips, as illustrated in Figure 5.48. The edges of the boards may be square edged or rebated. The boards, which form a rigid, fairly robust casing to steelwork, can be self-finished or plastered.

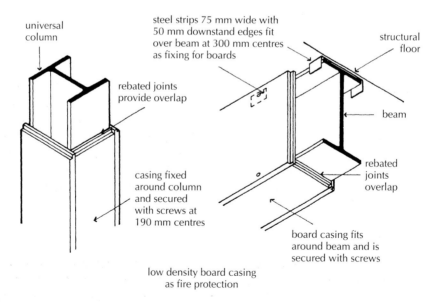

Figure 5.48   Vermiculite/gypsum board.

Plasterboard casings

Plasterboard casings can be formed from standard thickness plasterboard or from a board with a gypsum/vermiculite core for improved fire resistance. The boards are cut to size and fixed to metal straps around steel sections. The boards may be self-finished or plastered. This is a moderately robust casing. Figure 5.49 is an illustration of a board casing.

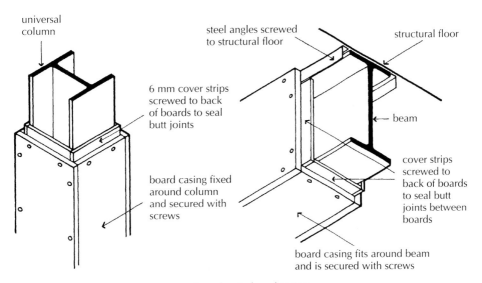

universal column

steel angles screwed to structural floor

structural floor

6 mm cover strips screwed to back of boards to seal butt joints

beam

board casing fixed around column and secured with screws

cover strips screwed to back of boards to seal butt joints between boards

board casing fits around beam and is secured with screws

medium density board casing

Figure 5.49   Board casing.

Preformed casings

These casings are made in preformed 'L' or 'U' shapes ready for fixing around the range of standard column or beam sections, respectively. The boards are made of vermiculite and gypsum, or with a sheet steel finish on a fire resisting lining, as illustrated in Figure 5.50. The vermiculite and gypsum boards are screwed to steel straps fixed around the steel sections and the sheet metal faced casings by interlocking joints and screws. These preformed casings provide a neat, ready finished surface with good resistance to knocks and abrasions in the case of the metal faced casings.

Plaster and lath

Plaster on metal lath casing is one of the traditional methods of fire protection for structural steelwork. Expanded metal lath is stretched and fixed to stainless steel straps fixed around steel sections with metal angle beads at arrises, as illustrated in Figure 5.51. The lath is covered with vermiculite gypsum plaster

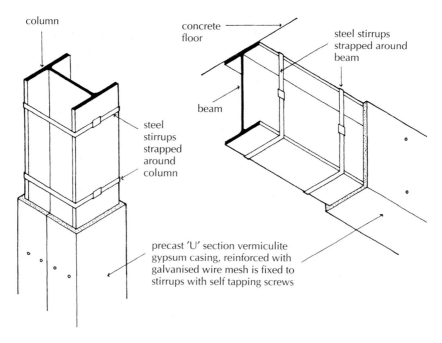

column

concrete
floor

steel stirrups
strapped around
beam

beam

steel
stirrups
strapped
around
column

precast 'U' section vermiculite
gypsum casing, reinforced with
galvanised wire mesh is fixed to
stirrups with self tapping screws

Figure 5.50   Preformed casings.

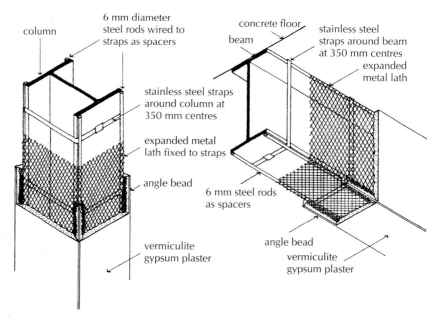

column

6 mm diameter
steel rods wired to
straps as spacers

concrete floor

beam

stainless steel
straps around beam
at 350 mm centres

expanded
metal lath

stainless steel straps
around column at
350 mm centres

expanded metal
lath fixed to straps

angle bead

6 mm steel rods
as spacers

vermiculite
gypsum plaster

angle bead

vermiculite
gypsum plaster

Figure 5.51   Metal lath and plaster casing.

to provide an insulating fire protective casing that is trowelled smooth ready for decoration. This rigid, robust casing can suffer abrasion and knocks and is particularly suitable for use where a similar finish is used for ceilings and walls.

Concrete, brick or block casing

An insitu cast concrete casing is the traditional method of providing fire protection to structural steelwork and protection against corrosion. This solid casing is highly resistant to damage by knocks. To prevent the concrete spalling away from the steelwork during fires, it is lightly reinforced, as illustrated in Figure 5.52. The disadvantages of a concrete casing to steelwork are its mass, which considerably increases the dead weight of the frame, and the cost of on-site labour and materials in the formwork and falsework necessary to form and support the wet concrete.

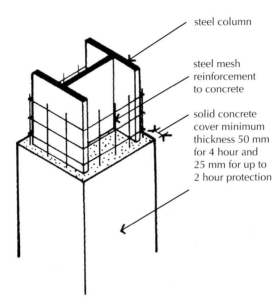

steel column

steel mesh reinforcement to concrete

solid concrete cover minimum thickness 50 mm for 4 hour and 25 mm for up to 2 hour protection

Figure 5.52   Concrete fire protection.

Brick casings to steelwork may be used where brickwork cladding or brick division or compartment walls are a permanent part of the building, or where a brick casing is used for appearance sake to match surrounding fairface brick. Otherwise a brick casing is an expensive, labour intensive operation in the necessary cutting and bonding of brick around columns.

Blockwork may be used as an economic means of casing columns, particularly where blockwork divisions or walls are built up to structural steelwork. The labour in cutting and bonding these larger units is considerably less than with bricks. The blocks encasing steelwork are reinforced in every horizontal joint with steel mesh or expanded metal lath.

5.7 Floor and roof construction

In the early days of iron and steel framed construction, floors were constructed with comparatively closely-spaced iron or steel filler beams, between main beams, giving support to shallow brick arches built between them on which concrete was spread to provide a level floor. This type of 'fire resisting' floor was in general use before the advent of reinforced concrete. The filler beam and brick arched floor gave way to the hollow clay block and insitu cast reinforced concrete floor described in *Barry's Introduction to Construction of Buildings*. This labour intensive form of construction was used for the advantage of the fire resisting property of the clay blocks and the reduction in dead weight afforded by the hollow blocks. With the development of the technique of mass production of precast reinforced concrete units, systems of precast reinforced concrete slabs, beams and infill concrete block floors became common for steel framed buildings. The advantage of these floor systems was a considerable reduction in site labour and speed of erection because the slabs and beams were 'self-centring', that is they required no support from below once in place, whereas previous floor systems required some support from below during construction.

Functional requirements

The functional requirements of floors and roofs are:

- ❑ Strength and stability
- ❑ Resistance to weather
- ❑ Durability and freedom from maintenance
- ❑ Fire safety
- ❑ Resistance to the passage of heat
- ❑ Resistance to the passage of sound

Strength and stability

The requirements from the Building Regulations are that buildings be constructed so that the loadbearing elements, foundations, walls, floors and roofs have adequate strength and stability to support the dead loads of the construction and anticipated loads on roofs, floors and walls without such undue deflection or deformation as might adversely affect the strength and stability of parts or the whole of the building. The strength and stability of floors and roofs depend on the nature of the materials used in the floor and roof elements, and the section of the materials used in resisting deflection (bending) under the dead and imposed loads. Under load, any horizontal element will deflect (bend) to an extent. Deflection under load is limited to about 1/300 of span to minimise cracking of rigid finishes to floors and ceilings and to limit the sense of insecurity the occupants might have, were the floor to deflect too obviously.

In general the strength and stability of a floor or roof is a product of the depth of the supporting members: the greater the depth, the greater the strength and stability.

Resistance to weather (roofs)

The requirements for resistance to the penetration of wind, rain and snow, and the construction and finishes necessary for both traditional and more recently-used roof coverings, are described in *Barry's Introduction to Construction of Buildings* and Chapter 3.

Durability and freedom from maintenance

The durability and freedom from maintenance of both traditional and the more recently-used roof coverings are described in *Barry's Introduction to Construction of Buildings* and Chapter 3. The durability and freedom from maintenance of floors constructed with steel beams, profiled steel decking and reinforced concrete depend on the internal conditions of the building. The majority of multi-storey framed buildings today are heated, so that there is little likelihood of moist internal conditions occurring, such as to cause progressive, destructive corrosion of steel during the useful life of the building.

Fire safety

The application of the Regulations, as set out in the practical guidance given in Approved Document B, is directed to the safe escape of people from buildings in case of fire rather than the protection of the building and its contents. Insurance companies that provide cover against the risks of damage to the buildings and contents by fire will generally require additional fire protection such as sprinklers and detection equipment.

Internal fire spread

Fire may spread within a building over the surface of materials covering walls and ceilings. The Regulations prohibit the use of materials that encourage spread of flame across their surface when subject to intense radiant heat and those which give off appreciable heat when burning. Limits are set on the use of thermoplastic materials used in rooflights and lighting diffusers. As a measure of ability to withstand the effects of fire, the elements of a structure are given notional fire resistance times, in minutes, based on tests. Elements are tested for the ability to withstand the effects of fire in relation to:

❑ Resistance to collapse (loadbearing capacity), which applies to loadbearing elements
❑ Resistance to fire penetration (integrity), which applies to fire separating elements (e.g. floors)
❑ Resistance to the transfer of excessive heat (insulation), which applies to fire separating elements

The notional fire resistance times, which depend on the size, height and use of the building, are chosen as being sufficient for the escape of occupants in the event of fire. The requirements for the fire resistance of elements of a structure do not apply to:

❑ A structure that supports only a roof unless:
   (a) the roof acts as a floor, e.g. car parking, or as a means of escape
   (b) the structure is essential for the stability of an external wall, which needs to have fire resistance
❑ The lowest floor of the building

To prevent rapid fire spread which could trap occupants, and to reduce the chances of a fire growing large, it is necessary to subdivide buildings into compartments separated by walls and/or floors of fire-resisting construction. The degree of subdivision into compartments depends on:

❑ The use and fire load (contents) of the building
❑ The height of the floor of the top storey as a measure of ease of escape and the ability of fire services to be effective
❑ The availability of a sprinkler system, which can slow the rate of growth of fire

The necessary compartment walls and/or floors should be of solid construction sufficient to resist the penetration of fire for the stated notional period of time in minutes. The requirements for compartment walls and floors do not apply to single-storey buildings.

Smoke and flame may spread through concealed spaces, such as voids above suspended ceilings, roof spaces and enclosed ducts and wall cavities in the construction of a building. To restrict the unseen spread of smoke and flames through such spaces, cavity barriers and stops should be fixed as a tight-fitting barrier to the spread of smoke and flames.

External fire spread
To limit the spread of fire between buildings, limits to the size of 'unprotected areas' of walls and also finishes to roofs, close to boundaries, are imposed by the Building Regulations. The term 'unprotected area' is used to include those parts of external walls that may contribute to the spread of fire between buildings. Windows are unprotected areas as glass offers negligible resistance to the spread of fire. The Regulations also limit the use of materials of roof coverings near a boundary that will not provide adequate protection against the spread of fire over their surfaces.

Resistance to the passage of heat
The requirements for the conservation of power and fuel by the provision of adequate insulation of roofs are described in *Barry's Introduction to Construction of Buildings* and Chapter 3.

Resistance to the passage of sound

A description of the transmission and perception of sound is given in *Barry's Introduction to Construction of Buildings*, Chapter 5. In multi-storey buildings the structural frame may provide a ready path for the transmission of impact sound over some considerable distance. The heavy slamming of a door, for example, can cause a sudden disturbing sound clearly heard some distance from the source of the sound by transmission through the frame members. Such unexpected sounds are often more disturbing than continuous background sounds such as external traffic noise. To provide resistance to the passage of such sounds it is necessary to provide a break in the path between potential sources of impact and continuous solid transmitters.

Precast hollow floor beams

The precast, hollow, reinforced concrete floor units illustrated in Figure 5.53 are from 400 mm to 1200 mm wide, 110 to 300 mm thick for spans of up to 10 m for floors and 13.5 m for the less heavily loaded roofs. The purpose of the voids in the units is to reduce deadweight without affecting strength. The reinforcement is cast into the webs between the hollows. The wide floor units are used where there is powered lifting equipment which can swing the units into place. These hollow floor units can be used as floor slabs with a non-structural levelling floor screed; alternatively they may be used with a structural reinforced concrete topping with tie bars over beams for composite action with the concrete casing to beams. Raised floor finishes can also be applied directly to the unfinished surface.

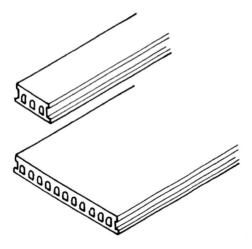

Figure 5.53   Hollow precast floor units.

The end bearing of these units is a minimum of 75 mm on steel shelf angles or beams and 100 mm on masonry and brick walls. The ends of these floor units are usually supported by steel shelf angles either welded or bolted to steel beams so that a part of the depth of the beam is inside the depth of the floor, as illustrated in Figure 5.54 and Photograph 5.5.

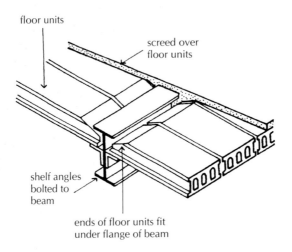

Figure 5.54   Hollow precast floor units on steel beam.

The ends of the floor units may be splayed to fit under the top flange of the beams. A disadvantage of the construction shown in Figure 5.54 is that the deep I section beam projects some distance below the floor units and increases the overall height of construction for a given minimum clear height between floor and underside of beam. To minimise the overall height of construction, it is recent practice in Sweden to use welded top hat profile beams with the floor units supported by the bottom flange, as illustrated in Figure 5.55.

The top hat section is preferred because of the difficulty of lowering and manoeuvring the units into the web of broad flange I section beams. This construction method is particularly suited to multi-storey residential flats where the comparatively small imposed loads on floors facilitate a combination of overall beam depth and floor units to minimise construction depth. A screed is spread over the floor for lightly loaded floors and roofs, and a reinforced concrete constructional topping for more heavily loaded floors.

Precast prestressed concrete floor units

These comparatively thin, prestressed solid plank, concrete floor units are designed as permanent centring (shuttering) for composite action with structural reinforced concrete topping, as illustrated in Figure 5.56. The units are 400 and 1200 mm wide, 65, 75 or 100 mm thick and up to 9.5 m long for floors and 10 m for roofs. It may be necessary to provide some temporary propping to the underside of these planks until the concrete topping has gained sufficient strength. A disadvantage of this construction is that as the planks are laid on top of the beams so that the floor spans continuously over beams, there

The hollow floor beams are lifted into position and simply placed on the steel beams. The whole working area has safety netting installed to prevent workers falling from the floor.

Raised floors can be installed directly on top of the beams.

Photograph 5.5 Installation of hollow precast concrete floor beams (courtesy of G. Throup).

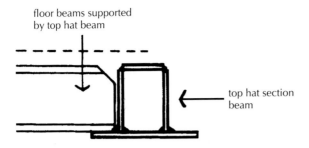

floor beams supported
by top hat beam

top hat section
beam

Figure 5.55   Top hat section beam.

is increase in overall depth of construction from top of floor to underside of beams.

Precast beam and filler block floor

This floor system of precast reinforced concrete beams or planks to support precast hollow concrete filler blocks is illustrated in Figure 5.57. For use with steel beams, the floor beams are laid between supports such as steel shelf angles fixed to the web of the beams or laid on the top flange of beams and the filler blocks are then laid between the floor beams. The reinforcement protruding from the top of the planks acts with the concrete topping to form a continuous floor system spanning across the structural beams. These small beams or planks and filler blocks can be manhandled into place without the need for heavy lifting equipment. This type of floor is most used in smaller scale buildings supporting the lighter imposed floor loads common in residential buildings, for example.

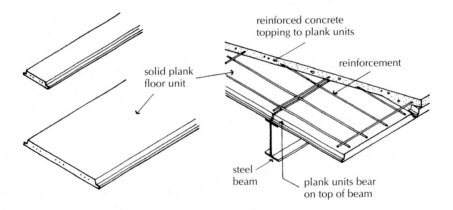

reinforced concrete
topping to plank units

reinforcement

solid plank
floor unit

steel
beam

plank units bear
on top of beam

Figure 5.56   Prestressed solid plank floor unit.

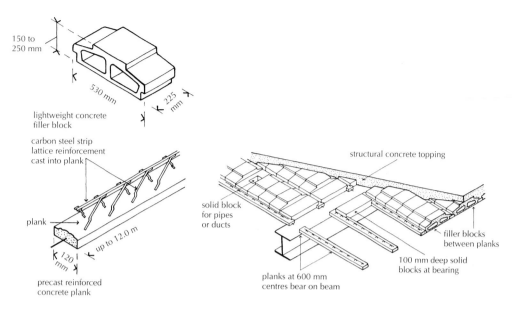

150 to
250 mm

530 mm

225 mm

lightweight concrete
filler block

carbon steel strip
lattice reinforcement
cast into plank

plank

up to 12.0 m

120 mm

precast reinforced
concrete plank

structural concrete topping

solid block
for pipes
or ducts

filler blocks
between planks

100 mm deep solid
blocks at bearing

planks at 600 mm
centres bear on beam

Figure 5.57   Precast beam and filter block floor.

Hollow clay block and concrete floor

This floor system, illustrated in *Barry's Introduction to Construction of Buildings*, consists of hollow clay blocks and insitu cast concrete reinforced as ribs between the blocks. This floor has to be laid on temporary centring to provide support until the insitu concrete has gained sufficient strength. This floor system is much less commonly used than it was due to the considerable labour required for setting up the temporary support and laying out the blocks and reinforcement.

Precast concrete T beams

Precast concrete T beam floors are mostly used for long span floors and particularly roofs of such buildings as stores, supermarkets, swimming pools and multi-storey car parks where there is a need for wide span floors and roofs, and the depth of the floor is no disadvantage. The floor units are cast in the form of a double T, as illustrated in Figure 5.58. The strength of these units is in the depth of the tail of the T, which supports and acts with the comparatively thin top web. A structural reinforced concrete topping is cast on top of the floor units.

Cold rolled steel deck and concrete floor

The traditional concrete floor to a structural steel frame consisted of reinforced concrete, cast insitu with the concrete casing to beams, cast on timber centring and falsework supported at each floor level until the concrete had sufficient strength to be self-supporting. The very considerable material and labour costs

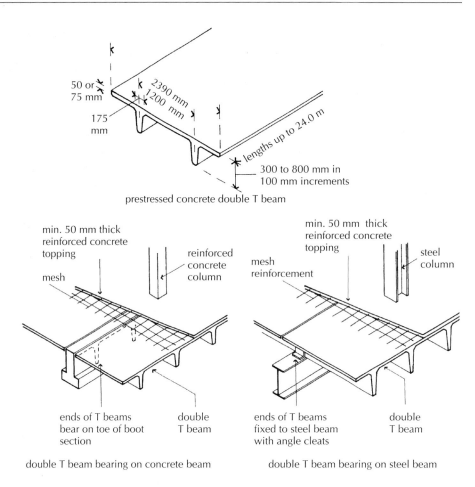

50 or 75 mm

175 mm

2390 mm

1200 mm

lengths up to 24.0 m

300 to 800 mm in 100 mm increments

prestressed concrete double T beam

min. 50 mm thick reinforced concrete topping

mesh

reinforced concrete column

ends of T beams bear on toe of boot section

double T beam

double T beam bearing on concrete beam

min. 50 mm thick reinforced concrete topping

mesh reinforcement

steel column

ends of T beams fixed to steel beam with angle cleats

double T beam

double T beam bearing on steel beam

Figure 5.58   Prestressed concrete double T beam.

in erecting and striking the support for the concrete floor led to the adoption of the precast concrete self-centring systems such as the hollow beam and plank, and beam and infill block floors. The term 'self-centring' derives from the word 'centring' used to describe the temporary platform of wood or steel on which insitu cast concrete is formed. The precast concrete beam, plank and beam, and block floors do not require temporary support, hence the term self-centring. A disadvantage of the precast concrete beam and plank floors for use with a structural steel frame is that it is practice to erect the steel frame in one operation. Raising the heavy, long precast concrete floor units and moving them into position is, to an extent, impeded by the skeleton steel frame.

Of recent years profiled cold rolled steel decking, as permanent formwork, acting as the whole or a part of the reinforcement to concrete, has become the principal floor system for structural steel frames. The profiled steel deck is easily handled and fixed in place as formwork (centring) for concrete. The profiled cold roll-formed, steel sheet decking, illustrated in Figure 5.59, is galvanised both sides as a protection against corrosion. The profile is shaped to bond to the concrete, using projections that taper in from the top of the deck. Another profile is of trapezoidal section with chevron embossing for key to concrete. The steel deck may be laid on the top flange of beams, as illustrated in Figure 5.59, or supported by shelf angles bolted to the web of the beam to reduce overall height and fixed in position on the steelwork with shot fired pins, self-tapping screws or by welding, with two fixings to each sheet. Side laps of deck are fixed at intervals of not more than 1 m with self-tapping screws or welding.

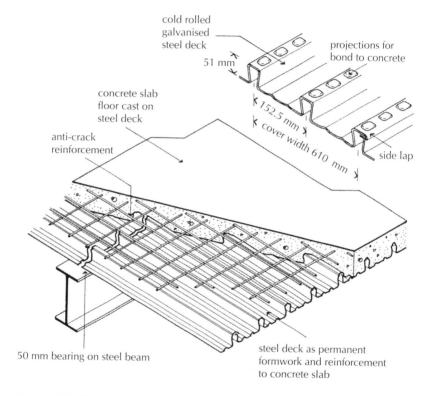

Figure 5.59   Steel deck and concrete floor.

For medium spans between structural steel beams, the profiled steel deck acts as both permanent formwork and as reinforcement for the concrete slab that is cast insitu on the deck. A mesh of anti-crack reinforcement is cast into the upper section of the slab, as illustrated in Figure 5.59. For long spans and heavy loads the steel deck can be used with additional reinforcement cast into the bottom of the concrete between the upstanding profiles and, for composite action between the floor and the beams, shear studs are welded to the beams and cast into the concrete. The steel mesh reinforcement cast into the concrete slab floor is sufficient to provide protection against damage by fire in most situations. For high fire rating the underside of the deck can be coated with sprayed-on protection or an intumescent coating. Where there is to be a flush ceiling for appearance and as a housing for services, a suspended ceiling is hung from hangers slotted into the profile or hangers bolted to the underside of the deck.

For particularly large spans or where cuts are made through the profile metal sheet for services, some temporary propping is required until the concrete has reached its seven-day maturity.

Slimfloor floor construction

'Slimfloor' is the name adopted by British Steel (now Corus) for a form of floor construction for skeleton steel framed buildings. This form of construction is an adaptation of a form of construction developed in Sweden where restrictions on the overall height of buildings dictated the development of a floor system with the least depth of floor construction to gain the maximum number of storeys within the height limitations. Slimfloor construction comprises beams fabricated from universal column sections to which flange plates are welded, as illustrated in Figure 5.60. The flange plates, which are wider overall than the flanges of the beams, provide support to profiled steel decking that acts in part as reinforcement and provides support for the reinforced concrete constructional topping. The galvanised, profiled steel deck units are 210 mm deep with ribs at 600 mm centres. The ribs and the top of the decking are ribbed to stiffen the plates and provide some bond to concrete. To seal the ends of the ribs in the decking, to contain the concrete that will be cast around beams, sheet steel stop ends are fixed through the decking to the flange plates, as illustrated in Figure 5.61. Constructional concrete topping is spread over the decking and into the

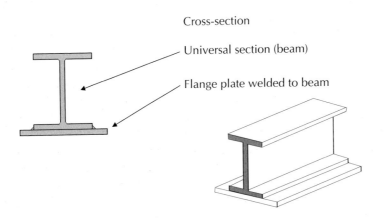

Cross-section

Universal section (beam)

Flange plate welded to beam

Figure 5.60   Slimfloor beam.

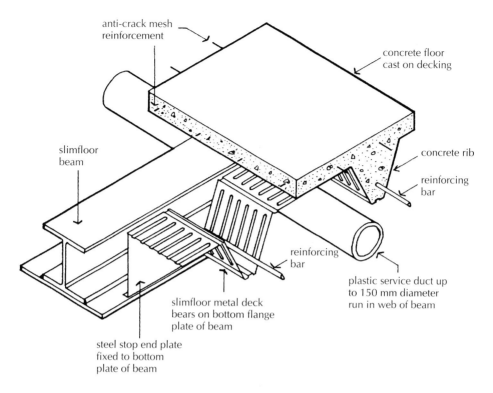

anti-crack mesh
reinforcement

concrete floor
cast on decking

slimfloor
beam

concrete rib

reinforcing
bar

reinforcing
bar

plastic service duct up
to 150 mm diameter
run in web of beam

slimfloor metal deck
bears on bottom flange
plate of beam

steel stop end plate
fixed to bottom
plate of beam

Figure 5.61   Slimfloor construction.

ribs around reinforcement in the base of the ribs and anti-crack reinforcement
in the floor slab.

   The galvanised pressed steel deck units are designed for spans of 6 m for
use with the typical grid of 9 m beam spans at 6 m centres. For spans of over
6 and up to 7.5 m the decking will need temporary propping at mid-span until
the concrete has developed adequate strength. The slimfloor may be designed
as a non-composite form of construction where the floor is assumed to have
no composite action with the beams, as illustrated in Figure 5.61. This non-
composite type of floor construction is usual where the imposed floor loads
are low, as in residential buildings, and the floor does not act as a form of
bracing to the structural frame. A particular advantage of the slimfloor is that
all or some of the various services, common to some modern buildings, may
be accommodated within the deck depth rather than being slung below the
structural floor over a false ceiling. Calculations and tests have shown that
150 mm diameter holes may be cut centrally through the web of the beams
at 600 mm spacing along the middle third of the length of the beam without
significantly affecting the load-carrying capacity of the beam. Figure 5.61 is an
illustration of the floor system showing a plastic tube sleeve run through the
web of a beam for service pipes and cables.

The ceiling finish may be fixed to the underside of the decking or hung from the decking to provide space for services such as ducting. Because of the concrete encasement to the beams, most slimfloor constructions achieve one hour's fire resistance rating without the need for applied fire protection to the underside of the beam. Where fire resistance requirement is over 60 minutes it is necessary to apply fire protection to the underside of the bottom flange plate.

The advantages of the slimfloor construction are:

❏ Speed of construction through ease of manhandling and ease of fixing the lightweight deck units, which provide a safe working platform
❏ Pumping of concrete obviates the need for mechanical lifting equipment
❏ The floor slab is lightweight as compared with insitu or precast concrete floors
❏ The deck profile provides space for both horizontal services in the depth of the floor and vertical services through the wide top flange of the profile
❏ Least overall depth of floor to provide minimum constructional depth consistent with robustness requirements dictated by design codes

Composite construction

Composite construction is the name given to structural systems in which the separate structural characteristics and advantages of structural steel sections and reinforced concrete are combined as, for example, in the T beam system. A steel frame, cased in concrete and designed to allow for the strength of the concrete in addition to that of the steel, is a form of composite construction. Where concrete encases steel sections, it is accepted that the stiffening and strengthening effect of the concrete on the steel can be allowed for in engineer's calculations. By reinforcing the concrete casing and allowing for its composite effect with the steel frame, a saving in steel and a reduction in the overall size of members can be achieved.

Shear stud connectors

A concrete floor slab bearing on a steel beam may be considered to act with the beam and serve as the beam's compressive flange, as a form of composite construction. This composite construction effect will only work if there is a sufficiently strong bond between the concrete and the steel, to make them act together in resisting shear stresses developed under load. The adhesion bond between the concrete and the top flange of the beam is not generally sufficient and it is usually necessary to fix shear studs or connectors to the top flange of the beam, which are then cast in the floor slab. The purpose of these studs and connectors is to provide a positive resistance to shear. Figure 5.62 is an illustration of typical shear stud connectors and Figure 5.63 is an illustration of composite floor and beam construction.

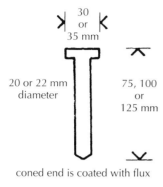

30
or
35 mm

20 or 22 mm
diameter

75, 100
or
125 mm

coned end is coated with flux

Figure 5.62   Shear stud connector.

Inverted T beam composite construction

The composite beam and floor construction described above employs the standard I section beams. The top flange of the beam is not a necessary part of the construction, as the concrete floor slab can be designed to carry the whole of the compressive stress, so that the steel in the top flange of the beam is wastefully deployed. By using an inverted T section member, steel is placed in the tension area and concrete in the compression area, where their characteristics are most useful. A cage of mild steel binders, cast into the beam casing and linked to the reinforcement in the floor slab, serves to make the slab and beam act as a form

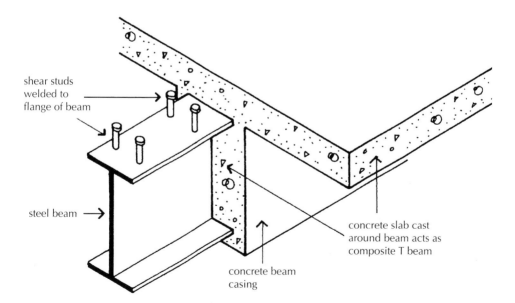

shear studs
welded to
flange of beam

steel beam →

concrete slab cast
around beam acts as
composite T beam

concrete beam
casing

Figure 5.63   Composite construction.

of composite construction by the adhesion bond of the concrete to the whole of the T section.

Preflex beams

The use of high tensile steel sections for long span beams has been limited owing to the deflection of the beams under load, which causes cracking of concrete casing, and possible damage to partitions and finishes. Preflex beams are made by applying and maintaining loads, which are greater than working loads, to pairs of steel beams. In this deflected position, reinforced concrete is cast around the tension flanges of the beams. When the concrete has developed sufficient strength the load is released and the beams tend to return to their former position. In so doing, the beams induce a compressive stress in the concrete around the tension flange, which prevents the beams wholly regaining their original shape. The beams now have a slight upward camber. Under loads, the deflection of these beams will be resisted by the compressive stress in the concrete around the bottom flange, which will also prevent cracking of concrete. Further stiffening of the beam to reduce deflection is gained by the concrete casing to the web of the beam. By linking the reinforcement in the concrete web casing to the floor slab, the concrete and steel can be made to act as a composite form of construction. These beams may be connected to steel columns, with end plates welded to beam ends and bolted to column flanges, or may be cast into reinforced concrete columns.

Preflex beams are considerably more expensive than standard mild steel beams and are designed, in the main, for use in long span heavily loaded floors. The slimfloor beam, which was developed in Sweden for use in the floors of multi-storey steel framed buildings to gain the maximum number of storeys within height limitations, may be used in composite construction.

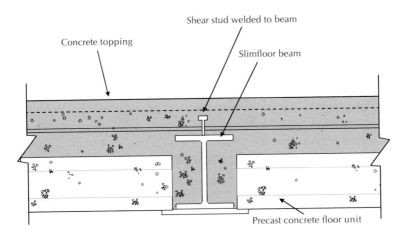

Figure 5.64   Composite floor construction.

For composite action, shear studs are welded to the top flange of a universal column section to which a wide bottom plate has been welded. This bottom plate serves as a bearing for hollow, precast reinforced concrete floor units. Structural concrete topping is spread and consolidated around the beam and as reinforced structural topping around transverse reinforcement, as illustrated in Figure 5.64. The result is a reinforced concrete floor acting compositely with the steel beam, the concrete casing tying to the beam and shear studs. The advantage of this construction is the least depth of floor of uniform depth. This type of floor is more expensive than a straightforward beam and slab floor.

6 Structural Concrete Frames

Reinforced concrete is one of the primary structural materials used in engineering and building works. Concrete has been used for a long time but it was the pioneering work of French engineer Joseph Monier that led to a patent in 1867 for the process of strengthening concrete by embedding steel in it. Since the early days of reinforcing concrete with steel to enhance its strength, reinforced concrete has become a common material on construction sites, used extensively for civil engineering works and widely applied in the construction of buildings.

6.1 Concrete

The three materials used in the production of concrete are cement, aggregate sand and water.

Cement

The cement used today was first developed by Joseph Aspdin, a Leeds builder, who took out a patent in 1824 for the manufacture of Portland cement. He developed the material for the production of artificial stone and named it Portland cement because, in its hardened state, it resembled natural Portland limestone in texture and colour. The materials of Aspdin's cement, limestone and clay, were later burned at a high temperature by Isaac Johnson in 1845 to produce a clinker which, ground to a fine powder, is what we now term Portland cement. The characteristics of cement depend on the proportions of the compounds of the raw materials used and the fineness of the grinding of the clinker, produced by burning the raw materials. A variety of Portland cements are produced, each with characteristics suited to a particular use. The more commonly used Portland cements are:

❏ Ordinary Portland cement
❏ Rapid hardening Portland cement
❏ Sulphate resisting Portland cement
❏ White Portland cement
❏ Low heat Portland cement
❏ Portland blastfurnace cement
❏ Water repellent cement
❏ High alumina (aluminous) cement

Ordinary Portland cement

Ordinary Portland cement is the cheapest and most commonly used cement, accounting for about 90% of all cement production. It is made by heating limestone and clay to a temperature of about 1300°C to form a clinker, rich in calcium silicates. The clinker is ground to a fine powder with a small proportion of gypsum, which regulates the rate of setting when the cement is mixed with water. This type of cement is affected by sulphates such as those present in ground water in some clay soils. The sulphates have a disintegrating effect on ordinary Portland cement. For this reason sulphate resisting cements are produced for use in concrete in sulphate bearing soils, marine works, sewage installations and manufacturing processes where soluble salts are present.

Rapid hardening Portland cement

Rapid hardening Portland cement is similar to ordinary Portland except that the cement powder is more finely ground. The effect of the finer grinding is that the constituents of the cement powder react more quickly with water, and the cement develops strength more rapidly. Rapid hardening cement develops in three days, a strength which is similar to that developed by ordinary Portland in seven days. With the advantage of the cement's early strength development, it is possible to speed up construction. With rapid hardening cement, the initial set is much shorter and formwork systems can be removed earlier. Although rapid hardening is more expensive than ordinary Portland cement, it is often used because of its early strength advantage. Rapid hardening Portland cement is not a quick setting cement. Several months after mixing there is little difference in the characteristics of ordinary and rapid hardening cements.

Sulphate resisting Portland cement

The aluminates within the cement, which are affected by sulphates, are reduced to provide increased resistance to the effect of sulphates. The effect of sulphates on ordinary cement is to combine with the constituents of the cement. As the sulphates react there is an increase in volume on crystallisation, which causes the concrete to disintegrate. Disintegration is severe where the concrete is alternately wet and dry, as in marine works. Because it is necessary to control, with some care, the composition of the raw materials of this cement, it is more expensive than ordinary cement. High alumina cement described later is also a sulphate resisting cement.

White Portland cement

White Portland cement is manufactured from china clay and pure chalk or limestone and is used to produce white concrete finishes. Due to the comparatively expensive raw material used (that is, china clay) and the process of manufacture, it is considerably more expensive than ordinary cement and is used in the main for the surface of exposed concrete and for cement renderings. Pigments may be added to the cement to produce pastel colours.

Low heat Portland cement

Low heat Portland cement is used mainly for mass concrete works in dams and other constructions where the heat developed by hydration of other cements would cause serious shrinkage cracking. The heat developed by the hydration of cement in concrete in construction works is dissipated to the surrounding air, whereas in large mass concrete works it dissipates slowly. Control of the constituents of low heat Portland causes it to harden more slowly and therefore develop less rapidly than other cements. The slow rate of hardening does not affect the ultimate strength of the cement yet allows the low heat of hydration to dissipate through the mass of concrete to the surrounding air.

Portland blastfurnace cement

Portland blastfurnace cement is manufactured by grinding Portland cement clinker with blast furnace slag, the proportion of slag being up to 65% by weight and the percentage of cement clinker no less than 35%. This cement develops heat more slowly than ordinary cement and is used in mass concrete works as a low heat cement. It has good resistance to the destructive effects of sulphates and is commonly used in marine works.

Water repellent cement

Water repellent cement is made by mixing a metallic soap with ordinary or white Portland cement. Concrete made with this cement is more water repellent and therefore absorbs less rainwater than concrete made with other cements and is thus less liable to dirt staining. This cement is used for cast concrete and cast stone for its water repellent property.

High alumina (aluminous) cement

High alumina (aluminous) cement is not one of the Portland cements. It is manufactured from bauxite and limestone or chalk in equal proportions. Bauxite is a mineral containing a higher proportion of alumina (aluminium oxide) than the clays used in the manufacture of Portland cements, hence the name given to this cement. The disadvantages of this cement are that there is a serious falling off in strength in hot moist atmospheres, and it is attacked by alkalis. This cement is little used for concrete in this country.

Aggregates

Concrete is a mix of particles of hard material, the aggregate, bound with a paste of cement and water with at least three quarters of the volume of concrete being occupied by aggregate. Volume for volume, cement is generally more costly than aggregate and it is advantageous, therefore, to use as little cement as necessary to produce a dense, durable concrete.

A wet concrete mix is spread in the form of foundation bases, slabs or inside formwork for beams and columns and compacted into a dense mass. There is a direct relation between the density and strength of finished concrete and the ease with which concrete can be compacted. The characteristics of the aggregate play a considerable part in the ease with which concrete can be compacted. The measure of the ease with which concrete can be compacted is described as the workability of the mix. Workability is affected by the characteristics of the particles of aggregate such as size and shape so that, for a given mix, workability can be improved by careful selection of aggregate.

The grading of the size and the shape of the particles of aggregate affects the amount of cement and water required to produce a mix of concrete that is sufficiently workable to be compacted to a dense mass. The more cement and water that are needed for the sake of workability, the greater the drying shrinkage there will be by loss of water as the concrete dries and hardens.

Characteristics of aggregate

Aggregate for concrete should be hard, durable and contain no materials that are likely to decompose or change in volume or affect reinforcement. Clay, coal or pyrites in aggregate may soften, swell, decompose and cause stains in concrete. Aggregate should be clean and free from organic impurities and coatings of dust or clay that would prevent the particles of aggregate from being adequately coated with cement and so lower the strength of the concrete.

Natural aggregates

Sand and gravel are the cheapest and most commonly used aggregate in the UK and consist of particles of broken stone deposited by the action of rivers and streams or from glacial action. Sand and gravel deposited by rivers and streams are generally more satisfactory than glacial deposits because the former comprise rounded particles in a wide range of sizes and weaker materials have been eroded by the washing and abrasive action of moving water. Glacial deposits tend to have angular particles of a wide variety of sizes, poorly graded, which adversely affect the workability of a concrete in which they are used.

Crushed rock aggregates are generally more expensive than sand and gravel, owing to the cost of quarrying and crushing the stone. Provided the stone is hard, inert and well graded, it serves as an admirable aggregate for concrete. The term 'granite aggregate' is used commercially to describe a wide range of crushed natural stones, some of which are not true igneous rocks. Natural granite is hard and dense and serves as an excellent aggregate. Hard sandstone and close-grained crystalline limestone, when crushed and graded, are commonly used as aggregate in areas where sand and gravel are not readily available.

Because of the depletion of inland deposits of sand and gravel, marine aggregates are used. They are obtained by dredging deposits of broken stone from the bed of the sea. Most of these deposits contain shells and salt. Though not normally harmful in reinforced concrete, limits should be set to the proportion of shells and salt in marine aggregates used for concrete. One of the disadvantages of marine fine aggregate is that it has a preponderance of one size of particle that can make design mix difficult. Sand from the beach is often of mainly single-sized particles and contains an accumulation of salts. Beach sands to be used as fine aggregate in concrete should be carefully washed to reduce the concentration of salts.

Artificial aggregates

Blastfurnace slag is the by-product of the conversion of iron ore to pig iron and consists of the non-ferrous constituents of iron ore. The molten slag is tapped from the blastfurnace and is cooled and crushed. In areas where there is a plentiful supply of blastfurnace slag, it is an economical and satisfactory aggregate for concrete.

Clean broken brick is used as an aggregate for concrete required to have a good resistance to damage by fire. The strength of the concrete produced with this aggregate depends on the strength and density of the bricks from which the aggregate is produced. Crushed engineering brick aggregate will produce a concrete of medium crushing strength. Porous brick aggregate should not be used for reinforced concrete work in exposed positions as the aggregate will absorb moisture and encourage the corrosion of the reinforcement.

Fine and coarse aggregate

'Fine aggregate' is the term used to describe natural sand, crushed rock and gravel, most of which passes through a number 5 British Standard (BS) sieve, and 'coarse aggregate' the term used to describe natural gravel, crushed gravel or crushed rock, most of which is retained on a 5 BS sieve. The differentiation of fine and coarse aggregate is made because in practice the fine and coarse aggregate are ordered separately for mixing to produce a determined mix for particular uses and strengths of concrete.

Grading of aggregate

The word 'grading' is used to describe the percentage of particles of a particular range of sizes in a given aggregate from fines (sand) to the largest particle size. A sound concrete is produced from a mix that can be readily placed and compacted in position, that is a mix that has good workability and after compaction is reasonably free of voids. This is affected by the grading of the aggregate and the water/cement ratio.

The grading of aggregate is usually given by the percentage by weight passing the various sieves used for grading. Continuously graded aggregate should contain particles graded in size from the largest to the smallest to produce a dense concrete. Sieve sizes from 75 to 5 mm (3 to $^3/_{16}$ inches) are used for coarse aggregate. An aggregate containing a large proportion of large particles is referred to as being 'coarsely' graded and one having a large proportion of small particles as 'finely' graded.

Particle shape and surface texture

The shape and surface texture of the particles of an aggregate affect the workability of a concrete mix. An aggregate with angular edges and a rough surface, such as crushed stone, requires more water in the mix to act as a lubricant to facilitate compaction than does one with rounded smooth faces to produce a concrete of the same workability. It is often necessary to increase the cement content of a mix made with crushed aggregate or irregularly shaped gravels to provide the optimum water/cement ratio to produce concrete of the necessary strength. This additional water, on evaporation, tends to leave small void spaces in the concrete, which will be less dense than concrete made with rounded particle aggregate. The addition of extra water, beyond that required in the chemical reaction (hydration), will weaken the concrete. Water that does not take part in the chemical reaction leaves voids as it evaporates out of the concrete.

The nature of the surface of the particles of an aggregate will affect workability. Gravel dredged from a river will have smooth surfaced particles, which will afford little frictional resistance to the arrangement of particles that takes place during compaction of concrete. A crushed granite aggregate will have coarse surfaced particles that will offer some resistance during compaction. The shape of particles of aggregate is measured by an angularity index, and the surface by a surface coefficient. Engineers use these to determine the true workability of a concrete mix, which cannot be judged solely from the grading of particles.

Water

Water for concrete should be reasonably free from such impurities as suspended solids, organic matter and dissolved salts, which may adversely affect the properties of concrete. Water that is fit for drinking is accepted as being satisfactory for mixing water for concrete.

6.2 Concrete mixes

The strength and durability of concrete are affected by the voids in concrete caused by poor grading of aggregate, incomplete compaction or excessive water in the mix.

Water / cement ratio

Workability

The materials used in concrete are mixed with water for two reasons, first to enable the reaction with the cement which causes setting and hardening to take place and second to act as a lubricant to render the mix sufficiently plastic for placing and compaction.

About a quarter part by weight of water to one part by weight of cement is required for the completion of the setting and hardening process. This proportion of water to cement would result in a concrete mix far too stiff (dry) to be adequately placed and compacted. About a half by weight of water to one part by weight of cement is required to make a concrete mix workable. It has been established that the greater the proportion of water to cement used in a concrete mix, the weaker will be the ultimate strength of the concrete. The principal reason for this is that the water, in excess of that required to complete the hardening of the cement, evaporates and leaves voids in the concrete, which reduce its strength. It is practice, therefore, to define a ratio of water to cement in concrete mixes to achieve a dense concrete. The water/cement ratio is expressed as the ratio of water to cement by weight, and the limits of this ratio for most concrete lie between 0.4 and 0.65. Outside these limits there is a great loss of workability below the lower figure and a loss of strength of concrete above the upper figure.

Water reducing admixtures

The addition of 0.2% by weight of calcium lignosulphonate, commonly known as 'lignin', to cement will reduce the amount of water required in concrete by 10% without loss of workability. This allows the cement content of a concrete mix to be reduced for a given water/cement ratio. Calcium lignosulphonate acts as a surface-active additive that disperses the cement particles, which then need less water to lubricate and disperse them in concrete. Water reducing admixtures such as lignin are promoted by suppliers as densifiers, hardeners, water proofers and plasticisers on the basis that the reduction of water content leads to a more dense concrete due to there being fewer voids after the evaporation of water. To ensure that the use of these admixtures does not adversely affect the durability of a concrete, it is practice to specify a minimum cement content.

Nominal mixes

Volume batching

The constituents of concrete may be measured by volume in batch boxes in which a nominal volume of aggregate and a nominal volume of cement are measured for a nominal mix, as for example in a mix of 1:2:4 of cement:fine:coarse aggregate. A batch box usually takes the form of an open top wooden box in which volumes of cement, fine and coarse aggregate are measured separately for the selected nominal volume mix. For a mix such as 1:2:4 one batch box will suffice, the mix proportions being gauged by the number of fillings of the box with each of the constituents of the mix.

Measuring the materials of concrete by volume is not an accurate way of proportioning and cannot be relied on to produce concrete with a uniformly high strength. Cement powder cannot be accurately proportioned by volume because while it may be poured into and fill a box, it can be readily compressed to occupy considerably less space. Proportioning aggregates by volume takes no account of the amount of water retained in the aggregate, which may affect the water/cement ratio of the mix and affect the proportioning, because wet sand occupies a greater volume than does the same amount of sand when dry. Volume batch mixing is mostly used for the concrete for the foundations and oversite concrete of small buildings such as houses. In these cases, the concrete is not required to suffer any large stresses, and the strength and uniformity of the mix is relatively unimportant. The scale of the building operation does not justify more exact methods of batching.

Weight batching
A more accurate method of proportioning the materials of concrete is by weight batching, by proportioning the fine and coarse aggregate by weight by reference to the weight of a standard bag of cement. Where nominal mixes are weight batched it is best to take samples of the aggregate and dry them to ascertain the weight of water retained in the aggregate and so adjust the proportion of water added to the mix to allow for the water retained in the aggregate. Water can be proportioned by volume or by weight.

Designed mixes
Designed mixes of concrete are those where strength is the main criterion of the specified mix, which is judged on the basis of strength tests. The position in which concrete is to be placed, the means used and the ease of compacting it, the nature of the aggregate and the water/cement ratio all affect the ultimate strength of concrete. A designed concrete mix is one where the variable factors are adjusted (selected) by the engineer to produce a concrete with the desired minimum compressive strength at the lowest possible cost. If, for example, the cheapest available local aggregate in a particular district will not produce a very workable mix, it would be necessary to use a wet mix to facilitate placing and compaction, and this in turn would necessitate the use of a cement-rich mix to maintain a reasonable water/cement ratio. In this example it might be cheaper to import a different aggregate, more expensive than the local one, which would produce a comparatively dry but workable mix requiring less cement. These are the considerations the engineer and the contractor have in designing a concrete mix.

Prescribed mixes and standard mixes
Prescribed mixes and standard mixes are mixes of concrete where the constituents are of fixed proportion by weight to produce a 'grade' of concrete with minimum characteristics strength.

Mixing, placing and compacting concrete

Mixing concrete
Concrete may be mixed by hand when the volume to be used does not warrant the use of mechanical mixing plant. The materials are measured out by volume

in timber gauge boxes, turned over on a clean surface several times dry and then water is added. The mix is turned over again until it has a suitable consistency and uniform colour. It is obviously difficult to produce mixes of uniform quality by hand mixing. A small hand-tilting mixer is often used. The mixing drum is rotated by a petrol or electric motor, the drum being tilted by hand to fill and empty it. This type of mixer takes over a deal of the backbreaking work of mixing, but does not control the quality of mixes as materials are measured by volume.

A concrete batch mixer mechanically feeds the materials into the drum where they are mixed and from which the wet concrete is poured. The materials are batched by either weight or volume. For extensive works, plant is installed on site that stores cement (delivered in bulk), measures the materials by weight and mechanically mixes them. Concrete for high strength reinforced concrete work can only be produced from batches (mixes) of uniform quality. Such mixes are produced by plant capable of accurately measuring and thoroughly mixing the materials.

Ready-mixed concrete

Ready-mixed concrete is extensively used today. It is prepared in mechanical, concrete mixing depots where the materials are stored, weight batched and mixed, and the wet concrete is transported to site in rotating drums mounted on lorries (cement mixers). The action of the rotating drum prevents aggregates from segregating and the concrete from setting and hardening for an hour or more. Once delivered it must be placed and compacted quickly as it rapidly hardens.

Placing and compacting concrete

The initial set of Portland cement takes place from half an hour to one hour after it is mixed with water. If a concrete mix is disturbed after the initial set has occurred, the strength of the concrete may be adversely affected. It is usual to specify that concrete be placed as soon after mixing as possible and not more than half an hour after mixing.

A concrete mix consists of particles varying in size from powder to coarse aggregate graded to, say, 40 mm. If a wet mix of concrete is poured from some height and allowed to fall freely, the larger particles tend to separate from the smaller. This action is termed 'segregation of particles'. Concrete should not, therefore, be tipped or poured into place from too great a height. It is usual to specify that concrete be placed from a height not greater than 1 m.

Once in place, concrete should be thoroughly consolidated or compacted. The purpose of compaction is to cause entrapped bubbles of air to rise to the surface in order to produce as dense and void-free concrete as possible. Compaction may be effected by agitating the mix with a spade or heavy iron bar. If the mix is dry and stiff this is a very laborious process and not very effective. A more satisfactory method is to employ a pneumatically-operated poker vibrator, which is inserted into the concrete and, by vibration, liberates air bubbles and compacts the concrete. As an alternative, the formwork of reinforced concrete may be vibrated by means of a motor attached to it.

Construction joints

Because it is not possible to place concrete continuously (on the vast majority of construction sites) it is necessary to form construction joints. A construction joint is the junction of freshly placed concrete with concrete that has been placed and set, for example concrete poured on the previous day. These construction joints are a potential source of weakness because there may not be a good bond between the two placings of concrete. When forming a construction joint, the previously placed concrete needs to be clean, with a sound surface exposed. The top surface of the concrete is usually broken away by means of a mechanical scabbler. This hammers the surface of the set concrete breaking away the loose surface, leaving a clean, surface for the new concrete to from a mechanical and chemical bond. There should be as few construction joints as practical and joints should be either vertical or horizontal. Joints in columns are made as near as possible to beam haunching, and those in beams at the centre or within the middle third of the span. Vertical joints are formed against a strip board. Water bars are fixed across or cast into construction joints where there is a need to provide a barrier to the movement of water through the joint (see Chapter 3).

Curing concrete

Concrete gradually hardens and gains strength after its initial set. For this hardening process to proceed and the concrete to develop its maximum strength there must be water present in the mix. If, during the early days after the initial set, there is too rapid a loss of water the concrete will not develop its maximum strength. The process of preventing a rapid loss of water is termed 'curing concrete'. Large exposed areas of concrete such as road surfaces are cured by covering the surface for at least a week after placing, with building paper, plastic sheets or wet sacks to retard evaporation of water. In very dry weather the surface of concrete may have to be sprayed with water in addition to covering it. The formwork around reinforced concrete is often kept in position for some days after the concrete is placed in order to give support until the concrete has gained sufficient strength to be self-supporting. This formwork also serves to prevent too rapid a loss of water and so helps to cure the concrete. In very dry weather it may be necessary to spray the formwork to compensate for too rapid a loss of water.

Specially designed curing agents can also be used. These are chemical liquids that are designed to be sprayed over the concrete. Once sprayed, the liquid produces a thin film that effectively seals the water, needed for hydration, in the concrete.

✕ *Deformation of concrete*

Hardened concrete will suffer deformation due to:

❑ Elastic deformation that occurs instantaneously and is dependent on applied stress
❑ Drying shrinkage that occurs over a long period and is independent of the stress in concrete
❑ Creep, which occurs over a long period and is dependent on stress in concrete
❑ Expansion and contraction due to changes in temperature and moisture
❑ ASR (alkali–silica reaction)

Elastic deformation
Under the stress of dead and applied loads of a building, hardened concrete deforms elastically. Vertical elements such as columns and walls are compressed and shorten in height and horizontal elements such as beams and floors lengthen due to bending. These comparatively small deformations, which are related to the strength of the concrete, are predictable and allowance is made in design.

Drying shrinkage
The drying shrinkage of concrete is affected principally by the amount of water in concrete at the time of mixing and to a lesser extent by the cement content of the concrete. It can also be affected by a porous aggregate losing water. Drying shrinkage is restrained by the amount of reinforcement in concrete. The rate of shrinkage is affected by the humidity and temperature of the surrounding air, the rate of airflow over the surface, and the proportion of surface area to volume of concrete. Where concrete dries in the open air in summer, small masses of concrete will suffer about a half of the total drying shrinkage a month after placing, and large masses about a half of the total shrinkage a year after placing. Shrinkage will not generally affect the strength or stability of a concrete structure but is sufficient to require the need for movement joints where solid materials such as brick and block are built up to the concrete frame.

Creep
Under sustained load, concrete deforms as a result of the mobility of absorbed water within the cement gel under the action of sustained stress. From the point of view of design, creep may be considered as an irrecoverable deformation that occurs with time at an ever-decreasing rate under the action of sustained load. Creep deformation continues over very long periods of time to the extent that measurable deformation can occur 30 years after concrete has been placed. The factors that affect creep of concrete are the concrete mix, relative humidity and temperature, size of member and applied stress.

Concrete is a mix of aggregate, water and cement. Most aggregates used in dense concrete are inert and do not suffer creep deformation under load. The hardened cement water paste surrounding the particles of aggregate is subject to creep deformation under stress due to movements of absorbed water. The relative volume of cement gel to aggregate therefore affects deformation due to creep. Changing from a 1:1:2 to a 1:2:4 cement, fine and coarse aggregate mix increases the volume of aggregate from 60 to 75%, yet causes a reduction in creep by as much as 50%.

Temperature, relative humidity and the size of members have an effect on the hydration of cement and migration of water around the cement gel towards the surface of concrete. In general, creep is greater the lower the relative humidity, and increases with a rise in temperature caused, for example, by solar heating. Small section members of concrete will lose water more rapidly than large members and will suffer greater creep deformation during the period of initial drying. The effect of creep deformation has the most serious effect through stress loss in prestressed concrete, deflection increase in large span beams, buckling of slender columns and buckling of cladding in tall buildings.

ASR (alkali–silica reaction)
The chemical reaction of high silica-content aggregate with alkaline cement causes a gel to form, which expands and causes concrete to crack. The expansion, cracking and damage to concrete is often most severe where there is an external source of water in large quantities. Foundations, motorway bridges and concrete subject to heavy condensation have suffered severe damage through ASR. The expansion caused by the gel formed by the reaction is not uniform in time or location. The reaction may develop slowly in some structures yet very rapidly in others and may affect one part of a structure but not another. Changes in the method of manufacture of cement, that has produced a cement with higher alkalinity, are thought to be one of the causes of some noted failures. To minimise the effect of ASR it is recommended that cement-rich mixes and high silica content aggregates be avoided.

6.3 Reinforcement

Concrete is strong in resisting compressive stress but comparatively weak in resisting tensile stress. The tensile strength of concrete is between one tenth and one twentieth of its compressive strength. Steel, which has good tensile strength, is cast into reinforced concrete members in the position or positions where maximum tensile stress occurs. To determine where tensile and compressive stresses occur in a structural member it is convenient to consider the behaviour of an elastic material under stress. A bar of rubber laid across, and not fixed to, two supports bends under load and the top surface shortens and becomes compressed under stress, and the bottom surface stretched under tensile stress, as illustrated in Figure 6.1.

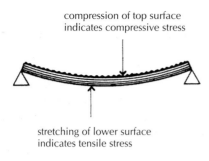

compression of top surface
indicates compressive stress

stretching of lower surface
indicates tensile stress

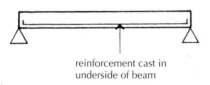

reinforcement cast in
underside of beam

Figure 6.1 Simply supported beam.

A member that is supported so that the supports do not restrain bending while under load is termed 'simply supported'. From Figure 6.1 it will be seen that maximum stretching due to tension occurs at the outwardly curved underside of the rubber bar. If the bar were of concrete it would seem logical to cast steel reinforcement in the underside of the bar. In that position the steel would be exposed to the surrounding air and it would rust and gradually lose strength. Further, if a fire occurred in the building, near to the beam, the steel might lose so much strength as to impair its reinforcing effect and the beam would collapse. It is practice, therefore, to cast the steel reinforcement into concrete so that there is at least 15 mm of concrete cover between the reinforcement and the surface of the concrete.

Concrete cover

For internal concrete structures 15 mm cover of concrete is considered sufficient to protect steel reinforcement from corrosion. External members need considerably more cover; in areas up to where reinforced concrete is exposed to seawater and abrasion, the concrete cover to the reinforcement should be 60 mm. From laboratory tests and experience of damage caused by fires in buildings, it has been established that various thicknesses of concrete cover will prevent an excessive loss of strength in steel reinforcement for particular periods of time. The presumption is that the concrete cover will protect the reinforcement for a period of time for the occupants to escape from the particular building during a fire. The statutory period of time for the concrete cover to provide protection against damage by fire varies with the size and type of building, from half an hour to four hours.

Bond and anchorage of reinforcement

The cement in concrete cast around steel reinforcement adheres to the steel just as it does to the particles of the aggregate and this adhesion plays its part in the transfer of tensile stress from the concrete to the steel. It is important, therefore, that the steel reinforcement be clean and free from scale, rust and oily or greasy coatings. Under load, tensile stress tends to cause the reinforcement to slip out of bond with the surrounding concrete due to the elongation of the member. This slip is partly resisted by the adhesion of the cement to the steel and partly by the frictional resistance between steel and concrete. To secure a firm anchorage of reinforcement to concrete and to prevent slip, it is usual practice to hook or bend the ends of bars, as illustrated in Figure 6.2.

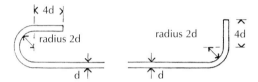

Figure 6.2   Hooked ends for reinforcing bars.

As an alternative to hooked or bent ends of reinforcing bars, deformed bars may be used. The simplest form of bar is the twisted, square bar illustrated in Figure 6.3, which through its twisted surface presents some resistance to slip. The round section, ribbed bar and the twisted, ribbed bars provide resistance to slip in concrete by the many projecting ribs illustrated in Figure 6.3. Deformed bars, which are more expensive than plain round bars, are used for heavily loaded structural concrete.

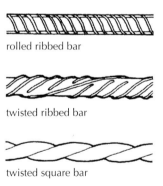

Figure 6.3   Deformed reinforcing bars.

Shear

Beams are subjected to shear stresses due to the shearing action of the supports, and the self-weight and imposed loads of beams. A pair of scissors does not cut paper, it shears it. The action of the blades, as they meet, is to force one side of the paper up and the other down and shear it into two pieces. The supports and the weight of the beam and its load act to shear a beam in the same way. Shear stress is greatest at the points of support and zero at mid-span in uniformly loaded beams. Shear failure occurs at an angle of 45°, as illustrated in Figure 6.4. Due to its poor tensile strength, concrete does not have great shear resistance and it is usual to introduce steel shear reinforcement in most beams of over, say, 2.5 m span.

To maintain the main reinforcing bars in place while concrete is being placed and until it has hardened, it is practice to use a system of stirrups or links, which are formed from light section reinforcing bars. These rectangular stirrups are attached by binding wire to the main reinforcement. To provide shear reinforcement at points of support in beams, the stirrups are more closely spaced, as illustrated in Figure 6.4.

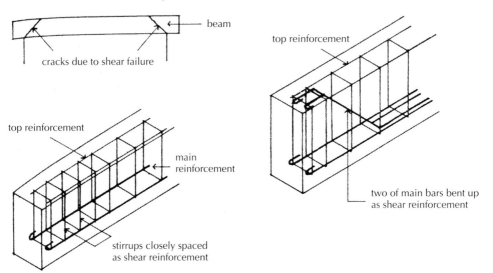

Figure 6.4   Shear reinforcement.

Fixed end support

A beam with pin jointed end support will suffer simple bending under load, whereas a beam with fixed end support is restrained from simple bending by the fixed ends, as illustrated in Figure 6.5. Because of the upward, negative bending close to the fixed ends, the top of the beam is in tension and the underside at mid-span is in tension due to positive bending. In a concrete beam with fixed ends it is not sufficient to cast reinforcement into the lower face of the beam only, as the concrete will not have sufficient tensile strength to resist tensile stresses in the top of the beam near points of support. Both top and bottom reinforcement are necessary, as illustrated in Figure 6.5, where the main bottom reinforcement is bent up at the support ends and continued as top reinforcement along the length of the main tensile end support. Top reinforcement has been omitted for the sake of clarity.

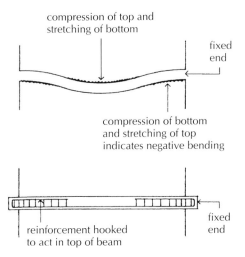

compression of top and
stretching of bottom

fixed
end

compression of bottom
and stretching of top
indicates negative bending

fixed
end

reinforcement hooked
to act in top of beam

Figure 6.5   Beams with fixed ends (top reinforcement omitted for clarity).

Because of the fixed end support, the upward negative bending at supports
will cause some appreciable deformation bending of columns around the connection of beam to column. Where a beam is designed to span continuously
over several supports, as illustrated in Figure 6.6, it will suffer negative, upward bending over the supports. At these points, the top of the beam will
suffer tensile stress and additional top reinforcement will be necessary. Here
additional top reinforcement is used against tensile stress, and the bottom reinforcement is cranked up over the support to provide shear reinforcement, as
illustrated in Figure 6.6.

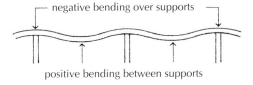

negative bending over supports

positive bending between supports

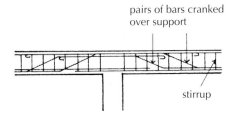

pairs of bars cranked
over support

stirrup

Figure 6.6   Reinforced concrete beam to span continuously over supports.

Cantilever beams

A cantilever beam projects from a wall or structural frame. The cantilever illustrated in Figure 6.7 may take the form of a reinforced concrete slab projecting from a building as a balcony or as several projecting cantilever beams projecting from a structural frame to support a reinforced concrete slab. As a simple explanation of the stress in a fixed end cantilever, assume it is made of rubber. Under load, the rubber will bend as illustrated in Figure 6.7. The top surface of the rubber will stretch and the bottom surface be compressed, indicating tensile stress in the top and compressive stress in the bottom. Under load, a reinforced concrete cantilever will suffer similar but less obvious bending with the main reinforcement in its top, as illustrated in Figure 6.7, with the necessary cover of reinforcement. Under appreciable load, shear reinforcement will also be necessary close to the point of support.

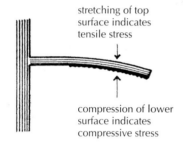

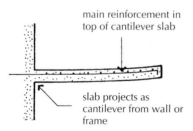

Figure 6.7   Cantilever slab.

Columns

Columns are designed to support the loads of roofs, floors and walls. If all these loads acted concentrically on the section of the column then it would suffer only compressive stress and it would be sufficient to construct the column of either concrete by itself or of reinforced concrete to reduce the required section area. In practice, the loads of floor and roof beams, and walls and wind pressure, act eccentrically, that is off the centre of the section of columns, and so cause some bending and tensile stress in columns. The steel reinforcement in columns is designed primarily to sustain compressive stress to reinforce the compressive strength of concrete, but also to reinforce the poor tensile strength of concrete against tensile stress due to bending from fixed end beams, eccentric loading and wind pressure.

Mild steel reinforcement

The cheapest and most commonly used reinforcement is round section mild steel rods of diameter from 6 to 40 mm. These rods are manufactured in long lengths and can be quickly cut and easily bent without damage. The disadvantages of ordinary mild steel reinforcement are that if the steel is stressed up to its yield point it suffers permanent elongation; exposed to moisture, it progressively corrodes; and on exposure to the heat generated by fires, it loses strength.

In tension, mild steel suffers elastic elongation, which is proportional to stress up to the yield stress, and it returns to its former length once stress is removed. At yield stress point, mild steel suffers permanent elongation and then, with further increase in stress again, suffers elastic elongation. If the permanent elongation of mild steel which occurs at yield stress were to occur in reinforcement in reinforced concrete, the loss of bond between the steel and the concrete and consequent cracking of concrete around reinforcement would be so pronounced as to seriously affect the strength of the member. For this reason maximum likely stresses in mild steel reinforcement are kept to a figure some two-thirds below yield stress. In consequence the mild steel reinforcement is working at stresses well below its ultimate strength.

Cold worked steel reinforcement

If mild steel bars are stressed up to yield point and permanent plastic elongation takes place and the stress is then released, subsequent stressing up to and beyond the former yield stress will not cause a repetition of the initial permanent elongation at yield stress. This change of behaviour is said to be due to a reorientation of the steel crystals during the initial stress at yield point. In the design of reinforced concrete members, using this type of reinforcement, maximum stress need not be limited to a figure below yield stress, to avoid loss of bond between concrete and reinforcement, and the calculated design stresses may be considerably higher than with ordinary mild steel. In practice it is convenient to simultaneously stress cold drawn steel bars up to yield point and to twist them axially to produce cold worked deformed bars with improved bond to concrete.

Deformed bars

To limit the cracks that may develop in reinforced concrete around mild steel bars, due to the stretching of the bars and some loss of bond under load, it is common to use deformed bars that have projecting ribs or are twisted to improve the bond to concrete. The types of deformed reinforcing bars generally used are ribbed bars that are rolled from mild steel and ribbed along their length, ribbed mild steel bars that are cold drawn as high yield ribbed bars, ribbed cold drawn and twisted bars, high tensile steel bars that are rolled with projecting ribs, and cold twisted square bars. Figure 6.3 is an illustration of some typical deformed bars.

Galvanised steel reinforcing bars

Where reinforced concrete is exposed externally or is exposed to corrosive industrial atmospheres it is sound practice to use galvanised reinforcement as a protection against corrosion of the steel to prevent rust staining of fairface finishes and inhibit rusting of reinforcement that might weaken the structure. The steel reinforcing bars are cut to length, bent and then coated with zinc by the hot dip galvanising process. The considerable increase in cost of the reinforcement is well worthwhile.

Stainless steel reinforcement

Stainless steel is an alloy of iron, chromium and nickel on which an invisible corrosion resistant film forms on exposure to air. Stainless steel is about ten times the cost of ordinary mild steel. It is used for reinforcing bars in concrete where the cover of concrete for corrosion protection would be much greater than that required for fire protection and the least section of reinforced concrete is a critical consideration.

Assembling and fixing reinforcement

Reinforcing steel for concrete is used in the main to provide resistance to tensile stresses in structural members. The steel reinforcing bars must therefore be placed and secured in the positions inside formwork where they will be most effective in reinforcing concrete that will be poured and compacted inside the formwork around the reinforcement. It is important, therefore, that the reinforcement is rigidly fixed in position so that it is not displaced when wet concrete is placed and compacted.

Reinforcement for structural beams and columns is usually assembled in the form of a cage, with the main and secondary reinforcement being fixed to links or stirrups that hold it in position. The principal purpose of these links is to secure the longitudinal reinforcing bars in position when concrete is being placed and compacted. They also serve to some extent in anchoring reinforcement in concrete and in addition provide some resistance to shear with closely spaced links at points of support in beams. Links are formed from small section reinforcing bars that are cut and bent to contain the longitudinal reinforcement. Stirrups or links are usually cold bent to contain top and bottom longitudinal reinforcement to beams and the main reinforcement to columns with the ends of each link overlapping, as illustrated in Figure 6.8. As an alternative, links may be formed from two lengths of bar, the main U shaped part of the link and a top section, as illustrated in Figure 6.8. The advantage of

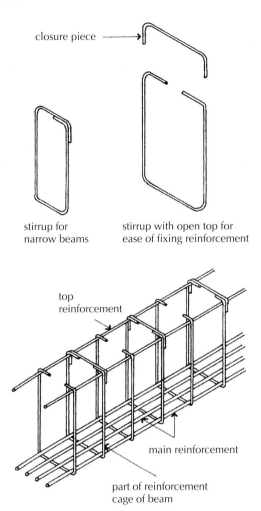

closure piece

stirrup for
narrow beams

stirrup with open top for
ease of fixing reinforcement

top
reinforcement

main reinforcement

part of reinforcement
cage of beam

Figure 6.8   Stirrups to form reinforcement cage of beams.

this arrangement of links is that where there are several longitudinal reinforcing bars in a cage they can be dropped in from the top of the links rather than being threaded through the links as the cage is wired up, thus saving time. Figure 6.8 is an illustration of part of a reinforcement cage for a reinforced concrete beam.

The separate cages of reinforcement for individual beams and column lengths are made up on site with the longitudinal reinforcement wired to the links with 1.6 mm soft iron binding wire that is cut to short lengths, bent in the form of a hair pin and looped and twisted around all intersections to secure reinforcing bars to links. The ends of binding wire must be flattened so that they do not protrude into the cover of concrete, where they might cause rust staining. Considerable skill, care and labour are required in accurately making up the reinforcing cages and assembling them in the formwork. This is one of the disadvantages of reinforced concrete where unit labour costs are high. At the junction of beams and columns there is a considerable confusion of reinforcement, compounded by large bars to provide structural continuity at the points of support and cranked bars for shear resistance.

Figure 6.9 is an illustration of the junction of the reinforcement for a main beam with an external beam and an external column. The longitudinal bars for the beams finish just short of the column reinforcement for ease of positioning the beam cages. Continuity bars are fixed through the column and wired to beam reinforcement. U bars fixed inside the column reinforcement and wired to the main beam serve to anchor beam to column against lateral forces.

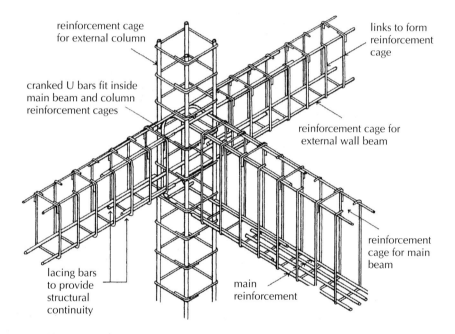

Figure 6.9   Reinforcement cages for reinforced concrete beam and column.

Figure 6.10A is an illustration of the reinforcement for the junction of four beams with a column. It can be seen that the reinforcement for intersecting beams is arranged to cross over at the intersection inside the columns. Figure 6.10B shows the next stage where the column, floor and beam have been cast and encased in concrete; starter bars are left protruding so that the lower column reinforcement cage can be tied to the next column cage. The correct term for linking the column starter bars to the next column cage is a 'column splice'. Figure 6.10B is an illustration of a column splice made in vertical cages for convenience in erecting formwork floor by floor and handling cages. In the reinforcement illustrated in Figures 6.9 and 6.10, the reinforcing bars are deformed to improve anchorage and obviate the necessity for hooked or bent ends of bars that considerably increase the labour of assembling reinforcing cages.

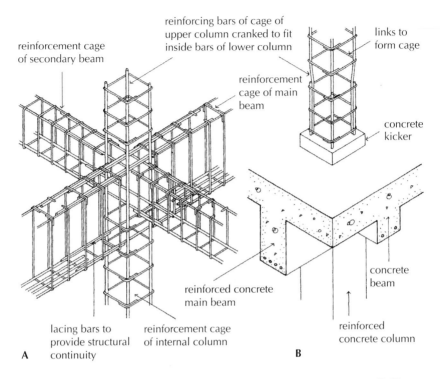

reinforcing bars of cage of upper column cranked to fit inside bars of lower column

reinforcement cage of secondary beam

links to form cage

reinforcement cage of main beam

concrete kicker

reinforced concrete main beam

concrete beam

lacing bars to provide structural continuity

reinforcement cage of internal column

reinforced concrete column

A

B

Figure 6.10   A, Reinforcement cages of internal columns and beams. B, Upper column cage spliced to lower cage.

Spacers for reinforcement

To ensure that there is the correct cover of concrete around reinforcement to protect the steel from corrosion and to provide adequate fire protection, it is necessary to fix spacers to reinforcing bars between the bars and the formwork. The spacers hold the reinforcement a set distance away from the face of the formwork, which will also be the face of the concrete. These spacers must be securely fixed so that they are not displaced during placing and compacting of concrete and are strong enough to maintain the required cover of concrete. Spacer blocks can be made from plastic, concrete or steel. Concrete spacer blocks are cast to the thickness of the required cover; they can be cast on site from sand and cement with a loop of binding wire protruding for binding to reinforcement, or ready prepared concrete spacers illustrated in Figure 6.11 may be used. The holes in the spacers are for binding wire. Plastic spacers are preferred to made-up cement and sand spacers for ease of use and security of fixing reinforcement in position.

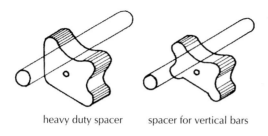

heavy duty spacer   spacer for vertical bars

Figure 6.11 Concrete spacers for reinforcement.

Plastic wheel spacers, as illustrated in Figure 6.12, are used with reinforcing bars to columns and to reinforcement in beams, with the spacers bearing on the inside face of formwork to provide the necessary cover for concrete around steel. The reinforcing bars clip into and are held firmly in place through the wheel spacers. The plastic pylon spacers illustrated in Figure 6.12 are designed to provide support and fixing to the bottom main reinforcement of beams. The reinforcement slips into and is held firmly in the spacer which bears on the inside face of the formwork to provide the necessary cover of concrete. These plastic spacers, which are not affected by concrete, are sufficiently rigid to provide accurate spacing and will not cause surface staining of concrete. They are commonly used in reinforced concrete.

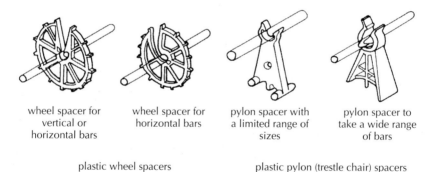

wheel spacer for vertical or horizontal bars | wheel spacer for horizontal bars | pylon spacer with a limited range of sizes | pylon spacer to take a wide range of bars

plastic wheel spacers   plastic pylon (trestle chair) spacers

Figure 6.12 Plastic wheel and pylon spacers.

To provide adequate support for top reinforcement which is cast into reinforced concrete floors, a system of chairs is used. The steel chairs are fabricated from round section mild steel rods to form a system of inverted Us which are linked by rods welded to them. Chairs are galvanised after fabrication. Chairs are either selected from a range of ready-made depths or purpose made to order. The steel chairs are placed on top of the main bottom reinforcement, which is supported by pylon spacers. The top bar of the chairs supports the top reinforcement, which is secured in place with binding wire. The chairs, illustrated in Figures 6.13 and 6.14, must be substantial enough to support the

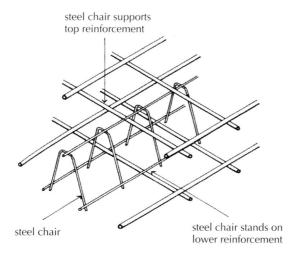

steel chair supports
top reinforcement

steel chair

steel chair stands on
lower reinforcement

Figure 6.13   Steel chair.

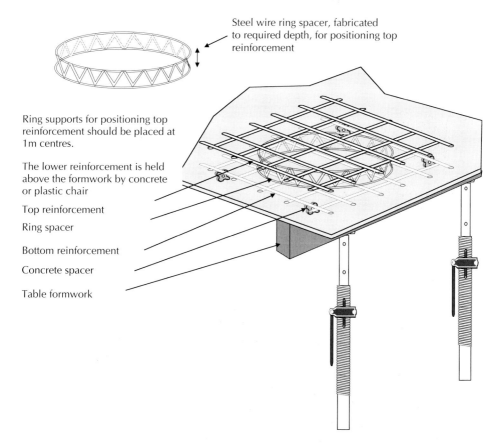

Steel wire ring spacer, fabricated
to required depth, for positioning top
reinforcement

Ring supports for positioning top
reinforcement should be placed at
1m centres.

The lower reinforcement is held
above the formwork by concrete
or plastic chair

Top reinforcement

Ring spacer

Bottom reinforcement

Concrete spacer

Table formwork

Figure 6.14   Steel ring spacer: floor reinforcement cage.

weight of those spreading and compacting the concrete. Reinforcement should be securely tied together; care should be taken to ensure that protruding ends of the tie wire do not intrude into the concrete cover.

6.4 Formwork and falsework

Formwork is the term used for the temporary timber, plywood, metal or other material used to contain, support and form wet concrete until it has gained sufficient strength to be self-supporting. Falsework is the term used to describe the temporary system or systems of support for formwork. Formwork and falsework should be strong enough to support the weight of wet concrete and pressure from placing and compacting the concrete inside the forms. Formwork should be sufficiently rigid to prevent any undue deflection of the forms out of true line and level and be sufficiently tight to prevent excessive loss of water and mortar from the concrete. The size and arrangement of the units of formwork should permit ease of handling, erection and striking. 'Striking' is the term used for dismantling formwork once concrete is sufficiently hard.

The traditional material for formwork was timber in the form of sawn, square edged boarding that is comparatively cheap and can be readily cut to size, fixed and struck. The material most used for lining formwork today is plywood (marine plywood), which provides a more watertight lining than sawn boards and a smoother finish. Joints between plywood are sealed with foamed plastic strips. Other materials used for formwork are steel sheet, glass reinforced plastics and hardboard. Where concrete is to be exposed as a finished surface, the texture of timber boards, carefully selected to provide a pattern from the joints between the boards and the texture of wood, may be used or any one of a variety of surface linings such as steel, rubber, thermoplastics or other material may be used to provide a finished textured surface to concrete.

Honeycombing and leaks
Formwork should be reasonably watertight to prevent small leaks causing unsightly stains on exposed concrete surfaces and large leaks causing honey-combing. Honeycombing is caused by the loss of water, fine aggregate and cement from concrete through large cracks, which results in a very coarse textured concrete finish which will reduce bond and encourage corrosion of reinforcement. To control leaks from formwork it is common to use foamed plastic strips in joints.

Release agents
To facilitate the removal of formwork and avoid damage to concrete as forms are struck, the surface of forms in contact with concrete should be coated with a release agent that prevents wet concrete adhering strongly to the forms. The

more commonly used release agents are neat oils with surfactants, mould cream emulsions and chemical release agents which are applied as a thin film to the inside faces of formwork before it is fixed in position.

Formwork support

The support for formwork is usually of timber in the form of bearers, ledgers, soldiers and struts (Photograph 6.2). For beams, formwork usually comprises bearers at fairly close centres, with soldiers and struts to the sides, and false-work ledgers and adjustable steel props, as illustrated in Figure 6.15. Form-work for columns is formed with plywood facings, vertical backing members and adjustable steel clamps, as illustrated in Figures 6.16 and 6.17. Falsework

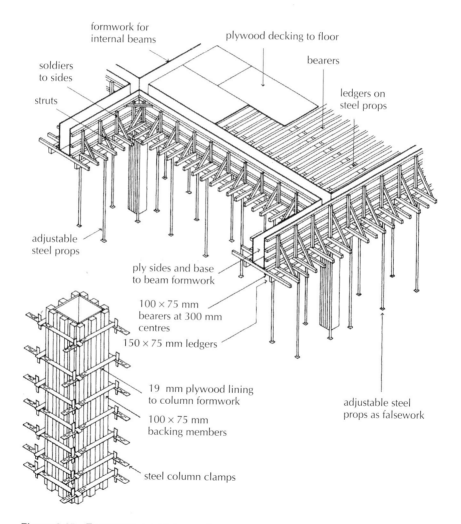

formwork for internal beams

plywood decking to floor

soldiers to sides

bearers

struts

ledgers on steel props

adjustable steel props

ply sides and base to beam formwork

100 × 75 mm bearers at 300 mm centres

150 × 75 mm ledgers

19 mm plywood lining to column formwork

100 × 75 mm backing members

steel column clamps

adjustable steel props as falsework

Figure 6.15   Formwork and falsework.

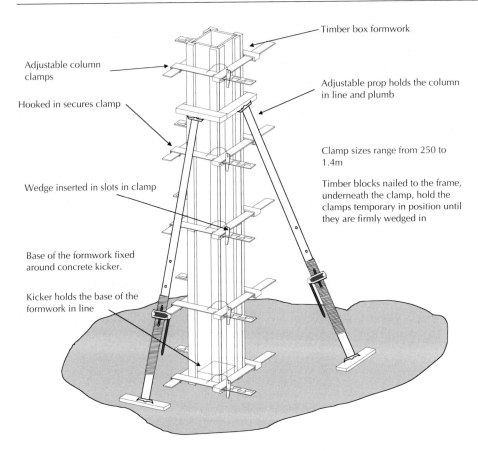

Timber box formwork

Adjustable column clamps

Hooked in secures clamp

Adjustable prop holds the column in line and plumb

Clamp sizes range from 250 to 1.4m

Wedge inserted in slots in clamp

Timber blocks nailed to the frame, underneath the clamp, hold the clamps temporary in position until they are firmly wedged in

Base of the formwork fixed around concrete kicker.

Kicker holds the base of the formwork in line

Figure 6.16   Column clamps and timber formwork.

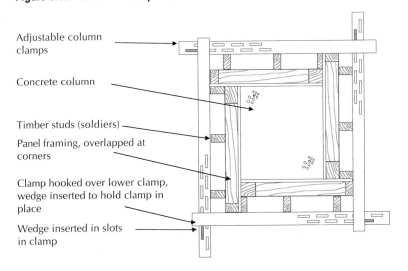

Adjustable column clamps

Concrete column

Timber studs (soldiers)

Panel framing, overlapped at corners

Clamp hooked over lower clamp, wedge inserted to hold clamp in place

Wedge inserted in slots in clamp

Figure 6.17   Plan of column clamps and timber formwork.

consists of adjustable steel props fixed as struts to the sides. Temporary false-work and formwork are struck and removed once the concrete they support and contain has developed sufficient strength to be self-supporting. In normal weather conditions the minimum period after placing ordinary Portland cement concrete that formwork can be struck is from 9 to 12 hours for columns, walls and sides of large beams, 11 to 14 days for the soffit of slabs and 15 to 21 days for the soffit of beams.

Column formwork

Column formwork can be constructed from short wall panels, with the ends of the panel overlapping (Figures 6.15, 6.16 and 6.17) or it can be constructed using specially designed and fabricated column formwork (Figures 6.18 and 6.19 and Photograph 6.1). Column formwork is made of timber, prefabricated steel units (Figure 6.18), or a combination of timber and steel. Glass reinforced plastic (GRP), expanded polystyrene, hardboard and plastic formers are also becoming more popular, as is cardboard. Where a circular column is required, it is possible to use disposable formwork, such as single-use cardboard formers. These are lightweight, easy to handle and can be quickly positioned. The cardboard tubes used to form the concrete column must be firmly held in position,

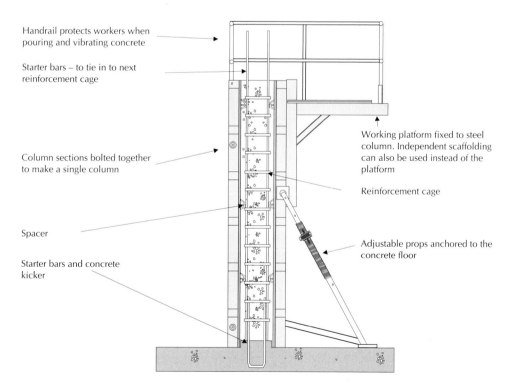

Handrail protects workers when pouring and vibrating concrete

Starter bars – to tie in to next reinforcement cage

Column sections bolted together to make a single column

Spacer

Starter bars and concrete kicker

Working platform fixed to steel column. Independent scaffolding can also be used instead of the platform

Reinforcement cage

Adjustable props anchored to the concrete floor

Figure 6.18   Proprietary steel column formwork (adapted from www.peri.ltd.uk).

Concrete kicker formed around
the starter bars

Concrete kicker formwork

Integral working
platform

Spacer

Reinforcement cage with
concrete spacer blocks to ensure
concrete cover is maintained

Steel column formwork erected
around concrete kicker and
reinforcement cage

Photograph 6.1   Circular steel column formwork.

adequately propped and clamped in place, before the concrete is poured. Extra care should be taken when pouring the concrete into lightweight formers because they can easily be knocked out of position.

With the sectional column formwork shown in Figure 6.16 the panels are erected around a concrete kicker. The concrete kicker is a small (40–50 mm) upstand cast in the concrete floor, which provides a firm object around which the column formwork can be erected. Prior to the formwork being erected, the concrete kicker is accurately cast onto the concrete floor. The small amount of formwork used to cast the kicker must be exactly the same cross-sectional

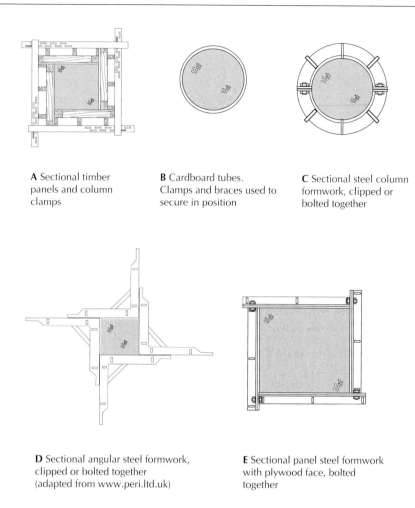

A Sectional timber panels and column clamps

B Cardboard tubes. Clamps and braces used to secure in position

C Sectional steel column formwork, clipped or bolted together

D Sectional angular steel formwork, clipped or bolted together (adapted from www.peri.ltd.uk)

E Sectional panel steel formwork with plywood face, bolted together

Figure 6.19   Types of column formwork – plan.

dimensions as the column. The kicker is accurately positioned so that the formwork can be accurately held in the correct position. Once the column is firmly clamped around the kicker, the formwork is checked for line and plumb. Adjustable props and clamps hold the columns firmly in position (Photograph 6.2). Having poured and vibrated the concrete, the columns should be checked to ensure that they are still line and plumb. If slight movement has occurred it is possible to adjust the props and bring the column back in line. The column should be cast 20 mm above the soffit of the underside of the next floor slab. This will provide a solid surface, which the table formwork for the floor slab can butt up against.

Horizontal formwork

It is becoming common to use patented formwork systems to cast large concrete floor slabs. These either take the form of plywood decks, fixed to steel or

Adjustable column
clamps

Reinforcement

Adjustable column
clamps

Adjustable prop

Photograph 6.2   Square column formwork systems.

aluminium bearers, which are held in position by interlinked and braced adjustable props (Photograph 6.3). Alternatively, table formwork can be used. The advantage of table formwork is that the supporting structure is already partly assembled as a braced table. The table can be hoisted by crane into the building and wheeled into position, before being properly secured. Where the concrete floor differs in soffit level, due to dropdown beams, additional props can be used to the formwork pieces that make up the beam (Figure 6.20).

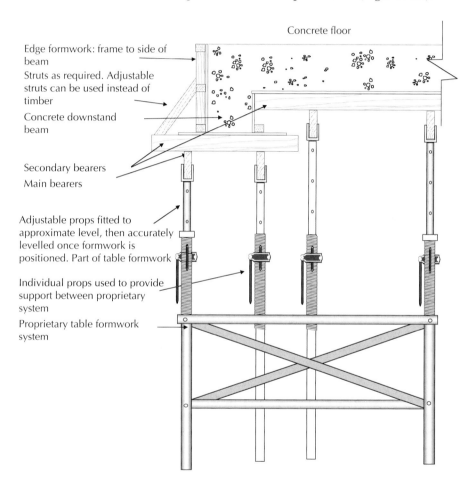

Figure 6.20   Beam and slab formwork.

To reduce the time that the main floor formwork is needed and the subsequent cost of hiring the formwork, intermediate props are used between the main formwork. The system of table and props is positioned, fixed in place and the concrete is then poured. Once the concrete has reached sufficient maturity, the majority of the horizontal support is removed. The intermediate props remain in position, continuing to provide support to the concrete slab. The table

Photograph 6.3   Prop and beam formwork.

formwork or the system of beams can be removed early, cleaned and quickly used on the next floor.

The props that are used to provide the longer-term support are usually designed to support both the concrete slab and the beams that support the main horizontal formwork. To ensure that the beams and the main horizontal formwork can be released early, the support mechanism is in two parts (Figure 6.21). At the head of the prop a collar or outer tube supports the horizontal beams and the main shaft of the prop is in direct contact with the concrete floor. Because the collar, sleeve or head that holds the horizontal bearers can be lowered and released independently of the main head, the prop is often called a drop head prop (Figure 6.21 and Photograph 6.4). Depending on the depth of the concrete floor and the properties of the concrete, using drop head props can allow the majority of the horizontal support to be struck after one day. This is a considerable saving in time, compared with the four to seven days' maturity normally required before removing the horizontal support.

Because it is easy to assemble, and can be quickly manoeuvred into position, table formwork is used to construct horizontal concrete surfaces. The tables are designed so that they can accommodate columns that would penetrate the decking (Figure 6.22). Table forms can be bolted or clipped together to provide the formwork for large horizontal surfaces (Figure 6.23 and Photograph 6.5).

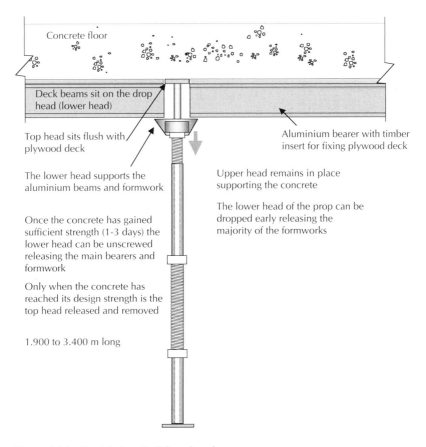

Figure 6.21   Double headed/drop head props.

Columns and other vertical concrete structures should be cast 20 mm above the soffit of the underside of the next floor slab. This provides a solid surface for the horizontal formwork to butt up against. The joint between vertical concrete surfaces (such as columns and walls) and horizontal formwork should be tight and sealed. Foam plastic sealing strips or gunned silicone rubber helps to seal the joint between the concrete and timber.

Prior to the concrete being poured it is vital that the surface of the formwork is cleaned. All joints should be taped and sealed, all loose debris and wire must be cleared away. A compressed air hose is usually used to blow away debris and dust (the reinforcement usually prevents brushes being used effectively). To make striking the formwork easier and reduce blowholes, the surface of the form should be coated with a release agent. Release agents are either chemical based, emulsion (mould cream) or neat oil with surfactant. Mould cream should not be used on steel forms, and oil can lead to staining of the concrete surface; therefore a chemical release agent is used. Chemical agents help to produce high quality finishes, are easily applied and are the least messy of the release agents.

Main beams and
secondary beams
carried by drop head
prop

Adjustable prop

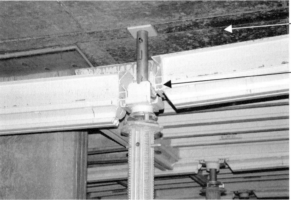

Concrete floor at partial
maturity

Drop head (outer
sleeve) can be released
and the main and
secondary beams
removed

Main head of the prop
remains in contact with
the concrete floor at all
times, continuing to
provide support until the
concrete reaches full
maturity

Photograph 6.4   Drop head prop.

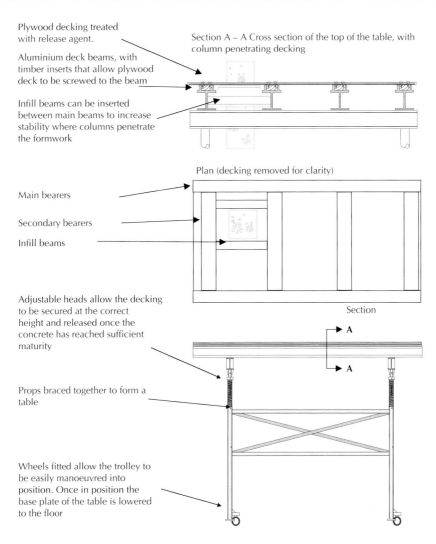

Plywood decking treated with release agent.

Aluminium deck beams, with timber inserts that allow plywood deck to be screwed to the beam

Infill beams can be inserted between main beams to increase stability where columns penetrate the formwork

Section A – A Cross section of the top of the table, with column penetrating decking

Plan (decking removed for clarity)

Main bearers

Secondary bearers

Infill beams

Adjustable heads allow the decking to be secured at the correct height and released once the concrete has reached sufficient maturity

Section

A

A

Props braced together to form a table

Wheels fitted allow the trolley to be easily manoeuvred into position. Once in position the base plate of the table is lowered to the floor

Figure 6.22　Table formwork.

Flying formwork

Table formwork comes in a range of sizes. It is not uncommon to see tables in excess of 5 m long. Large systems of prefabricated formwork reduce the assembly time. Where the building structure is designed in regular grid patterns, large tables prove economical. The large formwork systems are manoeuvred on trolleys, which wheel the forms to the edge of the building where they are lifted from the building using 'C hooks' fitted to cranes (Figure 6.24). Because the large frames are easily manoeuvred from the edge of the building and 'slung' through the air to their next position, they tend to be termed 'flying forms'. Using such large tables considerably reduces the cycle time to complete

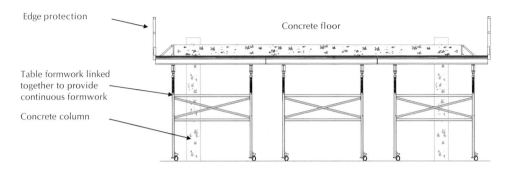

Edge protection

Concrete floor

Table formwork linked together to provide continuous formwork

Concrete column

Figure 6.23   Table formwork system.

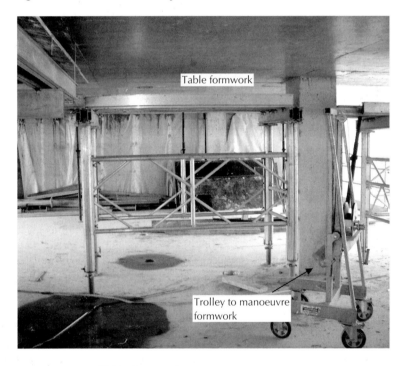

Table formwork

Trolley to manoeuvre formwork

Photograph 6.5   Table formwork.

the formwork for the next floor. Large flying forms can halve the time required to assemble smaller tables or beam and prop systems. Horizontal formwork must be struck gradually to avoid shock overloading in the concrete slab, which could cause the slab to fail. To eliminate any unnecessary vibration and shock, large table forms must be released slowly and the jacks slightly lowered first.

Wall formwork
It is common to see prefabricated steel or aluminium patient systems used on large concrete structures. Figure 6.25 shows a typical steel wall system.

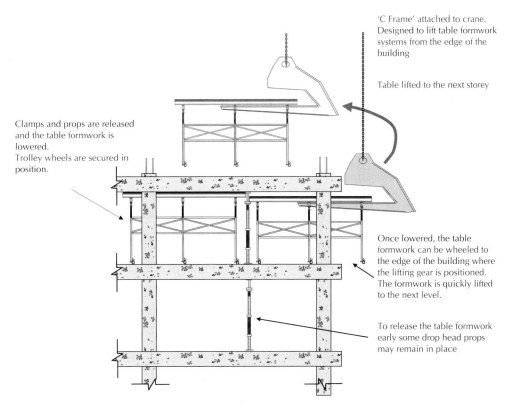

Figure 6.24   Flying formwork.

Photographs 6.6 and 6.7 show the formwork and steel reinforcement within the wall being assembled. As the working platform is already an integral part of the formwork, extra scaffolding around the formwork is not required. The formwork is manoeuvred into position by crane and quickly bolted together and propped (Photograph 6.7). Threaded steel ties are used to hold the two faces of formwork together. A steel or plastic sleeve, through which the tie runs, acts as a spacer holding the faces of the formwork the required distance apart. The sleeve is cut to the width of the wall. Various proprietary ties are available and not all require the use of a sleeve.

6.5 Prestressed concrete

Because concrete has poor tensile strength, a large part of the area of an ordinary reinforced concrete beam plays little part in the flexural strength of the beam under load. In the calculation of stresses in a simply supported beam, the strength of the concrete in the lower part of the beam is usually ignored. When reinforcement is stretched before or after the concrete is cast and the stretched reinforcement is anchored to the concrete, it causes a compressive prestress

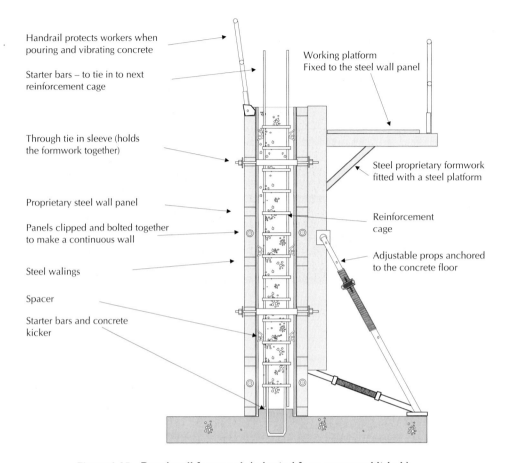

Handrail protects workers when pouring and vibrating concrete

Starter bars – to tie in to next reinforcement cage

Through tie in sleeve (holds the formwork together)

Proprietary steel wall panel

Panels clipped and bolted together to make a continuous wall

Steel walings

Spacer

Starter bars and concrete kicker

Working platform Fixed to the steel wall panel

Steel proprietary formwork fitted with a steel platform

Reinforcement cage

Adjustable props anchored to the concrete floor

Figure 6.25   Panel wall formwork (adapted from www.peri.ltd.uk).

in the concrete as it resists the tendency of the reinforcement to return to its original length. This compressive prestress makes more economical use of the concrete by allowing all of the section of concrete to play some part in supporting load. In prestressed concrete the whole or part of the concrete section is compressed before the load is applied, so that when the load is applied the compressive prestress is reduced by flexural tension.

In ordinary reinforced concrete, the concrete around reinforcement is bonded to it and must, therefore, take some part in resisting tensile stress. Because the tensile strength of concrete is low it will crack around the reinforcement under load, and hair cracks on the surface of concrete are not only unsightly, they also reduce the protection against fire and corrosion. The effectiveness of the concrete cover will be reduced when cracking occurs. In designing reinforced concrete members it is usual to limit the anticipated tensile stress in order to limit deflection and the extent of cracking of concrete around reinforcement.

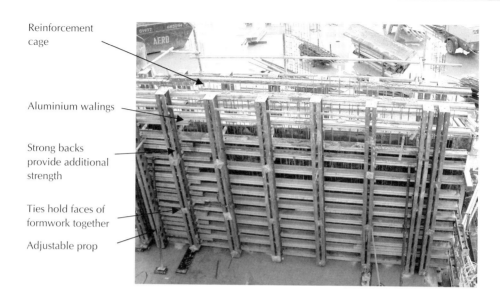

Reinforcement cage

Aluminium walings

Strong backs provide additional strength

Ties hold faces of formwork together

Adjustable prop

Photograph 6.6   Wall formwork.

This is a serious limitation in the most efficient use of reinforced concrete, particularly in long-span beams.

When reinforcement is stretched and put under tensile stress and then fixed in the concrete, once the prestress is released, the tendency of the reinforcement to return to its original length induces a compressive stress in concrete. The stretching of reinforcement before it is cast into concrete is described as pre-tensioning and stretching reinforcement after the concrete has been cast as post-tensioning. The advantage of the induced compressive prestress caused either by pre- or post-tensioning is that under load the tensile stress developed by bending is acting against the compressive stress induced in the concrete and in consequence cracking is minimised. If cracking of the concrete surface does occur and the load is reduced or removed, then the cracks close up due to the compressive prestress. Another advantage of the prestress is that the compressive strength of the whole of the section of concrete is utilised and the resistance to shear is considerably improved, so obviating the necessity for shear reinforcement.

For the prestress to be maintained, the steel reinforcement must not suffer permanent elongation or creep under load. High tensile wire is used in prestressed concrete to maintain the prestress under load. Under load, a prestressed concrete member will bend or deflect and compressive and tensile stresses will be developed in opposite faces, as previously explained. Concrete in parts of the member will therefore have to resist compressive stress induced by the prestress as well as compressive stress developed during bending. For this reason high compressive strength concrete is used in prestressed work to

One side of the wall formwork is erected against the concrete kicker

The reinforcement is then tied to the starter bars

Void formers are installed to allow for door openings

Finally the other side of the wall formwork is placed in position

Ties hold faces of formwork together

Tower scaffolding erected to pour concrete

Photograph 6.7   Wall formwork erection sequence.

gain the maximum advantage of the prestress. A consequence of the need to use high strength concrete is that prestressed members are generally smaller in section than comparable reinforced concrete ones.

Pre-tensioning

High tensile steel reinforcing wires are stretched between anchorages at each end of a casting bed and concrete is cast around the wires inside timber or steel moulds. The tension in the wires is maintained until the concrete around them has attained sufficient strength to take up the prestress caused by releasing the wires from the anchorages. The bond between the stretched wires and the concrete is maintained by the adhesion of the cement to the wires, by frictional resistance and the tendency of the wires to shorten on release and wedge into the concrete. To improve frictional resistance the wires may be crimped or indented, as illustrated in Figure 6.26. When stressing wires are cut and released from the anchorages in the stressing frame, the wires tend to shorten, and this shortening is accompanied by an increase in diameter of the wires which wedge into the ends of the member, as illustrated in Figure 6.27. Pre-tensioning of concrete is mainly confined to the manufacture of precast large-span members such as floor beams, slabs and piles. The stressing beds required for this work are too bulky for use on site.

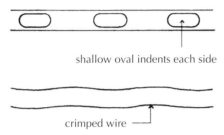

shallow oval indents each side

crimped wire

Figure 6.26   Prestressing wires.

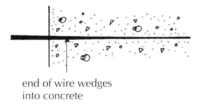

end of wire wedges
into concrete

Figure 6.27   Anchorage to stressing wire.

Post-tensioning

After the concrete has been cast inside moulds or formwork and has developed sufficient strength to resist the stress, stressing wires are threaded through ducts or sheaths cast in along the length of the member. These prestressing wires are anchored at one end of the member and are then stretched and anchored at the opposite end to induce the compressive stress. The advantage of post-tensioning is that the stressing wires or rods are stressed against the concrete and there is no loss of stress as there is in pre-tensioning due to the shortening of the wires when they are cut from the stressing bed. The major part of the drying shrinkage of concrete will have taken place before it is post-tensioned and this minimises loss of stress due to shrinkage of concrete. The systems of post-tensioning used are: Freyssinet, Gifford–Udall–CCL, Lee–McCall, Magnel–Blaton and the PSC one wire system.

Post tension cables and ducts are positioned within the reinforcement

The tension cables are simply anchored into the concrete at one end and tension exerted at the other

Reinforcement spacer chair

Photograph 6.8A   Post-tensioned floors.

The Freyssinet system

A duct is formed along the length of the concrete beam as it is cast. The duct is formed by casting concrete around an inflatable tube which is withdrawn when the concrete has hardened. At each end of the beam a high tension, concrete anchor cylinder is cast into the beam. The purpose of the anchor cylinder is to provide a firm base for the head of the jack used to tension wires. A 7 mm diameter cable of high tensile wires arranged around a core of fine-coiled wire is threaded through the duct. The wires are held at one end between the cast

Grouting tube, used
once tension is exerted

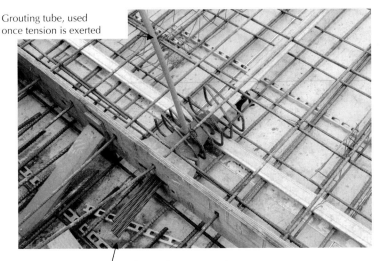

Steel cables are left protruding at the access end.

Once the concrete is poured tension is exerted and a report is
prepared on each tension cable reporting the level of force
exerted.

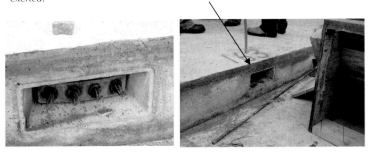

Photograph 6.8B   Post-tensioned floors.

in concrete anchor cylinder and a loose concrete anchor cone that is hammered
in tightly to secure the wires. A hydraulically operated ram is anchored to the
stressing wires, and the ram of the jack is positioned to bear on a loose ring
that bears on the concrete anchor cone around the wires. A piston on the ram
applies stress to the wires, which are anchored by the ram, forcing the anchor
cone into the anchor cylinder as illustrated in Figure 6.28. The stressing wires
are released from the jack and the protruding ends of wire cut off. A grout of
cement and water is forced under pressure into the cable duct to protect the
wires from corrosion (Photographs 6.8A and 6.8B).

The Gifford–Udall–CCL system
A duct is formed along the length of a concrete beam around an inflatable
former which is withdrawn when the concrete has hardened. Steel thrust rings

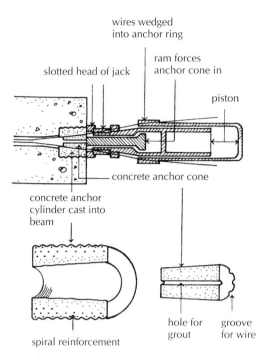

wires wedged
into anchor ring

ram forces
anchor cone in

slotted head of jack

piston

concrete anchor cone

concrete anchor
cylinder cast into
beam

hole for
grout

groove
for wire

spiral reinforcement

Figure 6.28 Freyssinet system.

are fixed to the ends of the beam, through which the stressing wires are threaded the length of the beam. The individual stressing wires are threaded through holes in a steel anchor plate at one end of the beam and firmly secured with steel grip barrels and wedges illustrated in Figure 6.29. The stressing wires are threaded through a steel anchor plate at the other end of the beam. Each wire is separately stressed by a jack and secured by ramming in split tapered wedges into a grip barrel. When the stressing operation is complete, the duct for the wires is filled under pressure with cement grout. The advantage of this system is that the precise stress in each wire is controlled, whereas in the Freyssinet system all wires are jacked together and, if one wire were to break, the remaining wires would take up their share of the total stress and might be overstressed.

The Lee–McCall system

An alloy bar is threaded through a duct in the concrete member and stressed by locking a nut to one end and stressing the rod the other end with a jack and anchoring it with a nut. The simplicity of this system is self-evident.

The Magnel–Blaton system

High tensile wires are arranged in layers of four wires each and are held in position by metal spacers. The layers of wire are threaded through a duct in

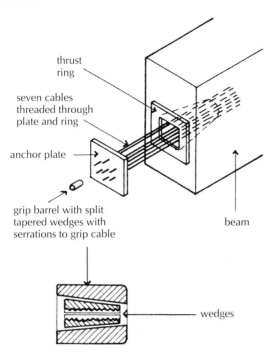

thrust ring

seven cables threaded through plate and ring

anchor plate

grip barrel with split tapered wedges with serrations to grip cable

beam

wedges

Figure 6.29   CCL system.

the concrete member. One end of the wires is fixed in metal sandwich plates against an anchor plate cast into the concrete. Pairs of wires are stressed in turn and wedged in position. The stressed wires are grouted in position in the duct by introducing cement grout through a hole in the top of the member leading to the duct.

The PSC one wire system

A duct is formed along the length of a concrete beam by an inflatable former. Guide bushes are cast into the ends of the beam, as illustrated in Figure 6.30. One, two or four high tensile wires are threaded through holes in anchor blocks at each end of the beam. The wires at one end are secured in the anchor block by ramming in split, tapered wedges around the wires in the holes in the anchor block. At the other end of the beam each wire is separately stressed by a jack and then secured by ramming in split tapered wedges. The cable duct is then filled with cement grout through the centre hole in the anchor blocks.

The advantage of the Gifford–Udall–CCL, the Lee–McCall, the Magnel–Blaton and the PSC systems over the Freyssinet system is that each wire or pair of wires is stressed individually so that the stress can be controlled and measured, whereas with the Freyssinet system there is no such control.

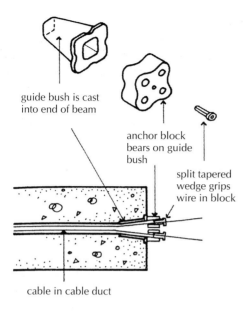

guide bush is cast
into end of beam

anchor block
bears on guide
bush

split tapered
wedge grips
wire in block

cable in cable duct

Figure 6.30   PSC one wire system.

6.6 Lightweight concrete

It may be advantageous to employ lightweight concrete such as no fines concrete for the monolithic loadbearing walls of buildings, and aerated concrete for structural members such as roof slabs supporting comparatively light loads to combine the advantage of reduced deadweight and improved thermal insulation. The various methods of producing lightweight concrete depend on:

❑ The presence of voids in the aggregate
❑ Air voids in the concrete
❑ Omitting fine aggregate, or
❑ The formation of air voids by the addition of a foaming agent to the concrete mix

The aggregates described for use in lightweight concrete building blocks are also used for mass concrete or reinforced concrete structural members, where improved thermal insulation is necessary and where the members, such as roof slabs, do not sustain large loads.

No fines concrete
No fines concrete consists of concrete made from a mix containing only coarse aggregate, cement and water. The coarse aggregate may be gravel, crushed brick or one of the lightweight aggregates. The coarse aggregate used in no

fines concrete should be as near one size as practicable to produce a uniform distribution of voids throughout the concrete. To ensure a uniform coating of the aggregate particles with cement/water paste it is important that the aggregate be wetted before mixing and the maximum possible water/cement ratio, consistent with strength, be used to prevent separation of the aggregate and cement paste.

Construction joints should be as few as possible and vertical construction joints are to be avoided if practicable because successive placings of no fines concrete do not bond together firmly as do those of ordinary concrete. Because of the porous nature of this concrete it must be rendered externally or covered with some protective coating or cladding material and the no fines concrete plastered or covered internally. A no fines concrete wall provides similar insulation to a sealed brick cavity wall of similar thickness.

Aerated and foamed concretes

An addition of one part of powdered zinc or aluminium to every thousand parts of cement causes hydrogen to evolve when mixed with water. As the cement hardens, a great number of small sealed voids form in the cement to produce aerated concrete, which usually consists of a mix of sand, cement and water. Foamed concrete is produced by adding a foaming agent, such as resin soap, to the concrete mix. The foam is produced by mixing in a high speed mixer or by passing compressed air through the mix to encourage foaming. As the concrete hardens, many sealed voids are entrained. Aerated and foamed concretes are used for building blocks and lightweight roofing slabs.

Surface finishes of concrete

When first used, the dull grey colour and coarse texture of dry concrete was not considered an attractive external finish. The inevitable marks caused by the joints in the formwork in which the wet material was cast and the inevitable irregular stains caused by rain and dust were considered unattractive, particularly in wet, northern European climates. From the middle of the 20th century the solid, rugged appearance of concrete finishes became fashionable. A variety of finishes to the exposed face of both insitu cast and precast concrete were used to enhance the appearance and mask weather staining of concrete surfaces. The principal finishes used were surfaces left untreated, either smooth or textured from formwork, textures formed by hammering or tooling using masonry techniques, and surfaces finished with aggregate exposed. The surface finish to concrete is dictated by architectural fashion.

Plain concrete finishes

Concrete is generally placed inside formwork in stages and, when the formwork is removed, variations in colour and texture and fine hair cracks usually clearly indicate the different placings of concrete. On drying, concrete shrinks and fine irregular shrinkage cracks appear in the surface in addition to the cracks and variations due to successive placings. One school of thought is to accept the cracks and variations in texture and colour as a fundamental of the material and make no attempt to control or mask them. Another school of thought is at pains to mask cracks and variations by means of designed joints and profiles on the surface.

Board marked concrete finishes are produced by compacting concrete by vibration against the surface of the timber formwork so that the finish is a mirror of the grain of the timber boards and the joints between them. This type of finish varies from the regular shallow profile of planed boards to the irregular marks of rough sawn boards and the deeper profile of boards that have been sand blasted to pronounce the grain of the wood. A necessary requirement of this type of finish is that the formwork be absolutely rigid to allow dense compaction of concrete to it and that the boards be non-absorbent.

One method of masking construction joints is to form a horizontal indentation or protrusion in the surface of the concrete where construction joints occur by nailing a fillet of wood to the inside face of the timber forms or by making a groove in the boards so that the groove or protrusion in the concrete masks the construction joint. Various plain concrete finishes can be produced by casting against plywood, hardboard or sheet metal to produce a flat finish or against corrugated sheets or crepe rubber to produce a profiled finish.

Tooled surface finishes

One way of masking construction joints, surface crazing of concrete and variations in colour is to tool the surface with hand or power operated tools. The action of tooling the surface is to break up the fine particles of cement and fine aggregate which find their way to the surface when wet concrete is compacted inside formwork, and also to expose the coarse texture of aggregate.

Bush hammering

A round headed hammer with several hammer points on it is vibrated by a power driven tool which is held against the surface and moved successively over small areas of the surface of the concrete. The hammer crushes and breaks off the smooth cement film to expose a coarse surface. This coarse texture effectively masks the less obvious construction joints and shrinkage cracks.

Point tooling

A sharp pointed power vibrated tool is held on the surface and causes irregular indentations and at the same time spalls off the fine cement paste finish. By moving the tool over the surface, a coarse pitted finish is obtained, the depth of pitting and the pattern of the pits being controlled by the pressure exerted and the movement of the tool over the surface. For best effect with this finish, as large an aggregate size as possible should be used to maintain an adequate cover of concrete to reinforcement. The depth of the pitting should be allowed for in determining the cover required.

Dragged finish
A series of parallel furrows is tooled across the surface by means of a power operated chisel pointed tool. The depth and spacing of the furrows depend on the type of aggregate used in the concrete and the size of the member to be treated. This highly skilled operation should be performed by an experienced mason.

Margins to tooled finishes
Bush hammered and point tooled finishes should not extend to the edges or arrises of members as the hammering operation required would cause irregular and unsightly spalling at angles. A margin of at least 50 mm should be left untreated at all angles. As an alternative, a dragged finish margin may be used with the furrows of the dragging at right angles to the angle.

Exposed aggregate finish
This type of finish is produced by exposing the aggregate used in the concrete or by exposing a specially selected aggregate applied to the face or faces of the concrete. In order to expose the aggregate, it is necessary either to wash or brush away the cement paste on the face of the concrete or to ensure that the cement paste does not find its way to the face of the aggregate to be exposed. Because of the difficulties of achieving this with insitu cast concrete, exposed aggregate finishes are confined in the main to precast concrete members and cladding panels. One method of exposing the aggregate in concrete is to spray the surface with water, while the concrete is still green, to remove cement paste on the surface. The same effect can be achieved by brushing and washing the surface of green concrete. The pattern and disposition of the aggregate exposed this way is dictated by the proportioning of the mix and placing and compaction of the concrete, and the finish cannot be closely controlled.

To produce a distinct pattern or texture of exposed aggregate particles, it is necessary to select and place the particles of aggregate in the bed of a mould or alternatively to press them into the surface of green concrete. This is carried out by precasting concrete. Members cast face down are prepared by covering the bed of the mould with selected aggregate placed at random or in some pattern. Concrete is then carefully cast and compacted on top of the aggregate so as not to disturb the face aggregate in the bed of the mould. If the aggregate is to be exposed in some definite pattern it is necessary to bed it in water-soluble glue in the bed of the mould on sheets of brown paper that are washed off later. Once the concrete member has gained sufficient strength it is lifted from the mould and the face is washed to remove cement paste. Large aggregate particles which are to be exposed are pressed into a bed of sand in the bed of the mould and the concrete is then cast on the large aggregate. When the member is removed from the mould after curing, the sand around the exposed aggregate is washed off. Alternatively, large particles may be pressed into the surface of green concrete and rolled, to bed them firmly and evenly.

6.7 Concrete structural frames

François Hennebique was chiefly responsible for the development of reinforced concrete for use in buildings, first as reinforced concrete piles and later as reinforced concrete beams and columns. In 1930 Freyssinet began development work that led to the use of prestressed concrete in building. The first reinforced concrete framed building to be built in the UK was the General Post Office building in London, which was completed in 1910. Subsequently comparatively little use was made of reinforced concrete in the UK until the end of the Second World War (1945). Steel had been the traditional material used for structural frames and engineers regarded reinforced concrete with some suspicion. The great shortage of steel that followed the end of the Second World War prompted engineers to use reinforced concrete as a substitute for steel in structural building frames. The shortage of steel continued for some years after the end of the war. At the time the conventional method of providing fire protection to structural steel frames was to encase beams and columns in concrete that was cast insitu in formwork around the steel. This concrete casing added nothing to the strength of the steel members, added considerably to the dead weight of the frame and was costly in the formwork and falsework necessary for casting concrete. With first a shortage and later the comparatively high cost of steel, it was common to use a reinforced concrete structural frame with the concrete providing compressive strength and fire protection, with the small section steel rods cast in to provide tensile strength where it was most needed. Up to the early 1980s the majority of framed buildings in the UK were constructed with reinforced concrete frames. More recently steel has become a more economic alternative for some building types, such as multi-storey framed structures and wide span single-storey shed buildings.

The conventional steel frame consists of I section columns and beams that have greater strength on the axis parallel to the web of the section than at right angles to it. It is logical, therefore, to connect the main beams to the flanges of columns and span one-way slabs between main beams with ties at right angles to main beams. The floor slabs bear on the beams in simple bending and do not act monolithically with the beams supporting them. In these conditions the rectangular column grid is the most economical arrangement.

The members of a reinforced concrete frame can be moulded to any required shape so that they can be designed to use concrete where compressive strength is required and steel reinforcement where tensile strength is required, and the members do not need to be of uniform section along their length or height. The singular characteristics of concrete are that it is initially a wet plastic material that can be formed to any shape inside formwork, for economy in section as a structural material or for reasons of appearance, and when it is cast insitu it will act monolithically as a rigid structure.

These characteristics are at once an advantage and a disadvantage. Unlimited choice of shape is an advantage structurally and aesthetically but may well be a disadvantage economically in the complication of formwork and falsework necessary to form irregular shapes. A monolithically cast reinforced concrete frame has advantageous rigidity of connections in a frame and in a solid wall or shell structure but this rigidity is a disadvantage in that it is less able to accommodate movements due to settlement, wind pressure, and temperature and moisture changes than is a more flexible structure.

The cost of formwork for concrete can be considerably reduced by repetitive casting in the same mould in the production of precast concrete cladding and structural frames, and the rigidity of the concrete frame can be of advantage on subsoils of poor or irregular bearing capacity and where severe earth movements occur as in areas subject to earthquakes. In spite of considerable publicity emphasising the advantages of steel as a structural frame material, the insitu cast reinforced concrete frame is still extensively used for both single- and multi-storey buildings as a convenient and economic skeleton frame within which or on which a variety of wall envelopes may be supported or hung to provide the appearance of traditional loadbearing walling, panel cladding, infill framing and thin sheet finishes.

Insitu cast frames

Structural frame construction
The principal use of reinforced insitu cast concrete as a structural material for building is as a skeleton frame of columns and beams with reinforced concrete floors and roof. In this use, reinforced concrete differs little from structural steel skeleton frames cased in concrete. In those countries where unit labour costs are low and structural steel is comparatively expensive, a reinforced concrete frame is widely used as a frame for both single- and multi-storey buildings such as the small framed building, with solid end walls and projecting balconies with upstands, illustrated in Figure 6.31.

The insitu cast, reinforced concrete structural frame is much used for multi-storey buildings such as flats and offices. Repetitive floor plans can be formed inside a skeleton frame of continuous columns and floors. To use the same formwork and falsework, floor by floor, variations in the reinforcement and/or mix of concrete in columns, to support variations in loads, can provide a uniform column section. The uniformity of column section and formwork makes for a speedily erected and economic structural frame. An advantage of the reinforced concrete structural frame is that the columns, beams and floor slabs

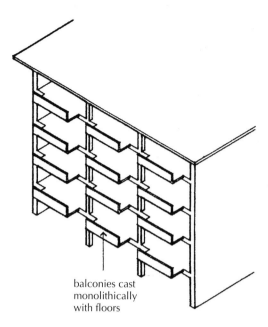

balconies cast
monolithically
with floors

Figure 6.31   Insitu cast concrete frames.

provide a level, solid surface on which walls and partitions can be built and between which walls, partitions and framing may be built and secured by bolting directly to a solid concrete backing.

A reinforced concrete structural frame with one-way spanning floors is generally designed on a rectangular grid for economy in the use of materials in the same way as a structural steel frame. Where floors are cast monolithically with a reinforced concrete frame, the tie beams that are a necessary part of a structural steel frame may be omitted as the monolithically cast floors will act as ties. The insitu reinforced concrete floors, illustrated in Figure 6.32, span one way between the upstand beams in external walls and the pair of internal beams supported by internal columns. This arrangement provides open plan floor areas each side of a central access corridor. An advantage of the upstand beams in the external walls is that the head of windows may be level with or just below the soffit of the floor above for the maximum penetration of daylight.

Cross wall and box frame construction

Multi-storey structures, such as blocks of flats and hotels with identical compartments planned on successive floors one above the other, require permanent, solid, vertical divisions between compartments for privacy, and sound and fire resistance. In this type of building it is illogical to construct a frame and then build solid heavy walls within the frame to provide vertical separation,

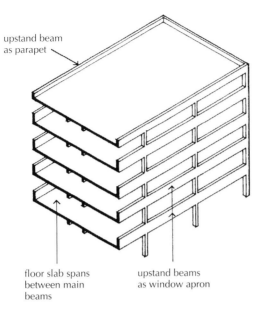

upstand beam
as parapet

floor slab spans
between main
beams

upstand beams
as window apron

Figure 6.32   Insitu cast frame.

with the walls taking no part in loadbearing. A system of reinforced concrete cross walls at once provides sound and fire separation and acts as a structural frame supporting floors, as illustrated in Figure 6.33.

Between the internal cross walls, reinforced concrete beam and slab or plate floors may be used. Where flats are planned on two floors as maisonettes the

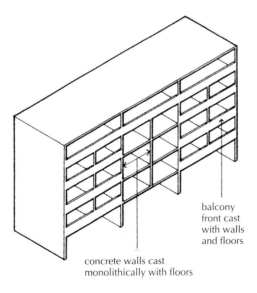

balcony
front cast
with walls
and floors

concrete walls cast
monolithically with floors

Figure 6.33   Cross wall construction.

intermediate floor of the maisonette may be of timber joist and concrete beam construction to reduce cost and dead weight. The intermediate timber floor inside maisonettes is possible where Building Regulations require vertical and horizontal separation between adjacent maisonettes.

A system of box frame, insitu cast external and internal walls and floors may be used where identical floor plans are used for a multi-storey building without columns or beams. The inherent strength and stability of the rigidly-connected walls and floors is used to advantage with both internal and external walls perforated for door and window openings as required, as illustrated in Figure 6.34. The structural logic of this system does not necessarily result in the most economical form of building because of the considerable labour cost in the extensive formwork and falsework needed. A straightforward system of skeleton frame with external cladding and solid internal division walls may often be cheaper.

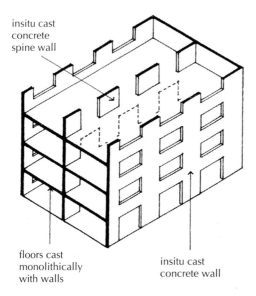

insitu cast concrete spine wall

floors cast monolithically with walls

insitu cast concrete wall

Figure 6.34　Box frame construction.

Wind bracing

A steel frame depends on the use of continuously rolled, comparatively slender steel sections that are put together in the form of a skeleton frame. Where a part or parts of a building have to be enclosed, as for example lift shafts, stairs and lavatories, it is usual to construct the steel frame around these parts and then enclose them with brickwork carried at each floor by the frame. Initially concrete is a wet, plastic material that can be moulded to any desired shape, and the shape is not dependent on the reinforcement, which can be disposed to suit the shape of the concrete.

In a reinforced concrete structure it is not logical to cast beams and columns to support permanent brick walls when a monolithically cast concrete wall will serve the dual function of frame and enclosing wall. In multi-storey reinforced concrete framed buildings it is usual to contain the lifts, stairs and lavatories within a service core, contained in reinforced concrete walls, as part of the frame. The hollow reinforced concrete column containing the services and stairs is immensely stiff and will strengthen the attached skeleton frame against wind pressure. In addition to stiffening the whole building, such a service core may also carry a considerable part of floor loads by cantilevering floors from the core and using props in the form of slender columns on the face of the building. Similarly, monolithically cast reinforced concrete flank end walls of slab blocks may be used to stiffen a skeleton frame structure against wind pressure on its long facade.

Floor construction

Insitu cast concrete floors
The principal types of reinforced insitu cast concrete floor construction are:

- ❑ Beam and slab
- ❑ Waffle grid slab
- ❑ Drop beam and slab
- ❑ Flat slab

Beam and slab floor
A beam and slab floor is generally the most economic and therefore most usual form of floor construction for reinforced concrete frames. When a reinforced concrete frame is cast monolithically with reinforced concrete floors it is logical to design the floor slabs to span in both directions so that all the beams around a floor slab can bear part of the load. This two-way span of floor slabs effects some reduction in the overall depth of floors as compared to a one-way spanning floor slab construction. Since the most economical shape for a two-way spanning slab is square, the best column grid for a reinforced concrete frame with monolithically cast floors is a square one, as illustrated in Figure 6.35.

The insitu cast reinforced concrete floor illustrated in Figure 6.35 combines main and secondary beams as a grid to provide the least thickness of slab for economy in the mass of concrete in construction, and comparatively widely spaced columns. This square grid results in the minimum thickness of floor slab and minimum depth of beams, and therefore the minimum dead weight of construction. Departure from the square column grid, because of user requirements and circulation needs in a building, will increase the overall depth, weight and therefore cost of construction of a reinforced concrete frame.

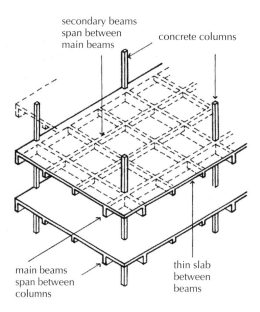

Figure 6.35   Square grid beam and slab floor.

The rectangular column grid, illustrated in Figure 6.36, supports main beams between columns that support one-way spanning floors with the beams between columns. The floor slab can be cast insitu on centering and falsework, or precast concrete floor beams or planks may be used. This arrangement involves closely spaced columns and the least mass of concrete in floors.

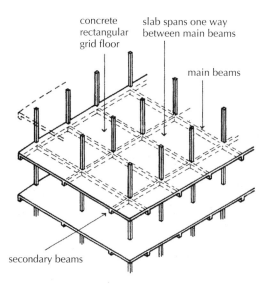

Figure 6.36   Rectangular grid beam and slab floor.

In a steel frame the skeleton of columns and beams is designed to carry the total weight of the building. The floors, which span between beams, act independently of the frame. With an insitu cast reinforced concrete frame and floor construction, columns, beams and floors are cast and act monolithically. The floor construction, therefore, acts with and affects the frame, and should be considered as part of it.

Waffle grid slab floor

If the column grid is increased from about 6.0 to about 12.0 m² or near square it becomes economical to use a floor with intermediate cross beams supporting thin floor slabs, as illustrated in Figure 6.37. The intermediate cross beams are cast on a regular square grid that gives the underside of the floor the appearance of a waffle, hence the name. The advantage of the intermediate beams of the waffle is that they support a thin floor slab and so reduce the dead weight of the floor as compared to a flush slab of similar span. This type of floor is used where a widely spaced square column grid is necessary and floors support comparatively heavy loads. The economic span of floor slabs between intermediate beams lies between 900 mm and 3.5 m. The waffle grid form of the floor may be cast around plastic or metal formers (as illustrated in Figure 6.37) laid on timber centering so that the smooth finish of the soffit may be left exposed.

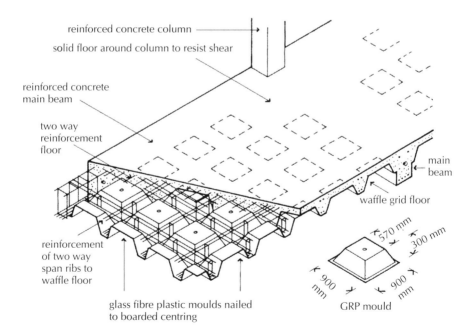

Figure 6.37   Waffle grid insitu cast reinforced concrete floor.

Drop slab floor

This floor construction consists of a floor slab, which is thickened between columns in the form of a shallow but wide beam, as illustrated in Figure 6.38. A drop slab floor is of about the same dead weight as a comparable slab and beam floor and will have up to half the depth of floor construction from top of slab to soffit of beams. On a 12.0 m² column grid the overall depth of a slab and beam floor would be about 1200 mm whereas the depth of a drop slab floor would be about 600 mm. This difference would cause a significant reduction in overall height of construction of a multi-storey building. This form of construction is best suited to a square grid of comparatively widely spaced columns selected for large, unobstructed areas of floor. Because of the additional reinforcement required for shallow depth, wide span beams, this type of floor is more expensive than a traditional rectangular grid beam, and slab floor.

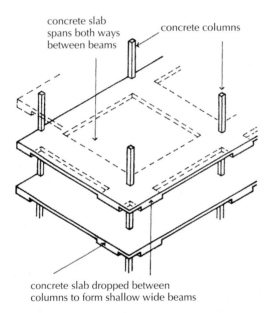

concrete slab spans both ways between beams

concrete columns

concrete slab dropped between columns to form shallow wide beams

Figure 6.38   Drop slab floor.

Flat slab (plate) floor

In this floor construction the slab is of uniform thickness throughout, without downstand beams and with the reinforcement more closely spaced between the points of support from columns. To provide sufficient resistance to shear at the junction of columns and floor, haunched or square headed columns are often formed. Figure 6.39 is an illustration of this floor. The dead weight of this floor and its cost are greater than for the floor systems previously described but its depth is less and this latter advantage provides the least overall depth

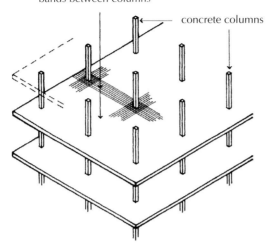

concrete flat slab floor
heavily reinforced in wide,
bands between columns

concrete columns

Figure 6.39   Flat slab (plate) floor.

of construction in multi-storey buildings. The floor slabs in the floor systems described above may be of solid reinforced construction or constructed with one of the hollow, or beam or plank floor systems.

In modern buildings it is common to run air conditioning, heating, lighting and fire fighting services on the soffit of floors above a false ceiling and these services occupy some depth below which minimum floor heights have to be provided. Even though the beam and slab or waffle grid floors are the most economic forms of construction in themselves, they may well not be the most advantageous where the services have to be fixed below and so increase the overall depth of the floor from the top of the slab to the soffit of the false ceiling below, because the services will have to be run below beams and so increase the depth between false ceiling and soffit of slab. Here it may be economic to bear the cost of a flat slab or drop slab floor in order to achieve the least overall height of construction and its attendant saving in cost. Up to about a third of the cost of an insitu cast reinforced concrete frame goes to providing, erecting and striking the formwork and falsework for the frame and the centering for the floors. It is important, therefore, to maintain a uniform section of column up the height of the building and repetitive floor and beam design as far as possible, so that the same formwork may be used at each succeeding floor. Alteration of floor design and column section involves extravagant use of formwork. Uniformity of column section is maintained by using high strength concrete with a comparatively large percentage of reinforcement in the lower, more heavily loaded storey heights of the columns, and progressively less strong concrete and less reinforcement up the height of the building.

Precast reinforced concrete floor systems

Precast reinforced concrete floor beams, planks, tee beams or beam and infill blocks that require little or no temporary support and on which a screed or structural concrete topping is spread are commonly used with structural steel frames and may be used for insitu cast concrete frames instead of insitu cast floors. Precast beams and plank floors that require no temporary support in the form of centering are sometimes referred to as self-centering floors. The use of these floor systems with skeleton frame reinforced concrete multi-storey buildings is limited by the difficulty of hoisting and placing them in position and the degree to which the operation would interrupt the normal floor by floor casting of slabs and columns.

Precast hollow floor units

These large precast reinforced concrete, hollow floor units are usually 400 or 1200 mm wide, 110, 150, 200, 250 or 300 mm thick and up to 10 m long for floors, and 13.5 m long for roofs. The purpose of the voids or hollows in the floor units is to reduce dead weight without affecting strength. The reinforcement is cast into the webs between hollows. Hollow precast reinforced concrete floor units can be used by themselves as floor slab with a levelling floor screed or they may be used with a structural reinforced concrete topping with tie bars over beams for composite action with the beams. When used for composite action it is usual to fix the reinforcing tie bars into slots in the ends of units. These tie bars are wired to loops of reinforcement cast in and protruding from the top of beams for the purpose of continuity of structural action. End bearing of these units should be a minimum of 75 mm on steel and concrete beams, and 100 mm on masonry and brick walls. Figure 6.40 is an illustration of precast hollow floor units bearing on an insitu cast beam.

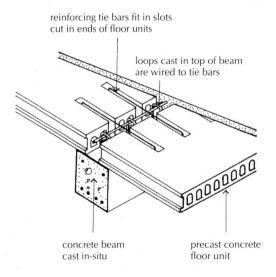

reinforcing tie bars fit in slots
cut in ends of floor units

loops cast in top of beam
are wired to tie bars

concrete beam
cast in-situ

precast concrete
floor unit

Figure 6.40 Precast concrete floor units.

Precast concrete plank floor units

These comparatively thin, prestressed solid plank, concrete floor units which are little used with skeleton frame concrete structures are designed as permanent centering and for composite action with reinforced concrete topping. The units are 400 or 1200 mm wide, 65, 75 or 100 mm thick and up to 9.5 m long for floors, and 10 m long for roofs. It may be necessary to provide some temporary propping to the underside of these planks until the concrete topping has gained sufficient strength.

Precast concrete tee beams

Precast prestressed concrete T beam floors are mostly used for long span floors in such buildings as stores, supermarkets, swimming pools and multi-storey car parks where there is a need for wide span floors and the depth of this type of floor is not a disadvantage. The floor units are cast in the form of a double tee. The strength of these units is in the depth of the ribs which support and act with the comparatively thin top web. A structural reinforced concrete topping is cast on top of the floor units, which bear on the toe of a boot section concrete beam.

Precast beam and filler block floor

This floor system consists of precast reinforced concrete planks or beams that support precast hollow concrete filler blocks, as illustrated in Figure 6.41. The planks or beams are laid between supports with the filler blocks between them, and a concrete topping is spread over the planks and filler blocks. The reinforcement protruding from the top of the plank acts with the concrete topping to form a reinforced concrete beam. The advantage of this system is that the lightweight planks or beams and filler blocks can be lifted and placed in position much more easily than the much larger hollow concrete floor units.

Hollow clay block and concrete floor

A floor system of hollow clay blocks and insitu cast reinforced concrete beams between the blocks and concrete topping, cast on centering and falsework, was for many years extensively used for the fire resisting properties of the blocks. This floor system is much less used because of the very considerable labour in laying the floor.

6.8 Precast reinforced concrete frames

The development of the structural building frame depended on the use of steel and, later, reinforced concrete as a skeleton frame to support floors and walls of the traditional materials, stone and brick. This form of construction was facilitated by a plentiful supply of cheap labour to construct the frame and build walls using the traditional skills of masonry and bricklaying. The supply of skilled and unskilled labour was adequate at the time for the construction

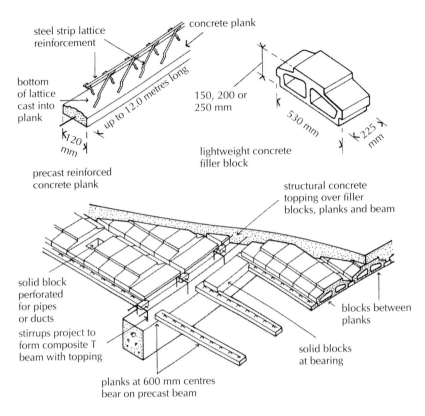

steel strip lattice reinforcement

concrete plank

bottom of lattice cast into plank

precast reinforced concrete plank

up to 12.0 metres long

120 mm

150, 200 or 250 mm

lightweight concrete filler block

530 mm

225 mm

structural concrete topping over filler blocks, planks and beam

solid block perforated for pipes or ducts

stirrups project to form composite T beam with topping

planks at 600 mm centres bear on precast beam

blocks between planks

solid blocks at bearing

Figure 6.41   Precast beam and filler block floor.

of the comparatively few large buildings in cities and towns. The extensive programmes of rebuilding and rehousing that followed the end of the Second World War (1945) coincided with a shortage of the traditional building materials, i.e. brick, stone, timber and steel, and a depleted labour force wholly inadequate to the scale of the projected work. In the event, a substantial part of the building programmes was met by the use of concrete in the form of precast frames, cladding units and wall frames. The combination of the use of precast concrete units and the introduction of the tower crane made it possible to produce standard components and assemble them on site with the minimum of skilled labour.

The advantage of precast concrete was that the materials – cement, sand, gravel and crushed rock – were readily available and could be combined with the least amount of steel to produce a material that was structurally sound and could serve both as a frame and as a wall material for buildings. Mechanisation of the production of standard, repetitively cast units off the site and their assembly on site required the least amount of labour, either skilled or unskilled. The two basic forms of precast concrete frames are:

❑ The frame members precast either as separate lengths of column and beam or combined as beam and column units, which are assembled and joined on site ready for infill panels or a cladding of brick or precast concrete units
❑ The wall frame units which serve both as structure and walling combined in large wall units, usually storey height, assembled and joined on site to precast concrete floor units

The advantage of the precast frame is that it allows the use of various wall and cladding finishes, either as infill panels fixed inside the frame or as a covering of cladding. The wall frame, however, is a large, solid, storey height panel of concrete that can be varied only in the surface finish of concrete and the size and disposition of windows.

In the years from 1950 to 1970 precast concrete frames, wall frames and cladding were extensively used as an accepted form of construction for multi-storey housing and other large buildings where standard units of construction could be used economically. Since the early 1970s, precast concrete has lost favour as a material for framing and cladding buildings. This loss of favour is principally a change of fashion as glass reinforced cement, then plastics and more recently sheet metal cladding panels were used instead of precast concrete, and for reasons of economy steel was used for structural frames in place of concrete. Publicised failures of precast concrete wall units, such as that at Ronan Point, have done much to add to the disfavour of concrete as a building material. These failures were not failures of the material itself but failures of the systems of jointing due to poor design or faulty workmanship or a combination of both.

Precast concrete has been established as a sound, durable material for framing and cladding buildings where repetitive casting of units is an acceptable and economic form of construction. Fashions change and it is likely that precast concrete will once again find favour as a building material. The chief problem in precast concrete framework is joining the members on site, particularly if the frame is to be exposed, to provide a solid, rigid bearing in column joints and a strong, rigid bearing of beams to columns that adequately ties beams to columns for structural rigidity. Where the frame is made up of separate precast column and beam units there is a proliferation of joints to be made on site. The number of site joints is reduced by the use of precast units that combine two or more column lengths with beams, as illustrated in Figure 6.42. The number of columns and beams that can be combined in one precast unit depends on the particular design of the building and the facilities for casting, transporting, hoisting and fixing units on site.

The general arrangement of precast structural units is as separate columns, often two-storey height and as cruciform, H or M frames. The H frame unit is often combined with under window walling, as illustrated in Figure 6.42.

The two basic systems of jointing used for connections of column to column are by direct end bearing or by connection to a bearing plate welded to

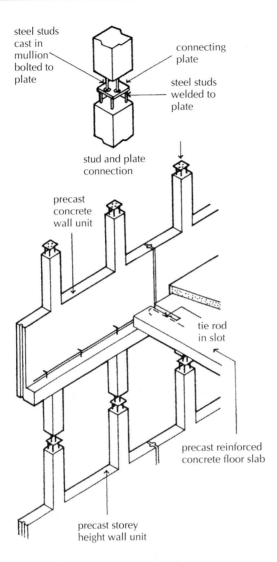

steel studs
cast in
mullion
bolted to
plate

connecting
plate

steel studs
welded to
plate

stud and plate
connection

precast
concrete
wall unit

tie rod
in slot

precast reinforced
concrete floor slab

precast storey
height wall unit

Figure 6.42   Precast concrete wall units.

protruding studs. Direct bearing of ends is effected through a locating dowel which can also be used as a post tensioning connection, as illustrated in Figure 6.43. A coupling plate connection is made by welding a plate to studs protruding from the end of one column and bolting studs protruding from the other to the plate, as illustrated in Figure 6.43. Plainly the studs and plate must be accurately located or else there will be an excessive amount of site labour in making this connection. The completed joint is usually finished by casting concrete around the joint. Alternatively the joint may be made with bronze studs and plate, and left exposed as a feature of an externally exposed frame.

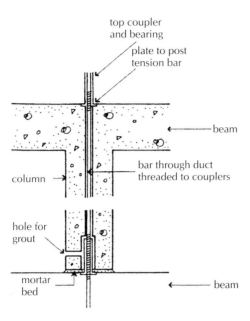

top coupler
and bearing

plate to post
tension bar

beam

bar through duct
threaded to couplers

column

hole for
grout

mortar
bed

beam

Figure 6.43   Precast concrete frame to frame joint.

One method of joining beams to columns is by bearing on a haunch cast in the columns and by connecting a steel box, cast in the end of beams, to an angle or plate set in a housing in columns, as illustrated in Figure 6.44.

A steel box is cast into the protruding ends of beams, which bear on to a steel angle plate cast into a housing in the column. A bolt is threaded through a hole in the end of the steel box end of one beam, a hole in the column and a hole in the box in the end of the next beam, and secured with a nut. This firmly clamps the beams to the columns.

The precast, hollow floor units bear on the rebate in beams. Slots cast in the floor units accommodate steel tie bars, which are hooked over peripheral tie bars. The tie bars are wired to loops, which are cast into beams. Structural concrete topping is spread, compacted and levelled over the floor slabs and into the space between the ends of slabs and beams to form an insitu reinforced concrete floor. Precast floor units bear either directly on concrete beams or, more usually, on supporting nibs cast for the purpose. Ends of floor units are tied to beams through protruding studs or insitu cast reinforcement so that the floor units serve to transfer wind pressure back to an insitu cast service and access core.

The precast reinforced concrete wall frames, illustrated in Figure 6.45, that combine four columns with a beam were used with drop in beams as the structural wall frame system for a 22-storey block of flats in Westminster. The precast framework is tied to the central core through the precast concrete floor units at each floor level, which are dowel fixed to the precast frame and tied with reinforcement to the insitu core; the precast framework is vertically tensioned

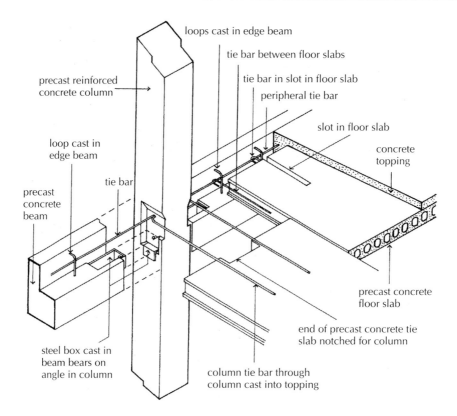

Figure 6.44   Precast reinforced concrete structural frame.

by couplers through columns, as illustrated in Figure 6.46, so that column ends are compressed to the dry mortar bed. Storey height frames are linked by short lengths of beam, which are dropped in and tied to the frames. The precast framework was designed for rapid assembly through precasting and direct bearing of beams on columns and end bearing of columns, to avoid the use of insitu cast joints that are laborious to make and which necessitate support of beams while the insitu concrete hardens. The top hung, exposed aggregate, precast concrete cladding panels have deep rebate horizontal joints and open vertical joints with mastic seals to columns, as illustrated in Figures 6.45 and 6.46.

Precast concrete wall frames
Precast concrete wall frames were used extensively in Russia and northern European countries in the construction of multi-storey housing where repetitive units of accommodation were framed and enclosed by large precast reinforced concrete wall panels that served as both external and internal walls and as a structural frame. The advantages of this system of building are that large, standard, precast concrete wall units can be cast off site and rapidly assembled

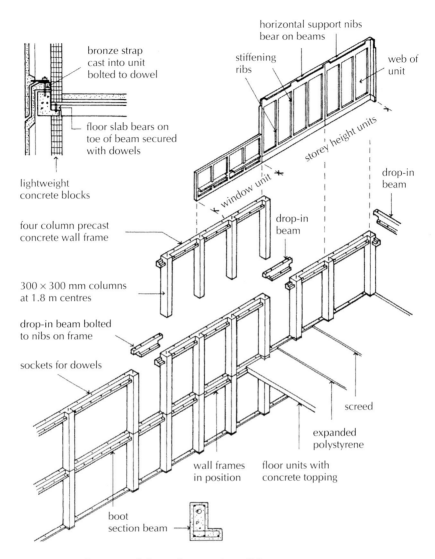

bronze strap
cast into unit
bolted to dowel

floor slab bears on
toe of beam secured
with dowels

lightweight
concrete blocks

four column precast
concrete wall frame

300 × 300 mm columns
at 1.8 m centres

drop-in beam bolted
to nibs on frame

sockets for dowels

horizontal support nibs
bear on beams

stiffening
ribs

web of
unit

storey height units

drop-in
beam

window unit

drop-in
beam

screed

expanded
polystyrene

wall frames
in position

floor units with
concrete topping

boot
section beam

Figure 6.45   Precast reinforced concrete wall frames.

on site largely independently of weather conditions, a prime consideration in countries where temperatures are below freezing for many months of the year.

Reinforced concrete wall frames can support the loads of a multi-storey building, can be given an external finish of exposed aggregate or textured finish that requires no maintenance, can incorporate insulation either as a sandwich or lining and have an internal finish ready for decoration. Window and door openings are incorporated in the panels so that the panels can be delivered to site with windows and doors fixed in position. In this system of construction the prime consideration is the mass production of complete wall units off the site,

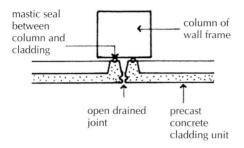

mastic seal between column and cladding

column of wall frame

open drained joint

precast concrete cladding unit

Figure 6.46 Vertical joint to cladding units.

under cover, by unskilled or semi-skilled labour assisted by mechanisation as far as practical towards the most efficient and speedy erection of a building. The appearance of the building is a consequence of the chosen system of production and erection rather than a prime consideration.

The concrete wall units will give adequate protection against wind and rain by the use of rebated horizontal joints and open drained vertical joints with back-up air seals similar to the joints used with precast concrete cladding panels. Some systems of wall frame incorporate a sandwich of insulation in the thickness of the panel, with the two skins of concrete tied together across the insulation with non-ferrous ties. This is not a very satisfactory method of providing insulation as a sufficient thickness of insulation for present day standards will require substantial ties between the two concrete skins, and the insulation may well absorb water from drying out of concrete and rain penetration, and so be less effective as an insulant. For best effect the insulation should be applied to the inside face of the wall as an inner lining to panels, or as a site fixed or built inner lining or skin.

The wall frame system of construction depends, for the structural stability of the building, on the solid, secure bearing of frames on each other, and the firm bearing and anchorage of floor units to the wall frames and back to some rigid component of the structure, such as insitu cast service and access cores. Figure 6.47 is an illustration of a typical precast concrete wall frame.

6.9 Lift slab construction

In this system of construction the flat roof and floor slabs are cast one on the other at ground level around columns or insitu cast service, stair and lift cores. Jacks operating from the columns or cores pull the roof and floor slabs up into position. This system of construction was first employed in America in 1950. Since then many buildings in America, Europe and Australia have been constructed by this method. The advantage of the system is that the only formwork required is to the edges of the slabs, and no centering whatever is required to

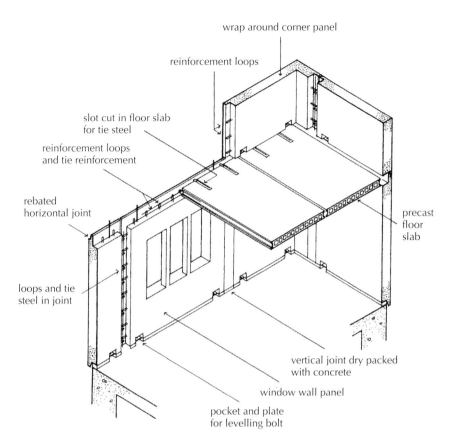

wrap around corner panel

reinforcement loops

slot cut in floor slab
for tie steel

reinforcement loops
and tie reinforcement

rebated
horizontal joint

precast
floor
slab

loops and tie
steel in joint

vertical joint dry packed
with concrete

window wall panel

pocket and plate
for levelling bolt

Figure 6.47   Precast concrete wall frame.

the soffit of roof or floors. The slabs are cast monolithically and can be designed to span continuously between and across points of support, and so employ the least thickness of slab. Where it is convenient to cantilever slabs beyond the edge columns and where cantilevers for balconies, for example, are required they can, without difficulty, be arranged as part of the slab. The advantages of this system are employed most fully in simple, isolated point block buildings of up to five storeys where the floor plans are the same throughout the height of the building and a flush slab floor may be an advantage. The system can be employed for beam and slab, and waffle grid floors, but the forms necessary between the floors to give the required soffit take most of the advantage of simplicity of casting on the ground.

Steel or concrete columns are first fixed in position and rigidly connected to the foundation, and the ground floor slab is then cast. When it has matured it is sprayed with two or three coats of a separating medium consisting of wax dissolved in a volatile spirit. As an alternative, polythene sheet or building paper may be used as a separating medium. The first floor slab is cast inside edge formwork on top of the ground floor slab, and when it is mature it is in turn coated or covered with the separating medium and the next floor slab is cast on top of it. The casting of successive slabs continues until all the floors and roof have been cast one on the other on the ground. Lifting collars are cast into each slab around each column. The slabs are lifted by jacks, operating on the top of each column, which lift a pair of steel rods attached to each lifting collar in the slab being raised. A central control synchronises the operation of the hydraulically operated reciprocating ram type jacks to ensure a uniform and regular lift.

The sequence of lifting the slabs depends on the height of the building, the weight of the slabs and extension columns, the lifting capacity of the jacks and the cross-sectional area of the columns during the initial lifting. The bases of the columns are rigidly fixed to the foundations so that when lifting commences the columns act as vertical cantilevers. The load that the columns can safely support at the beginning of the lift limits the length of the lower column height and the number of slabs that can be raised at one time. As the slabs are raised they serve as horizontal props to the vertical cantilever of the columns and so increasingly stiffen the columns, the higher the slabs are raised. The sequence of lifting illustrated in Figure 6.48 is adopted so that the roof slab, which is raised first, stiffens the columns, which are then capable of taking the load of the two slabs subsequently lifted, as illustrated. The steel lifting collars which are cast into each slab around each column provide a means of lifting the slabs and also act as shear reinforcement to the slabs around columns, and so may obviate the necessity for shear reinforcement to the slabs. Figure 6.49 is an illustration of a typical lifting collar fabricated from mild steel angle sections welded together and stiffened with plates welded in the angle of the sections.

The lifting collars are fixed to steel columns by welding shear blocks to plates welded between column flanges and to the collar after the slab has been raised into position, as illustrated in Figure 6.50. Connections to concrete columns are made by welding shear blocks to the ends of steel channels cast

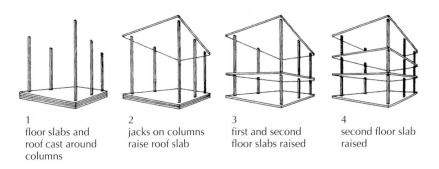

1
floor slabs and
roof cast around
columns

2
jacks on columns
raise roof slab

3
first and second
floor slabs raised

4
second floor slab
raised

Figure 6.48   Sequence of lifting slabs.

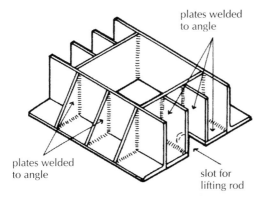

plates welded
to angle

plates welded
to angle

slot for
lifting rod

Figure 6.49   Lifting collar.

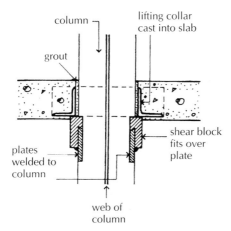

column

lifting collar
cast into slab

grout

plates
welded to
column

shear block
fits over
plate

web of
column

connection of slab to steel column

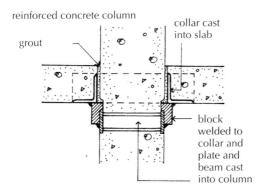

reinforced concrete column

grout

collar cast
into slab

block
welded to
collar and
plate and
beam cast
into column

connection of slab to concrete column

Figure 6.50   Connection of slab to columns.

into the column and by welding the collar to the wedges, as illustrated in Figure 6.50. With this connection it is necessary to cast concrete around the exposed steel wedges for fire protection. The connection of steel extension columns is made by welding, bolting or riveting splice plates to the flanges of columns at their junction. Concrete extension columns are connected either with studs protruding from column ends and bolted to a connection plate, or by means of a joggle connection.

7 Cladding and Curtain Wall Construction

The structural frame provides the possibility of endless variation in the form and appearance of buildings that no longer need be contained inside a load-bearing envelope. A large variety of walling materials are available to meet the changing needs of use, economy and fashion. The external walls of framed buildings differ from traditional loadbearing walls because the structural frame has an effect on, and hence influences the design of, the wall structure that it supports. To the extent that the structural frame may affect the functional requirements of an external wall, it should be considered as part of the wall structure. The use of the various materials for the external wall is, to an extent, influenced by the relative behaviour of the structural frame and the wall to accommodate differential structural, thermal and moisture movements that affect the functional requirements of a wall.

7.1 Functional requirements

Under load, both steel and concrete structural frames suffer elastic strain and consequent deflection (bending) of beams and floors, and shortening of columns. Deflection of beams and floors is generally limited to about one three hundredth of span, to avoid damage to supported facings and finishes. Shortening of columns by elastic strain under load can be in the order of 2.0 mm for each storey height of about 4 m, depending on load. Elastic shortening of steel columns may be of the order of 1 mm per storey height. The comparatively small deflection of beams and shortening of columns under load can be accommodated by the joints in materials such as brick, stone and block and the joints between panels, without adversely affecting the function of most wall structures.

Unlike steel, concrete suffers drying shrinkage and creep in addition to elastic strain. Drying shrinkage occurs as water, necessary for the placing of concrete and setting of cement, migrates to the surface of concrete members. The rate of loss of water and consequent shrinkage depend on the moisture content of the mix, the size of the concrete members and atmospheric conditions. Drying shrinkage of concrete will continue for some weeks after placing. For the small members of a structural frame, drying out of doors in summer, about half of the total shrinkage takes place in about one month and about three quarters in six months. For larger masses of concrete, about half the total shrinkage will take place one year after placing. The bond between the concrete and the reinforcement restrains drying shrinkage of concrete. Concrete in heavily reinforced members will shrink less than that in lightly reinforced sections. Drying shrinkage of the order of 2–3 mm for each 4 m of column length may well occur. The shrinkage occurs due to temperature variation and hydration. As the concrete sets it produces heat from the chemical reaction, which reduces over time. As the concrete hydrates, water is used in the chemical reaction and will also evaporate from the concrete – this will cause the concrete to shrink.

Creep of concrete is dependent on stress and is affected by humidity and by the cement content and the nature of the aggregate in concrete. The gradual creep of concrete may continue for some time; however, shortening of columns is minimal. Depending on the nature of the concrete, shrinkage due to creep could be of the order of 2.0 mm for each storey height of column over the long term. Like drying, shrinkage creep is restrained by reinforcement. Creep is much more of a problem in beams than it is in columns.

The combined effect of elastic strain, drying shrinkage and creep in concrete may well amount to a total reduction of up to 6 mm for each storey height of building. Because of these effects, it is necessary to make greater allowance for shortening in the design of wall structures supported by an insitu cast concrete frame than it is for a steel frame. Solid wall structures such as brick which are built within or supported by a concrete structural frame should be built with a 12 to 15 mm compression joint at each floor level to avoid damage to the wall by shortening of the frame and expansion of the wall materials due to thermal and moisture movements.

Experience shows that there are generally considerably greater inaccuracies in line and level with insitu cast concrete frames than there are with steel frames. There is an engineering tradition of accuracy of cutting and assembling steel that is not matched by the usual assembly of formwork for insitu cast concrete. Deflection of formwork under the load of wet concrete and some movement of formwork during the placing and compaction of concrete combine to create inaccuracies of line and level of both beams and columns in concrete frames that may be magnified up the height of multi-storey buildings. Allowances for these inaccuracies can be made where fixings for cladding are made by drilling for bolt fixings rather than relying on cast on or cast in supports and fixings. The advantage of the precast concrete frame is in the greater accuracy of casting of concrete in controlled factory conditions than on site.

Functional requirements

The functional requirements of a wall are:

❏ Strength and stability
❏ Resistance to weather

❑ Durability and freedom from maintenance
❑ Fire safety
❑ Resistance to the passage of heat
❑ Resistance to the passage of sound

Strength and stability

A wall structure should have adequate strength to support its own weight between points of support or fixing to the structural frame, and sufficient stability against lateral wind pressures. To allow for differential movements between the structural frame and the wall structure there has to be adequate support to carry the weight of the wall structure, and also restraint fixings that will maintain the wall in position and at the same time allow differential movements without damage to either the fixings or the wall material.

Brick and precast concrete cladding do not suffer the rapid changes of temperature between day and night that thin wall materials do because they act to store heat and lose and gain heat slowly. Thin sheet wall materials such as GRP, metal and glass suffer rapid changes in temperature and consequent expansion and contraction which may cause distortion and damage to fixings or the thin panel material or both.

In the design of wall structures faced with thin panel or sheet material the ideal arrangement is to provide only one rigid support fixing to each panel or sheet with one other flexible support fixing and two flexible restraint fixings. The need to provide support and restraint fixings with adequate flexibility to allow for thermal movement and at the same time adequately restrain the facing in place and maintain a weathertight joint has been the principal difficulty in the use of thin panel and sheet facings.

Resistance to weather

The traditional walling materials, brick, stone and block, serve to exclude rain from the inside of buildings by absorbing rainwater that evaporates to outside air during dry periods. The least thickness of solid wall material necessary to prevent penetration of rainwater to the inner face depends on the degree of exposure to driving rain. Common practice is to construct a wall as a cavity wall with an outer leaf of masonry as rain screen, a cavity and an inner leaf that provides adequate thermal resistance to the passage of heat, and an attractive internal finish.

Precast concrete wall panels act in much the same way as brick by absorbing rainwater. Because of the considerable size of these panels there have to be comparatively wide joints between panels to accommodate structural, thermal and moisture movements. The joints are designed with a generous overlap to horizontal and an open drained joint to vertical joints to exclude rain. Non-absorbent sheet materials, such as metal and glass, cause driven rain to flow under pressure in sheets across the face of the wall, so making the necessary joints between panels of the material highly vulnerable to penetration by rain.

These joints should at once be sufficiently wide to accommodate structural, thermal and moisture movements and serve as an effective seal against rain penetration. The materials that are used to seal joints are mostly short-lived as they harden on exposure to atmosphere and sun, and lose resilience in accommodating movement. The 'rain screen' principle is designed to provide a separate outer skin, to screen wall panels from scouring by wind and rain and deterioration by sunlight, and to improve the life and efficiency of joint seals.

Durability and freedom from maintenance

The durability of a wall structure is a measure of the frequency and extent of the work necessary to maintain minimum functional requirements and acceptable appearance.

Walls of brick and natural stone will very gradually change colour over the years. This slow change of colour, termed weathering, is generally accepted as one of the attractive features of these traditional materials. Walls of brick and stone facing require very little maintenance over the expected life of most buildings. Precast concrete wall panels which weather gradually may become dirt stained due to slow run off of water from open horizontal joints. This irregular and often unsightly staining, particularly around top edges of panels, is a consequence of the panel form of this type of cladding.

Panels of glass will maintain their lustrous fire glazed finish over the expected life of buildings but will require frequent cleaning of the surface if they are to maintain their initial appearance, and periodic attention to and renewal of seals. Of the sheet metal facings that can be used for wall structures, bronze and stainless steel, both expensive materials, will weather by the formation of a thin film of oxide that is impermeable and prevents further oxidation. Aluminium weathers with a light coloured, coarse textured, oxide film that considerably alters the appearance of the surface, although the material can be anodised to inhibit the formation of an oxide film or coated with a plastic film for the sake of appearance. Steel, which progressively corrodes to form a porous oxide, is coated with zinc, to inhibit the rust formation, and a plastic film as decoration. None of the plastic film coatings are durable as they lose colour over the course of a few years on exposure to sunlight and this irregular colour bleaching may well not be acceptable from the point of view of appearance to the extent that painting or replacement may be necessary in 10 to 25 years. In common with other thin panel materials, there will be a need for periodic maintenance and renewal of seals to joints between metal-faced panels.

Fire safety

The design of cladding and curtain wall construction must take into account fire safety. Primary concerns are the internal spread of fire across the surface materials of walls and ceilings, external fire spread over the fabric and fire spread within concealed spaces such as cavities.

Fire may spread within a building over the surface of materials covering walls and ceilings. The Building Regulations prohibit the use of materials that encourage spread of flame across their surface when subject to intense radiant heat and those which give off appreciable heat when burning. Limits are set on the use of thermoplastic materials used in rooflights and lighting diffusers.

To limit the spread of fire between buildings, limits to the size of 'unprotected areas' of walls and also finishes to roofs, close to boundaries, are imposed by the Building Regulations. The term 'unprotected area' is used to include those parts of external walls that may contribute to the spread of fire between buildings. Windows are unprotected areas as glass offers negligible resistance to the spread of fire. The Regulations also limit the use of materials of roof coverings near a boundary that will not provide adequate protection against the spread of fire over their surfaces.

Smoke and flames may spread through concealed spaces, such as voids above suspended ceilings, roof spaces, and enclosed ducts and wall cavities in the construction of a building. To restrict the unseen spread of smoke and flames through such spaces, cavity barriers and stops should be fixed as a tight-fitting barrier to the spread of smoke and flames.

Resistance to the passage of heat

The interiors of buildings built with insulated solid masonry walling and those clad with insulated panels of concrete, GRC, GRP, glass and metal, are heated by the transfer of heat from heaters and radiators to air (conduction), the circulation of heated air (convection) and the radiation of energy from heaters and radiators to surrounding colder surfaces (radiation). This internal heat is transferred to and through colder enclosing walls, roof and floors by conduction, convection and radiation to colder outside air.

The interiors of buildings clad with large areas of glass may gain a large part or the whole of their internal heat from a combination of solar heat gain through glass cladding and from internal artificial lighting to the extent that there may be little need for supplementary internal heating for parts of the year. Solar heat gain (and associated solar glare) can be controlled through the use of simple shading devices fixed externally and/or internally to the building fabric. As long as the interior of buildings is heated to a temperature above that of outside air, transfer of heat from heat sources to outside air will continue. For the sake of economy and to conserve limited supplies of fuel, it is sensible to seek to limit the rate of transfer of heat from inside to outside.

Ventilation

The use of sealed glazing and effective weatherseals to the joints of cladding panels and windows in the envelope of modern buildings has restricted, and to some extent controlled, the natural exchange of outside and inside air to provide ventilation of buildings. For comfort there should be a continuous change of air inside buildings to provide an adequate supply of oxygen, to limit the build-up of humidity, fumes, body odour and smells and provide a regular movement of air that is necessary for bodily comfort. The necessary movement of air inside sealed buildings may be induced artificially by mechanical systems of air conditioning which filter, dry and humidify air through a complex of inlet and extract ducting, connected to one or more air treatment plants. The pumps necessary to force air through the ducts may cause an unacceptable level of noise, and the air handling system is costly to install, maintain and run. To economise, it is often practice to install individual air conditioning heaters, which filter, dry and heat air that is recirculated from individual rooms with the effect that stale air is constantly circulated, so causing conditions of discomfort.

As an alternative, buildings may be constructed and finished with mainly open plan floor areas largely free of enclosed spaces, set around one or more central areas open from ground to roof level to provide facility for the natural movement of heated air to rise and so cause natural ventilation. This stack system of ventilation, so called by reference to the upward movement of air up a chimney stack, can be utilised by itself or with some small mechanical ventilation to provide comfort conditions with the least initial and running costs.

Thermal bridge

The members of a structural frame can act as a thermal bridge where the wall is built up to or between the frame, as illustrated later in Figure. 7.2 where the resistance to the thermal transfer through the brick slips and beam is appreciably less than through the rest of the wall. Similarly, there is a thermal bridge across a precast wall panel and a beam and column, as illustrated later in Figure 7.27. It may be difficult to provide an effective way of preventing the thermal bridge formed by the supporting structural frame. The effect of the bridge may be modified by the use of floor insulation and a suspended ceiling or by setting frame members, where possible, back from the outer face of the wall, as illustrated later in Figure 7.3. Wall panels of precast concrete, GRP and GRC have been used with a sandwich or inner lining of an insulating material. This arrangement is not entirely effective because the insulating material, if open pored as are many insulating materials, may absorb condensate water, which will reduce its thermal properties, and the edge finish to panels, necessary for rigidity and jointing, will act as a thermal bridge. Thin metal wall panel materials which are supported by a metal carrier system fixed across the face of the structural frame can provide thermal insulation more effectively by a sandwich, inner lining or inner skin of insulating material with the edge jointing material acting as a thermal break in the narrow thermal bridge of the edge metal, as illustrated later in Figure 7.45.

Resistance to the passage of sound

Manufacturers of cladding and curtain wall systems provide notional sound resistance figures for their products. The figures provide a useful guide to the expected noise reduction of a particular construction; however, the actual detailing at the cladding and curtain wall, especially at the junction with the structural frame, will affect the actual values. The most effective way of reducing impact sound is to isolate the potential source of impact from continuous solid transmitters such as structural frames. Resilient fixings to door frames and resilient bushes to supports for hard floor finishes effectively isolate the source of common impact sounds. The most effective barrier to airborne sound is an intervening mass such as a solid wall. The more dense and thick the material, the more effective it is as a barrier to airborne sound as the dense mass absorbs the energy generated by the sound source.

7.2 Terms and definitions

The term 'cladding' came into general use as a description of the external envelope of framed buildings, which clothed or clad the building in a protective coating that was hung, supported by or secured to the skeleton or structural frame. The word 'facings' has been used to describe materials used as a thin, non-structural, decorative, external finish such as the thin, natural stone facings applied to brick or concrete backing.

The words 'wall' or 'walling' will be used to describe the use of those materials such as stone, brick, concrete and blocks that are used as the external envelope of framed buildings where the appearance is of a continuous wall to the whole or part of several storeys or as walling between exposed, supporting beams and columns of the frame. The word 'cladding' will be used to describe panels of concrete, GRC, GRP, glass and metal fixed to and generally hung from the frame by supporting beams or inside light framing as a continuous outer skin to the frame. The external walls of framed buildings are broadly grouped as:

❏ Infill wall framing to a structural grid
❏ Solid and cavity walling of stone, block and brick
❏ Facings applied to solid and cavity background walls
❏ Cladding panels of precast concrete, GRC and GRP
❏ Thin sheet cladding of metal
❏ Glazed wall systems

7.3 Infill wall framing to a structural grid

Infill wall frames are fixed within the enclosing members of the structural frame or between projections of the frame, such as floors and roof slabs, which are exposed, as illustrated in Figure 7.1. The infill wall may be framed with timber

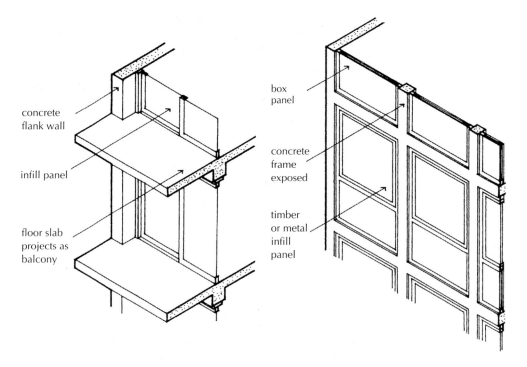

concrete
flank wall

infill panel

floor slab
projects as
balcony

box
panel

concrete
frame
exposed

timber
or metal
infill
panel

Figure 7.1   Infill panels.

or metal sections, with panels of an appropriate material secured within the frame.

The framing with its panels or sheet covering should have adequate strength and stability in itself to be self-supporting within the framing members and resist wind pressure and suction acting on it. Sufficient support and restraint fixings between the frame and the surrounding structural members are required. The framing, its panels and sheet covering must adequately resist the penetration of water to the inside face by a system of resilient mastic, drained and sealed joints. The joints between the framing and the structure should be filled with a resilient filler and weathersealed with mastic to accommodate structural, moisture and thermal movements. To enhance the thermal resistance of the lightweight framing and covering materials, double glazing and/or solar control glass should be used with double skin insulated panels, insulation between framing members or behind sheet covering materials.

In the 1950s and 1960s the infill wall frame system was much used in framed buildings, particularly for multi-storey housing. Many of the early infill wall frame systems suffered deterioration due to the use of steel framing poorly protected against corrosion, panel materials that absorbed water, and poor jointing materials that gave inadequate protection against rain penetration. These failures, coupled with the introduction of alternative walling materials such as concrete, GRC and GRP panels and glazed walls, led to loss of favour of wall infill framing. There were also problems with thermal bridging through the concrete frame, as would be the case with the building illustrated in Figure 7.1. Thermal bridging is difficult to design out of such structures, which are better suited to climates warmer than the UK. In countries where summer temperatures are high and shade from the sun is a necessity, many buildings are constructed with a reinforced concrete frame with projecting floors and roof for shade and as an outdoor balcony area in summer, as illustrated in Figure 7.1. Because of the protection afforded by the projecting floor slabs and roof against wind-driven rain and the diminution of daylight penetration caused by these projections, in winter months, it is common practice to form fully glazed infill panels in this form of construction.

7.4 Solid and cavity walling

Solid masonry walling

In the early days of the multi-storey structural frame, solid masonry was used for the external walling, built as a loadbearing structure off the supporting framework. Ashlar natural stone, brickwork and terracotta blocks were used. Masonry walling imposed considerable loads on the supporting frame and foundations, and was extravagant in the use of expensive materials. To improve thermal resistance and to provide a cavity as a barrier to the penetration of water to the inner face, it became practice to construct masonry walling as a cavity wall. With the introduction of the cavity wall, changing fashion and increasing expectation of thermal comfort in buildings, solid walling to framed buildings was, by and large, abandoned.

Cavity brick walling

With the use of cavity walling to framed buildings, it was considered necessary to provide support for at least two-thirds of the thickness of the outer leaf of the wall and the whole of the inner leaf at each floor level. This posed difficulties where the external face was to have the appearance of a traditional loadbearing wall. The solution was to fix special brick slips to mask the horizontal frame members at each floor level, as illustrated in Figure 7.2. A disadvantage of these brick slips is that even though they are cut or made from the same clay as the surrounding whole bricks, they may tend to weather to a somewhat different colour from that of the whole bricks and so form a distinct horizontal band that defeats the original objective. An alternative to the use of brick slips at each floor level is to build the external leaf of the cavity walling directly off a projection of the floor slab with the floor slab exposed as a horizontal band at each floor level, or to build the walling between floor beams and columns and so admit the frame as part of the facade. This technique has been largely

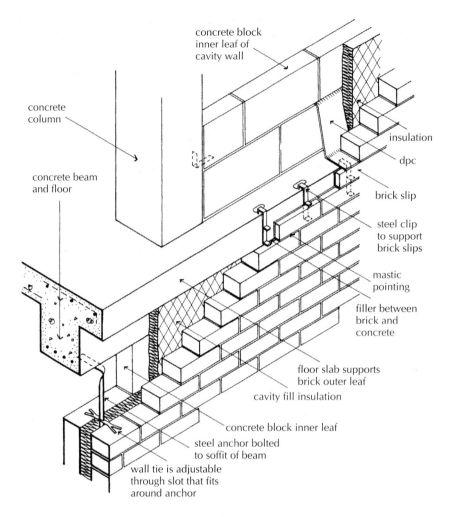

concrete block
inner leaf of
cavity wall

concrete
column

concrete beam
and floor

insulation

dpc

brick slip

steel clip
to support
brick slips

mastic
pointing

filler between
brick and
concrete

floor slab supports
brick outer leaf

cavity fill insulation

concrete block inner leaf

steel anchor bolted
to soffit of beam

wall tie is adjustable
through slot that fits
around anchor

Figure 7.2   Brick cladding to concrete frame. (Note: This form of construction is no longer used due to thermal bridge across the concrete floor and brick step.)

abandoned because of the problem of thermal bridging through the exposed floor slab.

The strength and stability of solid and cavity walling constructed as cladding to framed structures depends on the support afforded by the frame and the resistance of the wall itself to lateral wind pressure and suction. As a general principle, the slenderness ratio of walling is limited to 27:1, where the slenderness ratio is the ratio of the effective height or length to effective thickness. The effective thickness of a cavity wall may be taken as the combined thickness of the two leaves. To provide the appearance of a loadbearing wall to framed structures, without the use of brick slips, it is usual practice to provide support for the outer leaf by stainless steel brackets or angles built into horizontal brick joints, as illustrated in Figure 7.3.

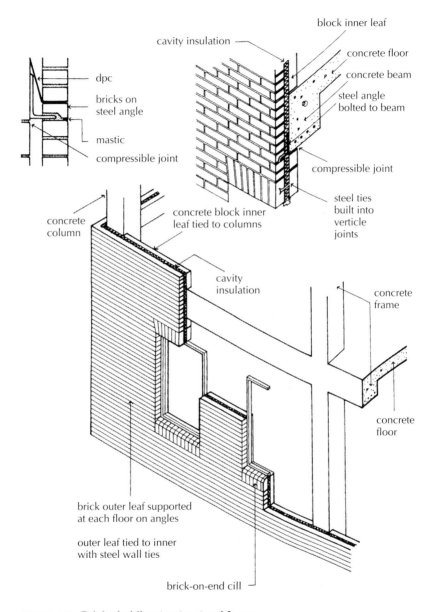

Figure 7.3   Brick cladding to structural frame.

A common support for the brick outer leaf of a cavity wall is a stainless steel angle secured with expanding bolts to a concrete beam, as illustrated in Figure 7.3. Depending on the relative thickness of the supporting flange of the angle and the thickness of the mortar joints, the angle may be bedded in the mortar joint or the bricks bearing on the angle may be cut to fit over the angle. To allow for relative movement between walling and the frame it is usual practice to form a horizontal movement joint at the level of the support angle by building in a compressible strip, which is pointed on the face with mastic to exclude water.

As an alternative to a continuous angle support, a system of support brackets may be used. These stainless steel brackets fit to a channel cast into the concrete. An adjusting bolt in each bracket allows some vertical adjustment and the slotted channel some horizontal adjustment so that the supporting brackets may be accurately set in position to support brickwork as it is raised. The brackets are bolted to the channel to support the ends of abutting bricks, as illustrated in Figure 7.4. A horizontal movement joint is formed at the level of the bracket support. Supporting angles or brackets may be used at intervals of not more than every third storey height of building or not more than 9 m, whichever is the less, except for four-storey buildings where the wall may be unsupported for its full height or 12 m, whichever is the less. Where support is provided at every third storey height, the necessary depth of the compressible movement joint may well be deeper than normal brick joints and be apparent on the face of the wall.

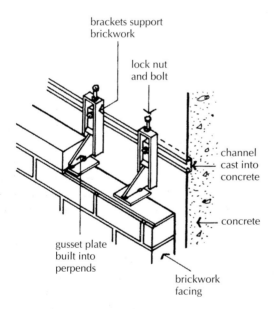

Figure 7.4 Loadbearing fixing for brickwork.

To provide support for the wall against lateral forces it is necessary to provide some vertical anchorage at intervals so that the slenderness ratio does not exceed 27:1. Fishtailed or flat anchors fitted to channels cast into columns are bedded in the face brickwork at the same intervals as wall ties, as illustrated in Figure 7.5, to provide lateral and vertical restraint. To provide horizontal, lateral restraint, anchors are fitted to slots in cast-in channels in beams or floor slabs at intervals of up to 450 mm. To provide anchorage to the top of the wall at each floor level where brick slips are used, it is usual to provide anchors that are bolted to the underside of the beam or slab and to fit stainless steel ties that are built into brickwork at 900 mm centres, as illustrated in Figure 7.2.

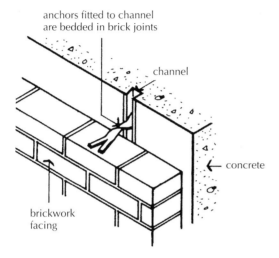

Figure 7.5   Restraint fixing for brickwork.

Where solid or cavity walling is supported on and built between the structural frame grid, some allowance should be made for movements of the frame, relative to that of the walling due to elastic shortening and creep of concrete, flexural movement of the frame, and thermal and moisture movements. Practice is to build in some form of compressible filler at the junction of the top of the walling and the frame members and the wall and columns as movement joints, with metal anchors set into cast-in channels in columns and bedded in brickwork and to both leaves of cavity walls at intervals similar to cavity wall ties.

Where cavity walling is built up to the face of columns of the structural frame and supported at every third storey, the support and restraint against lateral forces are provided by anchors. These anchors are fitted to cast-in channels and bedded in horizontal brick joints at intervals similar to cavity ties. To provide for movement along the length of walling, it is usual to form continuous vertical movement joints to coincide with vertical movement joints in the structural frame and at intervals of not more than 15 m along the length of continuous walling and at 7.5 m from bonded corners, with the joints filled with compressible strip and pointed with mastic. A wall of sound, well-burnt clay bricks should require no maintenance during the useful life of a building other than renewal of mastic pointing of movement joints at intervals of about 20 to 25 years.

Resistance to the penetration of wind-driven rain depends on the degree of exposure and the necessary thickness of the outer leaf of cavity walling and the cavity width. The use of cavity trays and a dpc at all horizontal stops to cavities is accepted practice. The purpose of these trays, illustrated in Figure 7.2, is to direct water that may collect inside the cavity away from the inner face of the wall. If the thickness of the outer leaf and the cavity is sufficient to resist penetration of water, there seems little logic in the use of these trays. To prevent water-soluble salts from the concrete of concrete frames finding their way to the face of brickwork and so causing unsightly efflorescence of salts, the face of concrete columns and beams that will be in contact with brickwork is painted with bitumen.

The requirements for resistance to the passage of heat usually necessitate the use of some material with comparatively good resistance to the transfer of heat, either in the cavity as cavity fill or partial fill with a lightweight block inner leaf, as illustrated in Figure 7.2. Where the cavity runs continuously across the face of the structural frame, as illustrated in Figure 7.3, the resistance to the transfer of heat of the wall is uninterrupted. Where a floor slab supports the outer leaf, as illustrated in Figure 7.2, there will be to an extent a cold bridge as the brick slips and the dense concrete of the floor slab will afford less resistance to the transfer of heat than the main cavity wall. The very small area of floor and ceiling may well be colder. Internal insulation around the floors, ceilings and columns may be used to reduce the impact of any cold bridges. Where internal insulation is used, vapour barriers should be used to prevent the warm, moisture-laden air reaching cold surfaces.

7.5 Facings applied to solid and cavity wall backing

The word 'facings' is used to describe comparatively thin, non-structural slabs of natural or reconstructed stone, faience, ceramic and glass tiles or mosaic which are fixed to the face of, and supported by, solid background walls or to structural frames as a decorative finish. Common to the use of these non-structural facings is the need for the background wall or frame to support the whole of the weight of the facing at each storey height of the building or at vertical intervals of about 3 m, by means of angles or corbel plates. In addition to the support fixings, restraint fixings are necessary to locate the facing units in true alignment and to resist wind pressure and suction forces acting on the wall. To allow for elastic and flexural movements of the structural frame and differential thermal and moisture movements, there must be flexible horizontal joints below support fixings and vertical movement joints at intervals along the length of the facings. Both horizontal and vertical movement joints must be sufficiently flexible to accommodate anticipated movements and be water resistant to prevent penetration of rainwater.

Natural and reconstructed stone facings

Natural and reconstructed stone facings are applied to the face of buildings to provide a decorative finish to simulate the effect of solidity and permanence traditionally associated with solid masonry. Because of the very considerable cost of preparation and fixing, this type of facing is mostly used for prestige buildings such as banks and offices in city centres. Granite is the natural stone much favoured for use as facing slabs for the hard, durable finish provided by polished granite and the wide range of colours and textures available from both native and imported stone. Polished granite slabs are used for the fine gloss surface that is maintained throughout the useful life of a building. To provide a more rugged appearance the surface of granite may be honed to provide a semi-polish, flame textured to provide random pitting of the surface or surface tooled to provide a more regular rough finish. Granite facing slabs are generally 40 mm thick for work more than 3.7 m above ground and 30 mm thick for work less than 3.7 m above ground.

Limestone, such as the native Portland, Bath or Clipsham, is used as facing, by and large, to resemble solid Ashlar masonry work, the slabs having a smooth finish to reveal the grain and texture of the material. These comparatively soft limestones suffer a gradual change of colour over the course of years, and this weathering is said to be an attractive feature of these stones. Limestone facing slabs are 75 mm thick for work more than 3.7 m above ground and 50 mm thick for work less than 3.7 mm above ground. Hard limestones, including Roman stone and a number of very dense stones from France and Germany, are much used as facings for the hardness and durability of the materials. This type of stone is generally used as flat, level finished, facing slabs in thicknesses of 40 mm for work more than 3.7 m above ground and 30 mm for work less than 3.7 m above ground.

Sandstones are used as facing slabs. Some care and experience are necessary in the selection of these native sandstones as the quality, and therefore the durability, of the stone may vary between stones taken from the same quarry. This type of stone is chosen for the colour and grain of the natural material whose colour will gradually change over some years of exposure. Because of the coarse grain of the material, it may stain due to irregular run-off of water down the face. Sandstone facing slabs are usually 75 mm thick for work 3.7 m above ground and 50 mm thick for work less than 3.7 m above ground.

Marble is less used for external facings in northern European climates, as polished marble finishes soon lose their shine. Coarser surfaces, such as honed or eggshell finishes, will generally maintain their finish, provided white or travertine marble is used. Marble facing slabs are 40 mm thick for work 3.7 m above ground and 30 mm thick for work below that level. Reconstructed stone made with an aggregate of crushed natural stone is used as facing slabs as if it were the natural material, in thicknesses the same as those for the natural stone.

Fixing natural and reconstructed stone facings

The size of stone facing slabs is generally limited to about 1.5 m in any one or both face dimensions or to such a size as is practical to win from the quarry. Stone facing slabs are fixed so that there is a cavity between the back of the slabs and the background wall or frame to allow room for fixings, tolerances in the sawn thickness of slabs and variations in background surfaces, and also to accommodate some little flexibility to allow for differential structural, thermal and moisture movements between the structure and the facing. The cavity or air space between the back of the facing slabs and the background walling or structure is usually from 10 to 20 mm and free from anything other than fixings so that the facing may suffer small movements without restraint by the background. Small differential movements are accommodated through the many joints between slabs and, more specifically, through vertical and horizontal control (movement) joints. The type of fixings used to support and secure facing slabs in position are:

❑ Loadbearing fixings
❑ Restraint fixings
❑ Combined loadbearing and restraint fixings
❑ Face fixings
❑ Soffit fixings

These fixings are made from one of the corrosion resistant metals such as stainless steel, aluminium bronze or phosphor bronze. Stainless steel is the general description for a group of steel alloys containing chromium and other elements. The type of stainless steel commonly used for structural fixings is austenitic stainless steel.

Loadbearing fixings

Corrosion resistant metal angles or corbel plates are used to carry the weight of the stone facing. These fixings are bolted to, built into or cramped to slots in the background wall or structure. The loadbearing fixings provide support at each floor level at not more than 3 m. The fixings bridge the cavity to provide support at the bottom or close to the bottom of slabs, with two fixings being used to each slab. Loadbearing fixings take the form of stainless steel angles or corbel plates that fit into slots cut in the bottom edge or into slots cut in the lower part of the back of slabs at each floor level or at vertical intervals of about 3 m.

Common practice is to support each facing slab on two supports, with the angle or corbel supports fixed centrally on vertical joints between slabs so that each supports two slabs. Angle and corbel plates should be at least 75 mm wide. At vertical movement joints two supports are used, one each side of the joint, to the lower edge or lower part of the two stone slabs each side of the joint. These separate supports should be at least 50 mm wide. Angle loadbearing fixings are bolted to the insitu concrete or brick background, with expanding bolts. Holes are drilled in the background into which the bolts make a tight fit so that, as the bolt is tightened, its end expands to make a secure fixing. Angles may be fixed to provide support to the bottom edge of slabs, with the supporting flange of the angle fitting into slots in the bottom edge of adjacent slabs so that a narrow horizontal joint between slabs may be maintained. Angle support to the thicker sedimentary stones is often made to grooves cut in the backs of adjacent stones, some little distance above the lower edge, into which the flange of the angle fits. This fixing is chosen where the edges of these laminated stones might spall where the lower edges were cut. Figure 7.6 is an illustration of loadbearing angle support fixings.

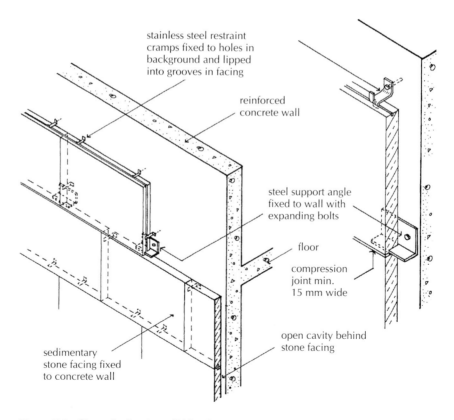

stainless steel restraint cramps fixed to holes in background and lipped into grooves in facing

reinforced concrete wall

steel support angle fixed to wall with expanding bolts

floor

compression joint min. 15 mm wide

open cavity behind stone facing

sedimentary stone facing fixed to concrete wall

Figure 7.6   Stone facing to solid background.

Corbel plate loadbearing fixings, which are the traditional means of providing support for stone facings, may be used as an alternative to angle supports, particularly for the thinner stones such as granite. Flat or fishtail corbel plates are from 6 to 16 mm thick, depending on the size of slab to be supported, 75 or 50 mm wide and from 125 mm long. The purpose of the fishtail end is to provide a more secure bond to the cement grout in which the corbel plate is set. A pocket is made in the concrete or brick background by drilling holes and chiselling to form a neat pocket into which the corbel plate is set in dry, rapid-hardening cement and sand, which is hammered in around the corbel. The one part cement to one part sand grout is left for at least 48 hours to harden. Corbel plate supports are usually fixed to provide support by fitting into slots cut in the back of adjacent stones some little distance above the lower edge of slabs, as illustrated in Figure 7.7.

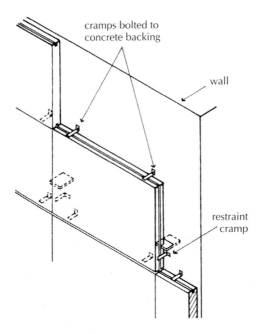

Figure 7.7   Corbel plate supports.

Two corbel plates are used to give support to the stone facing at each floor level or not more than 3 m. The thickness of the plates and the depth of their bed into the background have to be sufficient to give support for all the stone slabs between floors. A common alternative to flat corbel plates is to form the protruding ends of corbels, which fit into slots in the back of stones, to slope up at an angle of 158° to the horizontal. The upward slope provides a more positive seating for stones and to an extent serves to restrain and align the stones. The setting into the background of the corbel plates and the cutting of the slots in the back of the stones require a degree of skill to achieve both an intimate fit of stones to corbels and true alignment of stone faces. For setting in brick backgrounds, fishtail ended corbel plates are made for building in or grouting into pockets. Corbel plates may be shaped for bolting to backgrounds, as illustrated in Figure 7.8.

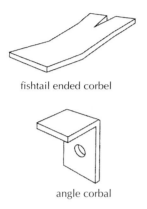

fishtail ended corbel

angle corbal

Figure 7.8   Corbel plates.

Restraint fixings

Restraint fixings take the form of strip metal or wire cramps shaped to hook into grooves or holes in the edges of slabs and formed for bolting to or being set into pockets or slot anchors in the background. The most straightforward form of restraint cramp consists of a narrow strip with one end bent up and holed for bolting to the background, with the other end double lipped to fit into grooves or slots in the top and bottom edges of slabs, as illustrated in Figure 7.6. A similar cramp has a fishtail end for grouting in pockets cut in the background. Strip metal restraint cramps may be shaped to fit into slot metal anchors that have been cast into concrete walls and frames. The anchors are cast in horizontally so that the cramp may be adjusted in the slot anchor to coincide with vertical joints between stones. The cramp may provide restraint through double lipped ends that fit to grooves in adjacent slabs or by a dowel that fits to holes in slab edges, as illustrated in Figure 7.9. Fishtail ended strip metal cramps are made for building or grouting into the horizontal joints of brick backgrounds. Dowels fit to holes in the cramps for setting into holes or grooves cut in the horizontal joints between stones.

Stainless steel wire restraint cramps are used for the thinner granite slab facings. These dense stones lend themselves to being accurately drilled to take the hooked ends of these cramps that fit into either the side or top and bottom edges of adjacent slabs. These stainless steel wire cramps are usually screwed or bolted to solid backgrounds and shaped to fit to holes or grooves cut in the horizontal joints of adjacent stones. As an alternative, a loose, wire cramp may

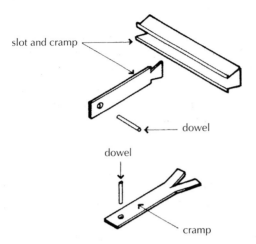

slot and cramp

dowel

dowel

cramp

Figure 7.9   Slot anchor and cramp.

be used to hook an upper stone to those below, as illustrated in Figure 7.10. For brick backgrounds a looped wire cramp is grouted into brick joints, with the double toed ends of the wire set into holes or grooves in adjacent stones. An advantage of wire cramps is that they can be bent on site to make an intimate fit to the holes or grooves cut in adjacent stones.

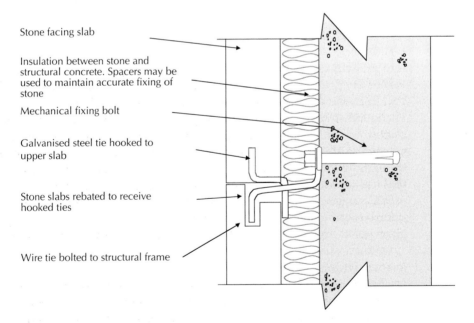

Stone facing slab

Insulation between stone and structural concrete. Spacers may be used to maintain accurate fixing of stone

Mechanical fixing bolt

Galvanised steel tie hooked to upper slab

Stone slabs rebated to receive hooked ties

Wire tie bolted to structural frame

Figure 7.10   Wire tie restraint fixing.

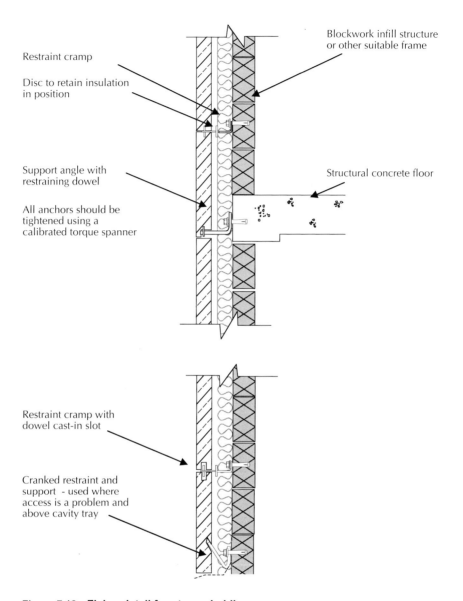

Restraint cramp

Disc to retain insulation
in position

Blockwork infill structure
or other suitable frame

Support angle with
restraining dowel

Structural concrete floor

All anchors should be
tightened using a
calibrated torque spanner

Restraint cramp with
dowel cast-in slot

Cranked restraint and
support - used where
access is a problem and
above cavity tray

Figure 7.12   Fixing detail for stone cladding.

slabs. Hangers fit into the lipped channel to allow for adjustment of the hanger to suit joints in stone soffit. With the system illustrated in Figure 7.14 the plates supported by the hangers may be cut to fit into grooves or slots cut in the edges between stones, or be made to fit into slots cut in the edges of four stones, depending on the thickness and weight of the stone slabs used and convenience in fixing. The number of hangers used for each soffit facing slab depends on the size and weight of each slab and the thickness of the lipped

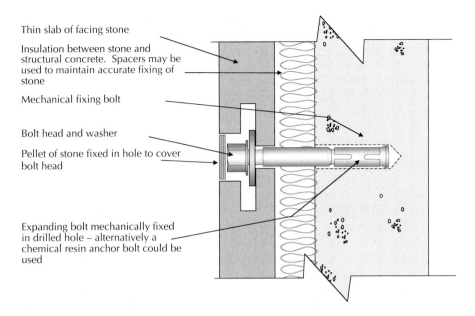

Thin slab of facing stone

Insulation between stone and structural concrete. Spacers may be used to maintain accurate fixing of stone

Mechanical fixing bolt

Bolt head and washer

Pellet of stone fixed in hole to cover bolt head

Expanding bolt mechanically fixed in drilled hole – alternatively a chemical resin anchor bolt could be used

Figure 7.13   Face fixing for stone cladding.

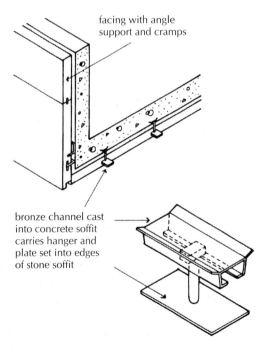

facing with angle support and cramps

bronze channel cast into concrete soffit carries hanger and plate set into edges of stone soffit

Figure 7.14   Soffit fixing for stone facing (insulation omitted for clarity).

edge of the slab bearing on, and therefore carrying, the slab. At the junction of soffit slab, and wall slabs, a hanger bolted to the wall background may have a lip or stud to fit to slots or holes in the edge of the soffit slabs.

Joints between stone slabs

Joints between stone facing slabs should be sealed as a barrier to the penetration of rainwater running off the face of the slabs. Where rainwater penetrates the joints between stone slabs it will be trapped in the cavity between the slabs and the background wall, will not evaporate to air during dry spells and may cause conditions of persistent damp. Open or butt joints between slabs should be avoided in external face work.

The joints between sedimentary stone slabs, such as limestone and sandstone, may be filled with a mortar of cement, lime and sand (or crushed natural stone) mix 1:1:6 and finished with either flush or slightly recessed pointing to a minimum depth of 5 mm. Joints between granite and hard limestone slabs are filled with a mortar of 1:2:8 cement, lime and sand (or stone dust) or 4:1 cement and sand to a minimum thickness of 3 mm. As an alternative to mortar filling the joints between stones, a sealant may be used. Sealants such as one part polysulphide, one part polyurethane/two parts polysulphide and two parts polyurethane are recommended for the majority of stones. These sealant joints should be not less than 5 mm wide. The jointing sealants will accommodate a degree of movement between stones without failing as a water seal for up to 15 to 20 years, when they may well need to be reformed. Mortar joints will take up some slight movement between stones but may in time not serve as an effective water seal as wind-driven rain may penetrate the fine cracks that open up. Some penetration of rainwater through joints between stones may well occur as sealants age and mortar cracks. There will generally be no large penetration of water into the cavity behind the stones so that unless obvious damp stains appear on the inside face of the background walling it is unlikely that anyone will be concerned to go to the considerable expense of hacking out and reforming the joints.

Movement joints

As a building is erected and loads increase there is an early measurable elastic shortening of columns of the order of about 2.5 mm per 4 m storey height of a reinforced concrete frame and a later gradual shortening due to creep of the same magnitude. A less pronounced shortening takes place as a steel frame is erected. Much of the early elastic shortening of the columns of a structure will have taken place before a wall cladding is fixed. The long-term shortening of reinforced concrete columns, through creep, has to be allowed for in horizontal movement joints. Differential temperature and moisture movements of a wall facing relative to the supporting structure will generally dictate the need to allow some movement of joints and fixings. There will be, for example, very considerable temperature differences between facing slabs on an exposed south facing wall and the structure behind so that differential thermal movement has to be allowed for both in joints and support and restraint fixings.

A general recommendation in the fixing of stone facing slabs is that there should be horizontal movement joints at each storey height below loadbearing support fixings or not more than 3 m. These joints are usually 10 to 15 mm deep and filled with one of the elastic sealants. Where so wide a joint would not be acceptable in facework finished with narrow joints, it is usual to accommodate movement in narrower sealant-filled horizontal joints to all the facework. Vertical movement joints are formed in facework where these joints occur in the structure, to allow for longitudinal structural, thermal and moisture movements. A continuous vertical joint is formed between stone facings and filled with sealant.

Faience slabwork

Faience is the term used to describe fire-glazed stoneware in the form of slabs that are used as a facing to a solid background wall. The best quality slabs are made from stoneware that shrinks and deforms less on firing than does earthenware. The fired slab is glazed and then refired to produce a fire glazed finish. The slabs are usually 300 × 200, 450 × 300 or 610 × 450 mm and 25 to 32 mm thick. They form a durable, decorative facing to solid walls. The glazed finish, which will retain its lustre and colour indefinitely, needs periodic cleaning, especially in polluted atmospheres. Faience slabwork was much used as a facing in the 1930s in the UK, as a facing to large buildings such as cinemas. This excellent facing material has since then lost favour. When first used, the slabs were fixed with cement mortar dabs to a keyed brick background. This unsatisfactory method of fixing made no allowance for differential movements and has been abandoned in favour of support fixings to each slab and restraint fixings and movement joints, in the same way that stone facings are fixed.

Terracotta

Terracotta (burnt earth) was much used in Victorian buildings as a facing because it is less affected by polluted atmospheres than natural limestone and sandstone facings. Fired blocks of terracotta, with a semi-glaze self-finish, were moulded in the form of natural stone blocks to replicate the form and detail of the stonework buildings of the time. The plain and ornamental blocks were made hollow to reduce and control shrinkage of the clay during firing. In use the hollows in the blocks were filled with concrete and the blocks were then laid as if they were natural stone. Well burned blocks of terracotta are durable even in heavily polluted atmospheres. This labour intensive system of facing is little used today.

Tiles and mosaic

'Tile' is the term used to describe comparatively thin, small slabs of burnt clay or cast concrete up to about 300 mm square and 12 mm thick. These small units of fired clay and cast concrete are used as a facing to structural frames and solid background walls. For many years practice has been to bond tiles directly to frames and walls with cement mortar dabs, which provides sufficient adhesion to maintain individual tiles in place. Unfortunately, this system of adhesion does not make any allowance for differential movements between the frame, background walls and tiles, other than in the joints between tiles, which can be considerable, particularly with insitu cast concrete work. To make allowance for movements in the structure and the facing, tiles should be supported and restrained by cramps that provide a degree of flexibility between the facing and the background. Plainly it would be both tedious and expensive to fix individual tiles in this way. For economy and ease of fixing, the tiles can be cast on to a slab of plain or reinforced concrete, which is then fixed in the same way as stone facing slabs. Mosaic is the term used to describe small squares of natural stone, tile or glass set out in some decorative pattern. The units of mosaic are usually no larger than 25 mm^2. A mosaic finish as an external facing should be used as a facing to a cast concrete slab in the same way as tiles.

7.6 Cladding panels

The word cladding is used in the sense of clothing or covering the building with a material to provide a protective or decorative cover. The cladding panels, usually storey (floor) height, serve the function of providing protection against wind and rain, and resistance to the transfer of heat from inside to outside, without providing structural support.

Precast concrete cladding panels

The aesthetic, economic and constructional advantages of the use of precast concrete cladding slabs or units as a facing and walling to framed structures were demonstrated by Le Corbusier in the multi-storey housing development, the Unit, at Marseilles, which was completed in 1952. The use of precast concrete and the design of this one building had a profound influence on the use of precast concrete for many years. The advantages of repetitive casting, speed of erection largely independent of weather, and the rugged appearance of the material that was a vogue of the 1950s and 1960s led to the extensive use of this system of wall cladding. Precast concrete or units are usually storey height, as illustrated in Figure 7.15 or column spacing wide as spandrel or undercill units for support and fixing to the structural frame. Precast concrete cladding units are hung on and attached to frames as a self-supporting facing and wall element which may combine all of the functional requirements of a wall element.

Precast concrete cladding units are cast with either the external face up or down in the moulds, depending on convenience in moulding and the type of finish. Where a finish of specially selected aggregate is to be exposed on the

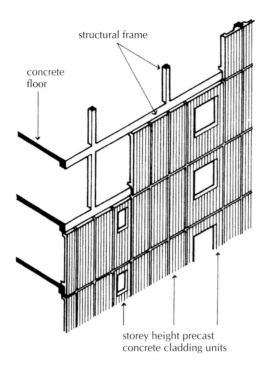

concrete
floor

structural frame

storey height precast
concrete cladding units

Figure 7.15 Storey height precast concrete cladding.

face, the face up method of casting is generally used for the convenience and accuracy in applying the special finish to the core concrete of the panel. Cladding units that are flat or profiled are generally cast face down for convenience in compacting concrete into the face of the mould bed. Strongly constructed moulds of timber, steel or glass fibre reinforced plastic are laid horizontal, the reinforcing cage and mesh are positioned in the mould, and concrete is placed and compacted. For economy in the use of the comparatively expensive moulds it is essential that there be a limited number of sizes, shapes and finishes to cladding units to obtain the economic advantage of repetitive casting. For strength and rigidity in handling, transport, lifting and support, and fixing, and to resist lateral wind pressures, cladding units are reinforced with a mesh of reinforcement to the solid web of units, and a cage of reinforcement to vertical stiffening ribs and horizontal support ribs. Figure 7.16 is an illustration of a storey height cladding unit.

The initial wet plastic nature of concrete facilitates the casting of a wide variety of shapes and profiles, from flat solid webs enclosing panels to the comparatively slender solid sections of precast concrete frames for windows. The limitation of width of cladding units is determined by facilities for casting, and size for transport and lifting. The width of the units cast face up is limited

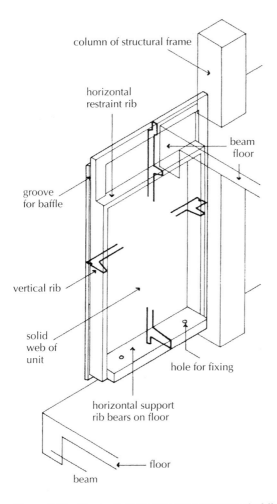

Figure 7.16 Storey height precast concrete cladding unit.

by ease of access to placing the face material in the moulds. The usual width of storey height panels is from 1200 to 1500 mm, or the width of one or two structural bays.

There is no theoretical limit to the size of precast units, provided they are sufficiently robust to be handled, lifted and fixed in place, other than limitations of the length of a unit that can be transported and lifted. In practice, cladding units are usually storey height for convenience in transport and lifting, and fixing in place. Cladding units two or more storeys in height have to be designed, hung and fixed to accommodate differential movements between the frame and the units, which are multiplied by the number of storeys they cover.

Storey height precast concrete cladding is supported by the structural frame, either by a horizontal support rib at the bottom of the units or by hanging on a horizontal support rib at the top of the units, as illustrated in Figure 7.17. Bottom support is preferred as the concrete of the unit is in compression and less likely to develop visible cracks and crazing than it is when top hung. Whichever system of support is used, the horizontal support rib must have an adequate projection for bearing on structural floor slabs or beams and for the fixings used to secure the units to the frame. At least two mechanical support and two restraint fixings are used for each unit. The usual method of fixing at supports is by the use of steel or non-ferrous dowels that are grouted into 50 mm square pockets in the floor slab. The dowel is then grouted into a 50 mm diameter hole in the support rib, as illustrated in Figure 7.18. The advantage of this dowel fixing is that it can readily be adjusted to inaccuracies in the structure and the panel. Dowel fixings serve to locate the units in position and act as restraint fixings against lateral wind pressures.

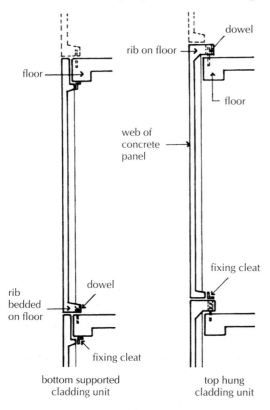

Figure 7.17   Storey height cladding panel.

The vertical stiffening ribs are designed for strength in resisting lateral wind pressures on the units between horizontal supports and strength in supporting

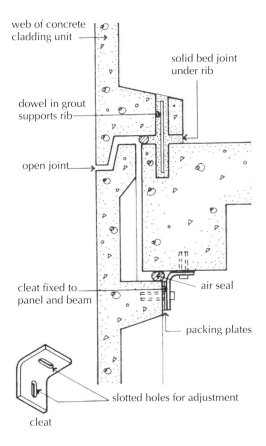

web of concrete
cladding unit

solid bed joint
under rib

dowel in grout
supports rib

open joint

cleat fixed to
panel and beam

air seal

packing plates

cleat

slotted holes for adjustment

Figure 7.18   Fixing for bottom supported cladding units.

the weight of the units that are either hung from or supported on the horizontal support ribs. The least thickness of concrete necessary for the web and the ribs is dictated largely by the cover of concrete necessary to protect the reinforcement from corrosion, for which a minimum web thickness of 85 or 100 mm is usual. The necessary cover of concrete to reinforcement makes this system of walling heavy, cumbersome to handle and fix, and bulky looking. Restraint fixings to the upper or lower horizontal ribs of cladding units, depending on whether they are top or bottom supported, must restrain the unit in place against movements and lateral wind pressure. The restraint fixing most used is a non-ferrous or stainless steel angle cleat that is either fixed to a slotted channel cast in the soffit of beams or slabs or, more usually, by expanding bolts fitted to holes drilled in the concrete. The cleat is bolted to a cast-in stud protruding from the horizontal rib of the unit, as illustrated in Figure 7.18. The slotted hole in the downstand flange of the cleat allows some vertical movement between the frame and the cladding.

Another system of fixing combines support fixing by dowels with restraint fixing by non-ferrous flexible straps that are cast into the units and fit over the dowel fixing. Support and restraint fixing may be provided by casting loops or hooked ends of reinforcement, protruding from the back of cladding units, into a small part of or the whole of an insitu cast concrete member of the structural frame. The disadvantage of this method is the site labour required in making a satisfactory joint, and the rigidity of the fixing that makes no allowance for differential movements between structure and cladding. At external angles on elevations, cladding units may be joined by a mitre joint or as a wrap around corner unit specially cast for the purpose, as illustrated in Figure 7.19.

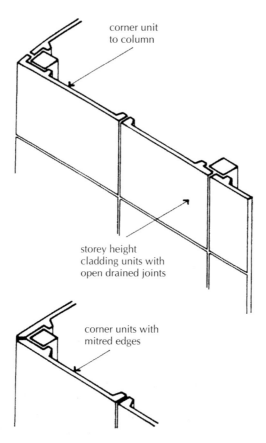

Figure 7.19   Corner units to concrete cladding.

The advantage of the wrap around corner unit is that the open drained joint may be formed against the solid background of a column, and the disadvantage of this unit is that the small, protruding lipped edge may be damaged in lifting and handling into place. It is not practical to make an invisible repair of a damaged edge of a unit. It is difficult to cast a neat, satisfactory drained mitre joint and maintain the mitre junction as a gap of uniform thickness. Whichever joint is used is a matter of choice, principally for reasons of appearance. A common use for precast concrete cladding units is as undercill cladding to continuous horizontal windows or as a spandrel unit to balcony fronts. Typical undercill units, illustrated in Figure 7.20, are designed for bottom rib support and top edge restraint at columns.

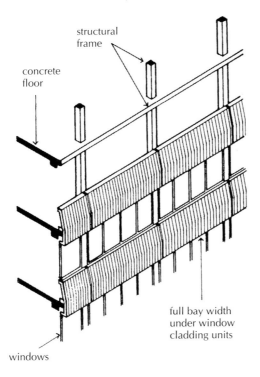

structural frame

concrete floor

full bay width under window cladding units

windows

Figure 7.20   Under window (spandrel) precast concrete cladding.

The units are designed to be supported by horizontal webs that bear on a structural beam or projection of the floor slab, as illustrated in Figure 7.21, which carries the weight of the spandrel unit. Support fixing is similar to that for bottom-supported storey height panels. Restraint fixing is by non-ferrous angle cleats with slotted holes to facilitate fixing to column face and unit. The reinforced concrete cill bearing and window head ribs are adequate to stiffen the web of the units against wind pressure and suction. Where the under window spandrel cladding is continued on a return face of the building, the regular grid of columns may be inset at corners to provide a small span

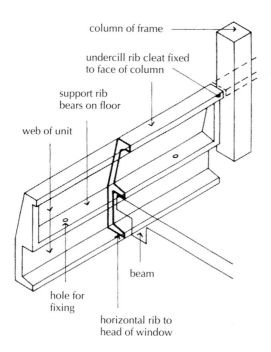

column of frame

undercill rib cleat fixed
to face of column

support rib
bears on floor

web of unit

beam

hole for
fixing

horizontal rib to
head of window

Figure 7.21   Under window concrete cladding unit.

cantilever to accommodate a wrap around spandrel unit supported on the
cantilever floor slab and restrained to the near to corner column.

As an alternative to bottom support, under window spandrel cladding units
may be designed and cast for top, cill level support. A system of separate cill
level beams is precast or may be insitu cast to provide support and fixing
specifically for the undercill ribs of the units, with restraint fixing through a rib
to the floor. The purpose of this arrangement is where structural columns are
widely spaced and it is convenient for casting, transport and handling to use
several units between columns. With the system of under window cladding,
the horizontal windows are framed for fixing to the underside and top of
the spandrel cladding units. The junction of the window framing is usually
weathered with gunned-in mastic as a weather seal. In this construction it is
probably more practical to use a sealed vertical joint between cladding units
than an open drained joint, to avoid the difficulty of making a watertight seal
at the junction of an open joint and window framing.

At the junction of a flat roof and a system of precast cladding units it is necessary to form an upstand parapet either in the structural frame or in purpose
cast parapet panels. For a parapet of any appreciable depth, an upstand beam
or a concrete upstand to a beam is formed and clad with purpose cast panels,
as illustrated in Figure 7.22. A non-ferrous metal capping weathers the top of

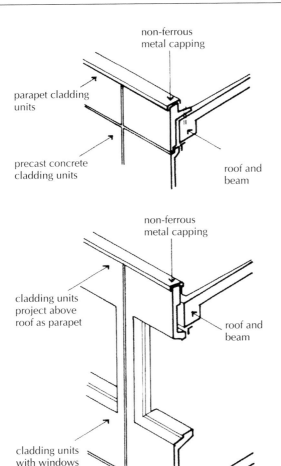

non-ferrous
metal capping

parapet cladding
units

precast concrete
cladding units

roof and
beam

non-ferrous
metal capping

cladding units
project above
roof as parapet

roof and
beam

cladding units
with windows

floor

Figure 7.22   Parapets to precast concrete cladding units.

the parapet. In any event, some form of upstand parapet is formed to avoid
a run-off of water from the roof down the face of the cladding, which would
cause irregular, unsightly staining. Where a system of parapet height cladding
panels is used, the panels are cast to provide a top rib that bears on and is
secured to the parapet, as illustrated in Figure 7.22, and restrained by angle
cleats between the soffit of the beam and the panels. A non-ferrous capping
weathers the top of the parapet and cladding panels.

As an alternative, a system of special storey height cladding panels designed
to extend over the face of, and be fixed to, the parapet may be used, as illustrated

in Figure 7.22. It is possible to cast both parapet cladding panels and storey height panels that extend up to and over the parapet and down the inside face of the parapet. This may be a perfectly satisfactory finish, provided the panels can be lifted and set in place without damage. To provide protection to the sealed or open vertical joints between panels, it is necessary to fix a non-ferrous capping over the parapet and down each side over the panels. Because of the plastic nature of wet concrete, it is possible to cast cladding units in a variety of profiled and textured finishes and to include openings for windows in individual cladding units, within the need for repetitive casting for economy. Because of the coarse grains of the materials used in concrete, the need for concrete cover to reinforcement and the limitations of casting, it is not possible to produce fine, sharp detail in cast concrete. In consequence, the profiles are heavy and rugged, which was one of the attractions of this form of wall cladding; it was at one time described as 'brutalism'. Figure 7.23 is an illustration of part of the wall of a framed structure finished with profiled, storey height window panels. The deep window profiles give a rugged appearance to the building. A disadvantage of deep profiles in a coarse grained material is that irregular run-off of rainwater down the face of the building will cause irregular staining. Some consider this staining unsightly; others claim it enhances the rugged nature of the building. The cladding panels are cast with ribs for bottom support at each floor, top ribs for restraint fixing and side ribs for open drained jointing. The panels may be delivered with the window frames fixed in place ready for site glazing.

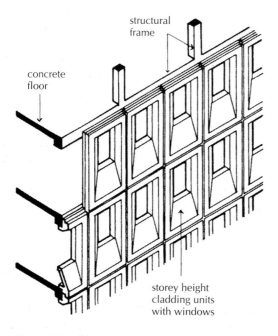

Figure 7.23 Storey height precast cladding units.

Surface finishes

Owing to compaction of wet concrete in the mould, the lower face of the concrete consists of a water-rich mix of the fine particles of cement and aggregate. On drying, this thin layer of cement-rich material shrinks and forms a surface of irregular fine cracks, and the surface may show marked colour differences due to variations in placing, compacting and mixing of the concrete. Because such a surface is not generally an acceptable finish, the exposed faces of precast concrete are usually treated to reveal the underlying aggregate. To provide an acceptable finish to the exposed faces of precast concrete panels it is practice to provide what is sometimes called an 'indirect finish' by abrasive blasting, surface grinding, acid washing or tooling to remove the fine surface layer and expose the aggregate and cement below. This surface treatment has the general effect of exposing a surface of reasonably uniform colour and texture. This form of surface treatment can produce a fine smooth finish by light abrasive sand or grit blasting or grinding, or a more coarse texture by heavy surface treatment.

It was the fashion for some years to use coarse finishes to precast concrete panels by exposing the surface aggregate by heavy abrasive blasting or tooling the surface to produce bush hammered, chisel or pointed tool finishes to emphasise the rugged, heavy nature of the panels. A variety of profiled finishes is produced by casting the panels face down in moulds against timber, etched metal or glass fibre formers to produce a distinct profiled finished face. The finished face of the panels is acid washed or abrasive blasted to remove the surface of fine particles of cement and expose the aggregate. Much favoured at one time was a board marked finish produced by casting on to the surface of boards, which had been grit blasted so that the finished concrete surface displayed the grain of the wood. Profiled finishes to precast concrete panels are not used as much as they were because the dull effect of these finishes, accentuated by surface staining, is not particularly attractive in the dull light of northern climates.

Exposed aggregate finishes are produced by casting the panels face up so that a selected aggregate of comparatively large stones may be spread over the wet compacted concrete and lightly compacted in place so that the aggregate is exposed. Once the concrete has hardened, the surface is washed or blasted to remove traces of cement to expose the colour and texture of the aggregate. This form of rugged finish, which at one time was highly fashionable, has since lost favour.

Applied finishes

It was common, where precast concrete cladding panels were used, to apply facing materials of brick or stone to the panels for the advantage of casting and fixing such traditional materials to minimise site labour, particularly in those climates where air temperature is below freezing for an appreciable part of the year.

Any type of brick that is reasonably frost resistant and durable may be used as a facing material and fixed to the precast concrete panels using the mould to produce the full range of brick construction features such as corbels, string courses, piers and arches. The facing brickwork is bonded to the concrete panel either with a mechanical key or by stainless steel or nylon filament ties. A mechanical key can be provided where bricks with holes in them are used and cut along the length of the brick, so that the resulting semicircular grooves may provide a bond to the concrete. To ensure a good bond to the bricks it is essential to saturate the bricks thoroughly before the backing concrete is placed. Where ties are used to retain the face bricks in place the nylon filament, stainless steel wire or bars are threaded through holes in the brickwork and turned up between bricks as loops to bond with the backing concrete.

The face brickwork may be pre-pointed or post-pointed. Fillets are suspended between the joints in brickwork, which is laid in the bed of the mould, and the pointing material of cement, lime and sand is then pumped into the space below the fillets. The face of the brickwork is protected from mortar by a cloth or paper impregnated with a retarder laid in the bed of the mould. The fillets are removed and the backing concrete placed and compacted over the bricks and around the tie loops. The face pointing, which is usually up to 25 mm deep, is allowed to harden before the concrete is placed. For post-pointing, the bricks are laid in the bed of the mould between Neoprene strips to provide the necessary recess of about 20 mm for pointing, and the concrete backing is then cast on the bricks and consolidated. Pointing is carried out on site when the cast panels are in place. This is a labour intensive way of providing a brick facing to concrete panels that at best will provide a simulation of traditional brickwork. Because there have to be both horizontal and vertical movement joints between the concrete panels, which will interrupt any attempt to copy loadbearing brickwork, it provides the possibility of setting the bricks in any pattern, other than the traditional bonded horizontal bonded pattern, for purely decorative purposes. Finishes of brick or stone to precast concrete cladding panels are less used than they were because of the considerable costs of casting, transporting and lifting into place these heavy, cumbersome wall units where brick walling and stone facings may be more economically applied on site to a structural frame.

Natural and reconstructed stone facing slabs are used as a decorative finish to precast concrete panels. Any of the natural or reconstructed stones used for stone facework to solid backgrounds may be used for facings to precast concrete panels. For ease of placing the stone facing slabs in the bed of the mould, it is usual to limit the size of the panels to not more than 1.5 m in any one dimension. Granite and hard limestone slabs not less than 30 mm thick and limestone, sandstone and reconstructed stone slabs not less than 50 mm thick are used. The facing slabs are secured to and supported by the precast panel through stainless steel corbel dowels at least 4.7 mm in diameter, that are set into holes in the back of the slabs and cast into the concrete panels at the rate of at least 11 per m^2 of panel and inclined at 45° or 60° to the face of the panel. Normal practice is that about half of the dowels are inclined up and half down, relative to the vertical position of the slab when in position on site. The dowels are set in epoxy resin in holes drilled in the back of the slabs. Flexible grommets are fitted around the dowels where they protrude from the back of the slab. These grommets, which are cast into the concrete of the panel, together with the epoxy resin bond of the dowel in the stone slab, provide a degree of flexibility to accommodate thermal and moisture movement of the slab relative to that of the supporting precast concrete cladding panel.

All joints between the stone facing slabs are packed with closed cell foam backing or dry sand, and all joints in the back of the stone slabs are sealed with plastic tape to prevent cement grout running in. When the precast panel is taken from the mould, the jointing material is removed for mortar or sealant jointing. To prevent the concrete of the precast panel bonding to the back of the stone slabs either polythene sheeting or a brushed on coating of clear silicone waterproofing liquid is applied to the whole of the back of the slabs. The purpose of this debonding layer is to allow the facing slabs free movement relative to the precast panel due to differential movements of the facing and the backing. The necessary joints between precast concrete cladding panels faced with stone facing slabs are usually sealed with a sealant to match those between the facing slabs.

Joints between precast concrete cladding panels

The joints between cladding panels must be sufficiently wide to allow for inaccuracies in both the structural frame and the cladding units, to allow unrestrained movements due to shortening of the frame and thermal and moisture movements, and at the same time to exclude rain. The two systems of making joints between units are the face sealed joint and the open drained and rebated joint. Sealed joints are made watertight with a sealant that is formed inside the joint over a backing strip of closed cell polyethylene, at or close to the face of the units, as illustrated in Figure 7.24. The purpose of the backing strip is to ensure a correct depth of sealant. Too great a depth or width of sealant will cause the plastic material of the sealant to move gradually out of the joint, due to its own weight.

Sealant material is applied by gun. The disadvantages of sealant joints is that there is a limitation to the width of joint in which the sealant material can successfully be retained, and that the useful life of the material is from 15 to 20 years, as it oxidises and hardens with exposure to sunlight and has to be raked out and renewed. Sealed joints are used in the main for the smaller cladding units. The sealants most used for joints between precast concrete cladding panels are two parts polysulphide, one part polyurethane, epoxy modified two parts polyurethane and low modulus silicone. Which of these sealants is used depends to an extent on experience in the use of a particular material and ease of application on site. The two part sealants require more skill in mixing the two components to make a successful seal than the one part material,

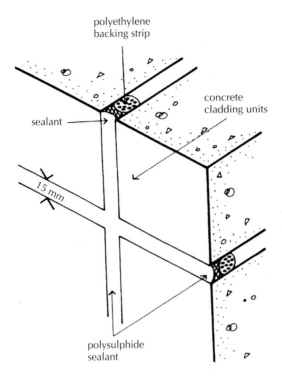

polyethylene
backing strip

concrete
cladding units

sealant

15 mm

polysulphide
sealant

Figure 7.24 Sealant joints to concrete cladding.

which is generally reflected in the relative cost of the sealants. A closed cell polyethylene backing strip is rammed into the joint and the sealant applied by power or hand pump gun, and compacted and levelled with a jointing tool.

Open drained joints between precast concrete cladding panels are more laborious to form than sealed joints and are mostly used for the larger precast panels where the width of the joint may be too wide to seal and where the visible open joint is used to emphasise the rugged, coarse textured finish to the panels. Open joints are the most effective system of making allowance for inaccuracies and differential movements and serving as a bar to rain penetration without the use of joint filling material.

Horizontal joints are formed as open overlapping joints with a sufficiently deep rebate as a bar to rain penetration, as illustrated in Figure 7.25. The rebate at the joint should be of sufficient section to avoid damage in transport, lifting and fixing in place. The thickness necessary for these rebates is provided by the depth of the horizontal ribs. The air seal formed at the back of horizontal joints is continuous in both horizontal and vertical joints as a seal against outside wind pressure and driving rain.

Vertical joints are designed as open drained joints in which a Neoprene baffle is suspended inside grooves formed in the edges of adjacent units, as illustrated in Figure 7.26. The open drained joint is designed to collect most of the rain in

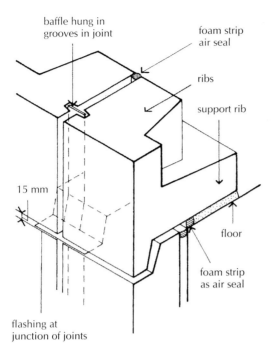

Figure 7.25   Horizontal open drained joint.

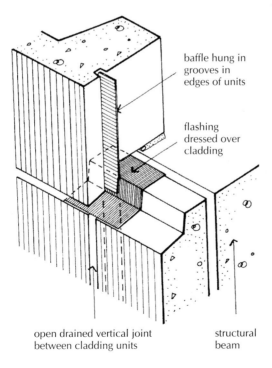

Figure 7.26   Vertical open drained joint.

the outer zone of the joint in front of the baffle, which acts as a barrier to rain that may run or be forced into the joint by wind pressure. The baffle is hung in the joint so that to an extent there is a degree of air pressure equalisation each side of the baffle due to the air seal at the back of the joint. This air pressure equalisation acts as a check to wind-driven rain that would otherwise be forced past the baffle if it were a close fit and there were no air seal at the back of the joint. At the base of each open drained joint there is a lead flashing, illustrated in Figures 7.25 and 7.26, which serves as a barrier to rain at the most vulnerable point of the intersection of horizontal and vertical joints. As cladding panels are fixed, the baffle in the upper joints is overlapped outside the baffle of the lower units.

Where there is a cavity between the back of the cladding units and an inner system of solid block walls or framing for insulation, air seals can be fitted between the frame and the cladding units. It is accepted that the system of open joints between units is not a complete barrier to rain. The effectiveness of the joint depends on the degree of exposure to driving rain, the degree of accuracy in the manufacture and assembly of the system of walling, and the surface finish of the cladding units. Smooth faced units will tend to encourage driven rain to sheet across and up the face of the units, and so cause a greater pressure of rain in joints than there would be with a coarse textured finish, which will disperse driven rain and wind, and so reduce pressure on joints. The backs of cladding panels will tend to collect moisture by possible penetration of rain through joints and from condensation of moisture-laden air from outside and warm moist air from inside by vapour pressure, which will condense on the inner face of panels. Condensation can be reduced by the use of a moisture vapour check on the warm side of insulation as a protection against interstitial condensation in the insulation and as a check to warm moist air penetrating to the cold inner face of panels. Precast concrete cladding panels are sometimes cast with narrow weepholes, from the top edge of the lower horizontal ribs out to the face, in the anticipation that condensate water from the back of the units will drain down and outside. The near certainty of these small holes becoming blocked by windblown debris makes their use questionable.

Attempts have been made to include insulating material in the construction of precast cladding, either as a sandwich with the insulation cast between two skins of concrete or as an inner lining fixed to the back of the cladding. These methods of improving the very poor thermal properties of concrete are not successful because of the considerable section of the thermal bridge of the dense concrete horizontal and vertical ribs that are unavoidable, and the likelihood of condensate water adversely affecting some insulating materials.

It has to be accepted that there will be a thermal bridge across the horizontal support rib of each cladding panel that has to be in contact with the structural frame. The most straightforward and effective method of improving the thermal properties of a wall structure clad with precast concrete panels is to accept the precast cladding as a solid, strong, durable barrier to rain with good acoustic and fire resistance properties and to build a separate system of inside finish with good thermal properties. Lightweight concrete blocks by themselves, or with the addition of an insulating lining, at once provide an acceptable internal finish and thermal properties. Block wall inner linings should be constructed independently of the cladding panels and structural members, as far as practical, to reduce interruptions of the inner lining, as illustrated in Figure 7.27.

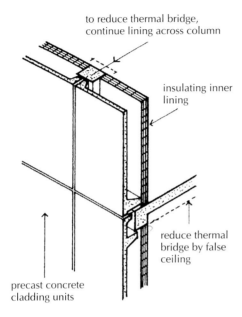

to reduce thermal bridge,
continue lining across column

insulating inner lining

reduce thermal bridge by false ceiling

precast concrete cladding units

Figure 7.27   Insulation lining to concrete cladding.

Glass fibre reinforced cement cladding panels (GRC)

Glass fibre reinforced cement as a wall panel material was first used in the early 1970s after studies at the Building Research Establishment and the production of an alkali resisting glass fibre. The material has since been used as a lightweight substitute for precast concrete in wall cladding in the UK, America, the Middle East and Japan. The principal advantage of GRC as a wall panel material is weight saving as compared to similar precast concrete panels. Much of the mass of concrete used in panels is required as protection of the steel reinforcement against atmospheric chemical attack, whereas alkali resisting glass fibre, which is not subject to attack, can be used in panels with a skin thickness of 10 to 15 mm and a weight saving of about 80% of that of a comparable concrete panel. This weight reduction will afford substantial savings in transport, handling and erection costs, and some small saving in structural frame members. Because of the fine grain of the material that is used in the manufacture of GRC and the freedom from the constraint of the need for steel reinforcement and its necessary cover against corrosion, this material can be formed in a wide variety of shapes, profiles and accurately finished mouldings such as that illustrated in Figure 7.28. The material has inherently good durability and chemical resistance, is non-combustible, not susceptible to rot and will not corrode or rust stain. The limiting factors in the use of this material arise from relatively large thermal and moisture movements, and the restricted ductility of the material.

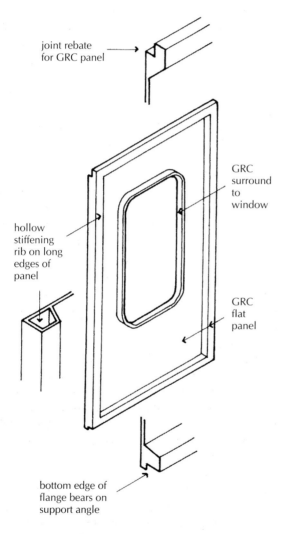

joint rebate for GRC panel

GRC surround to window

hollow stiffening rib on long edges of panel

GRC flat panel

bottom edge of flange bears on support angle

Figure 7.28　Ribbed single skin GRC panel.

The material is a composite of cement, sand and alkali resistant (AR) glass fibre in proportions of 40–60% cement, 20% water, up to 25% sand and 3.5–5%

glass fibre by weight. The glass fibre is chopped to lengths of about 35 mm before mixing. It is formed in moulds by spray application of the wet mix, which is built up gradually to the required thickness and compacted by roller. After the initial 3 mm thickness has been built up it is compacted by roller to ensure a compact surface finish. For effective hand spraying the maximum width of panel is about 2 m. For mass production runs of panel, a mechanised system is used with dual spray heads which spray fibre and cement, sand and water separately in the mould, which moves under the fixed spray heads. The mechanised spray results in a greater consistency of the mix and a more uniform thickness of panel than is usually possible with hand spraying. The moulds for GRC are either timber or the more durable GRP lined, timber framed types. Spray moulded GRC panels have developed sufficient strength 24 hours after moulding to be taken from moulds for curing. The size of GRC cladding panels is limited by the method of production as to width and to the storey height length for strength, transport and lifting purposes. It is also limited by the considerable moisture movement of the cement-rich material, which may fail if moisture movement is restrained by fixings. The usual thickness of GRC single skin panels is 10 to 15 mm.

As a consequence of moulding, the surface of a GRC panel is a cement-rich layer, which is liable to crazing due to drying shrinkage and to patchiness of the colour of the material due to curing. To remove the cement-rich layer on the surface and provide a more uniform surface, texture and colour, the surface can be acid etched, grit blasted or smooth ground. Alternatively, the panels can be formed in textured moulds so that the finished texture masks surface crazing and patchiness. Using ordinary or rapid hardening Portland cement, the natural colour of these panels is a light, dull grey. White or pigmented white cement can be used instead of Portland cement to produce a white or colour finish, which may well not be uniform, panel to panel. For a uniform colour finish that can be restored by repainting on site, coloured permeable coatings are used which have microscopic pores in their surface that allow a degree of penetration and evaporation of moisture that prevents blistering or flaking of the coating. Textured permeable finishes such as those used for external renderings, and microporous matt and glass finish paints are used. The thin single skin of GRC does not have sufficient strength or rigidity by itself to be used as a wall facing other than as a panel material of up to about 1 m^2 square, supported by a metal carrier system or bonded to an insulation core for larger panels, as illustrated in Figure 7.29.

The use of GRC panels formed around a core of insulation material has by and large been abandoned. In use the effect of the core of insulation is to cause an appreciably greater expansion of the external skin than that of the internal skin, with consequent deformation and bowing of the outer skin and possible failure. The thermal resistance of GRC is poor and it is now generally accepted that an inner skin of insulation should be provided separate from the GRC. A single skin GRC panel larger than 1 m^2 does not have sufficient strength by itself

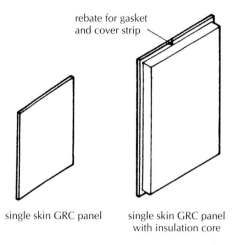

rebate for gasket
and cover strip

single skin GRC panel

single skin GRC panel
with insulation core

Figure 7.29 Single skin GRC panel.

and requires some form of stiffening. Stiffening is provided by solid flanges formed at the bottom and top edges of panels. The bottom flange provides support for the panel, and the top flange a means of restrain fixing. Stiffening ribs are formed in the vertical edges of panels. These stiffening ribs are usually hollow and formed around hollow or foam plastic formers to minimise weight and shrinkage of the cement-rich material as it dries. The four edges of panels are usually rebated to overlap at horizontal joints for weathering and to form weathered vertical joints. A flanged single skin, flat GRC panel is illustrated in Figure 7.30.

Storey height, spandrel and undercill GRC panels have been extensively used as both flat and shaped panels for the advantage of the fine grained, smooth surface of the material. The shaping of a panel adds a degree of stiffness to the thin material in addition to the flanges and stiffening ribs. Figure 7.31 is an illustration of a shaped, window panel of GRC.

The support and restraint fixings for GRC panels are designed to allow freedom of movement of the thin panel material to accommodate differential structural thermal and moisture movement of the panels relative to that of the structure. To this end a minimum of fixings is used. The weight of a GRC panel is supported by the structure or structural frame at two points near the base of the panel so that compressive stress acts on the bottom of the panel. Either one or both of the bottom supports is designed to allow some freedom of horizontal movement. To hold the panel in its correct position and allow freedom of movement, four restraint fixings are used near the four corners of the panel. These restraint fixings allow freedom of horizontal movement at the base and freedom of both horizontal and vertical movement at the two top corners. As an alternative to fixing GRC panels to the structure or structural frame, a separate stud frame is used as support for the panels, with the stud

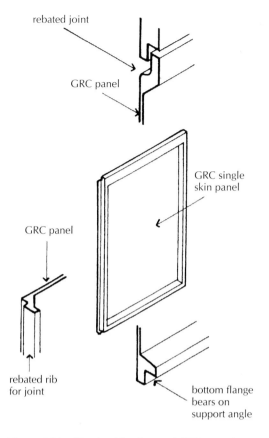

rebated joint

GRC panel

GRC single
skin panel

GRC panel

rebated rib
for joint

bottom flange
bears on
support angle

Figure 7.30   Single skin, flanged GRC panel.

frame fixed to the structural frame with fixings designed to allow freedom of movement of the stud frame relative to the structure.

The requirement for allowances for movement in fixings and the additional work and materials involved does add considerably to the cost of the use of this material as a facing, which is one reason for the loss of favour in the use of GRC panels. The common means of support for GRC panels is by stainless steel angles that are bolted to the solid structure or structural frame, with the horizontal flange of the angle providing support at the bottom of the panel, as illustrated in Figure 7.32.

The bottom flange of the GRC panel bears on the horizontal flange of the angle with metal packing pieces as necessary to level the panel. Either a separate restraint fixing is used or a stainless steel dowel is welded to the angle to fit into a tapered socket in the GRC flange. The socket is filled with a resilient filler to allow some freedom of movement. The dowel can serve to locate the panel in position and as a restraint fixing. This fixing allows for some horizontal movement through the resilient filler and the movement of the panel

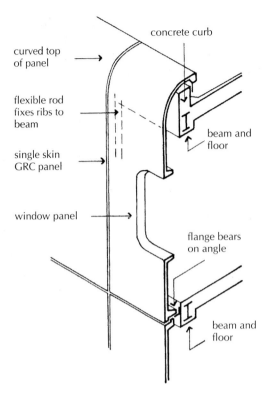

curved top of panel

concrete curb

flexible rod fixes ribs to beam

single skin GRC panel

window panel

beam and floor

flange bears on angle

beam and floor

Shaped GRC panels

Figure 7.31 Shaped GRC panels.

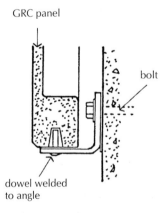

GRC panel

bolt

dowel welded to angle

Figure 7.32 Support angle for GRC panel.

on the angle. The edge of the seating angle may be masked by the joint filler, by setting the angle into a slot in the GRC or behind a rebated joint between panels. Support angles are secured with expanding bolts to insitu concrete or brickwork, and angles fixed to structural steel frames.

Restraint fixings

Restraint fixings should at once tie the panel back to the structure and allow for some horizontal and vertical movement of the panel relative to the fixing. Restraint fixing is usually provided by a stainless steel socket that is cut into the back of the solid flange of a GRC panel as it is being manufactured. The socket is threaded ready for a stainless steel bolt. The bolt is fitted to an oversize hole in an angle, which is bolted to the underside of a concrete or steel beam. Rotational movement of the GRC panel is allowed by a metal tube and plastic separating sleeve that fits around the stainless steel bolt in the oversize hole in the angle. The bolt is held in place in the angle by steel washers, and some movement is provided for by washers around the angle, as illustrated in Figure 7.33. Some care is required in the design to make allowance for access to make these somewhat complex fixings.

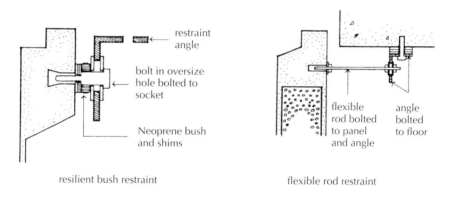

restraint angle

bolt in oversize hole bolted to socket

Neoprene bush and shims

resilient bush restraint

flexible rod bolted to panel and angle

angle bolted to floor

flexible rod restraint

Figure 7.33   Restraint fixings.

Stud frames

A large single skin GRC panel with top and bottom flanges and edge stiffening ribs may not have sufficient stiffness to adequately resist the stresses due to moisture movement of the cement-rich material and the considerable wind forces acing on a wall. To provide support for large single skin panels and at the same time make allowance for moisture movement, a system of stud frames was developed and has been extensively used. A frame of hollow and channel steel section is prefabricated with welded joints to the top, bottom and side members, and intermediate vertical sections spaced at about 600 mm, as illustrated in Figure 7.34. L section, 9.5 mm round steel section anchors are welded to the hollow section studs at about 600 mm centres to serve as flex anchors to the GRC panel. Near the base of the frame, T sections are welded to the sides of the edge studs to serve as gravity anchors. Angles are also welded to the back face of the studs as seating angles to support the stud frame.

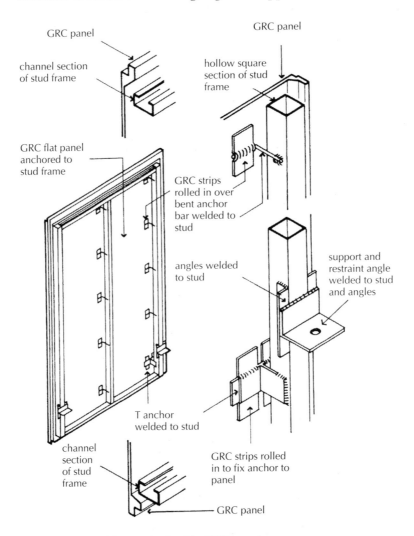

GRC panel

channel section of stud frame

GRC flat panel anchored to stud frame

GRC panel

hollow square section of stud frame

GRC strips rolled in over bent anchor bar welded to stud

angles welded to stud

support and restraint angle welded to stud and angles

T anchor welded to stud

channel section of stud frame

GRC strips rolled in to fix anchor to panel

GRC panel

Figure 7.34   Stud frame single skin GRC panel.

The fabricated stud frame is galvanised or powder coated to inhibit rusting. As the GRC panel is being manufactured, and the skin and flanges and ribs are formed, the stud frame is placed on the back of the compacted and still moist GRC, with the flex anchors and T bearing on the back of the panel. Moist strips of GRC are then rolled on to the back of the panel over the flex anchors and T anchors to secure them to the panel, as illustrated in Figure 7.34. The GRC strips, which are rolled over the flex anchors on to the back of the panel, provide a firm attachment to maintain the thin skin as a flat panel yet allow sufficient rotational and lateral movement of the panel to prevent failure.

Figure 7.35 is an illustration of the fixing of storey height stud frame panels of GRC to the beams of a structural steel frame. T section, steel beam brackets are welded to plates that are welded to the web of beams. These brackets support

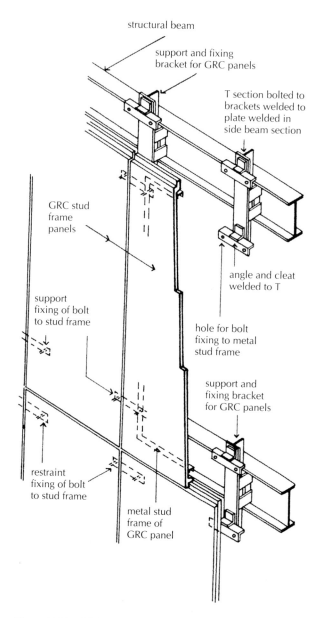

structural beam

support and fixing
bracket for GRC panels

T section bolted to
brackets welded to
plate welded in
side beam section

GRC stud
frame
panels

angle and cleat
welded to T

support
fixing of bolt
to stud frame

hole for bolt
fixing to metal
stud frame

support and
fixing bracket
for GRC panels

restraint
fixing of bolt
to stud frame

metal stud
frame of
GRC panel

Figure 7.35   GRC stud frame fixed to structural steel frame.

T sections welded to cleats. One beam bracket is welded to beams centrally on the junction of vertical and horizontal joints between GRC panels.

The weight of the stud frame is supported by the bearing of the bottom flange of the frame on the lower flange of the beam. Top and bottom restraint fixings, cast in the side ribs of the GRC panel, are bolted to the angles of the beam brackets. Open drained or mastic or gasket joints are made to joints between panels. Stud frames are best suited to flat storey height panels which are supported at floor levels. Curved and profiled panels may have sufficient stiffness in their shape and not justify the additional cost of a stud frame. Similarly, small under window and spandrel panels of GRC do not generally require a stud frame.

Joints between GRC panels

Mastic joint
The joints between GRC panels may be square for mastic sealant, rebated for gasket joints, channelled for open drained joints or rebated for overlap at joints. Which joint is used depends on the overall size of the panel and therefore the moisture movement that the joint will have to accommodate, the exposure of the panel to driving wind and rain, the jointing system chosen by the manufacturers and the convenience in making the joint. Appearance, as part of the overall design, may also be a consideration. For small GRC panels where the moisture movement of the cement-rich mix will be limited, a sealant joint is commonly used for the advantage of simplicity of application and renewal as necessary. The mastic sealant joint is made between the square edges of uniform width joints by ramming a backing strip of closed cell polyethylene into the joint as support and backing for the mastic sealant. The mastic sealant is run into the joint from a hand or power operated gun. The sealant is finished with a tool to provide a concave surface for the sake of appearance in keeping the mastic from the panel face, as illustrated in Figure 7.36.

A mastic sealant joint should be no wider than 15 mm as the sealant in a wider joint might sag and no longer seal the joint. The sealants most used are two part or one part polysulphide, one part polyurethane, epoxy modified polyurethane and low modulus silicones. Which is used depends on the skill of the operative and the ease of access and application. One part sealants are easier to use, but less effective than two part sealants. Silicones tend to be the most difficult to use and the most effective. In time, sealants may oxidise and harden and require renewal after a number of years.

Gasket joint
A gasket joint is used for larger GRC panels where accuracy of manufacture and fixing can provide a joint of uniform width. The elasticity of the gasket is capable of accommodating the moisture movements of the panel while maintaining a weather seal. A gasket joint may be used for vertical and horizontal joints or for vertical joints alone with overlapping rebated horizontal joints, as illustrated in Figure 7.37.

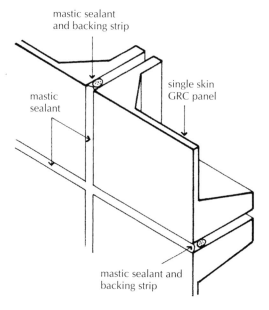

mastic sealant
and backing strip

mastic
sealant

single skin
GRC panel

mastic sealant and
backing strip

Figure 7.36   Mastic sealant joint.

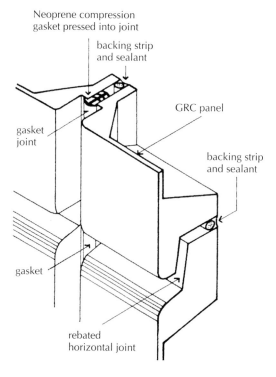

Neoprene compression
gasket pressed into joint

backing strip
and sealant

gasket
joint

GRC panel

backing strip
and sealant

gasket

rebated
horizontal joint

Figure 7.37   Gasket joint.

For gasket jointing, the edges of the panels are rebated to provide space for insertion of the gaskets. A backing strip and mastic seal is formed on the back face of the joint to exclude wind. The strip of Neoprene gasket is rammed into the joint to make a weather tight seal.

Where both vertical and horizontal joints are gasket-sealed, preformed cross over gaskets are heat welded to the ends of the four straight lengths at junctions of vertical and horizontal joints and then rammed into position. Being set in position some distance behind the GRC panel faces, the gaskets are protected against the scouring effect of wind and rain and also the hardening of the material due to oxidisation caused by direct sunlight. It is a reasonable expectation that these gasket joints will be effective during the life of most buildings. The advantage of the overlapping, rebated horizontal joint illustrated in Figure 7.37 is that the deep rebate will provide a degree of protection against rain running down the building face that a parallel flat face will not, where a similar vertical joint will provide protection from one direction only.

Open drained joint
Open drained, vertical joints are used for the larger GRC panels where it is not practical to form the narrow, accurately fitted joints necessary for both mastic and gasket jointing. Plastic channels are cast in a chase in the vertical edges of panels, as illustrated in Figure 7.38, inside which a plastic baffle is hung as

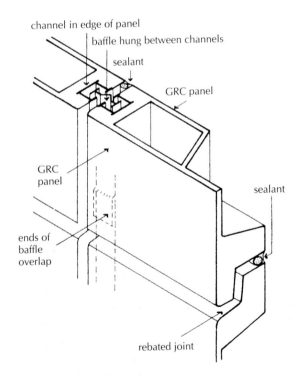

Figure 7.38 **Open drained joint.**

a loose fit. The inside joint between panels is sealed with a backing strip and mastic sealant joint.

The loose baffle acts as a first line of defence against wind driven rain, most of which will run down the baffle. Because the baffle is a loose fit, there will be a degree of wind pressure equalisation each side of the baffle, which will appreciably reduce wind pressure on rain that may find its way past the baffle. The advantage of this joint is that it allows for movement of the panel, inaccuracies in manufacture and assembly and acts as a reasonable barrier to wind driven rain. This open joint acts in conjunction with a rebated, overlapping horizontal joint which is mastic sealed on the inner face. It is accepted that this joint is not a complete barrier to rain. The effectiveness of the joint depends on the degree of exposure and the surface finish of the panels. Smooth faced panels will tend to encourage rain to sheet across panels and so cause greater pressure of rain than would coarse textured panels.

Glass fibre reinforced polyester cladding (GRP)

Glass fibre reinforced polyester laminate was first used as a thin skin wall cladding material in the mid-1950s and subsequently up to the late 1970s as a lightweight skin panel material for wall systems. This material has never been used as extensively as precast concrete and in recent years its use has declined. GRP is a composite of a durable, thermosetting polyester resin, reinforced with glass fibre, that is used as a thin laminate with high strength, low density, good corrosion and weather resistance but a low modulus of elasticity. It can be moulded without pressure or high temperature with comparatively simple, inexpensive equipment to produce an unlimited variety of shape and detail. The polyester resin is supplied as a viscous syrup-like material, which is polymerised or cured to a hard solid by the addition of chemical catalysts under controlled heat, shortly before it is laid up in the mould. Glass fibre is made by drawing molten glass to a filament of glass fibre, which has high tensile strength.

GRP cladding panels are made by 'laying up' or spraying the viscous GRP material in moulds lined with GRP. The surface of the mould is first waxed and polished and then coated with a release agent. In the 'laying up' process, the materials are laid in the mould by hand in layers of glass fibre mat and resin mixed with catalyst in successive layers, and consolidated by hand. In the spray process the materials are sprayed into the mould and consolidated by hand in layers. As a preliminary to laying up or spraying the GRP material in the mould, a thin gel coat of resin is spread on to the surface of the mould. The primary purpose of the gel coat is as a protection against moisture which might otherwise penetrate the surface of the GRP to the glass fibre and cause swelling, rupture and breakdown of the GRP laminate. Once the gel coat has hardened sufficiently to be tacky, the first layer of resin, catalyst and glass mat is spread in the mould and consolidated by roller, followed by successive layers up to the required thickness of the laminate. The moulded panel is then taken from the mould and cured in a box or chamber under controlled conditions of temperature and humidity to develop structural and dimensional stability.

Control of the process of manufacturing GRP panels has a most significant effect on the finished product in use as a wall panel material. Selection of the resin, catalyst, fillers and pigment for a particular purpose, the careful mixing of the materials, skill in application of the materials to form a sound laminate, control of curing and control of the conditions of temperature and humidity in the workshop all have a significant effect on the dimensional accuracy, stability, strength and durability of the finished product. One of the main reasons for the comparatively limited use of this material is the difficulty of control in manufacture and the failures that have been caused by faulty control and poor workmanship.

The natural colour of GRP is not generally accepted as an attractive finish for wall panels and it is usual to make panels that are coloured with a pigment added to the gel coat or the resin binder. The addition of pigment and the selection of colour can appreciably affect the weathering characteristics of GRP panels. Strong colours such as oranges and reds tend to fade through the effect of ultraviolet light, which causes the surface to chalk, and colour fastness may well be irregular between panels. Dark colours encourage high surface temperatures that increase the risk of separation of the laminae of the skin, a failure that is known as lamination. Variations in the surface of flat smooth faced panels, caused by mechanical and thermal distortion, may be obvious in surfaces more than about a metre wide. A matt texture or shallow ribbed finish will effectively mask distortions of the surface without noticeably affecting the smooth, sleek surface of this finish.

To enhance the poor resistance to damage and spread of flame characteristics of GRP it is practice to add fillers or chemical additives to the resin or to coat the surface. The addition of fillers to improve the fire retardancy of the material has the effect of weakening its capacity to resist weathering agents, and affects pigments which are added. The addition of fillers and pigment to improve fire resistance and colour appreciably reduces the weathering characteristics of this material. One method that has been used to provide protection against weathering and to improve fire resistance is to coat the surface with polyurethane to improve weathering and to modify the gel coat to improve fire resistance. Because GRP is an expensive material, it is used as a thin skin for all panels in thicknesses of from 3 to 6 mm and has to be stiffened with edge flanges, shaped profiles, stiffening ribs or a sandwich construction. To be effective as stiffening, shaped profiles must be deep in relation to the thickness of the skin. Because GRP is used for the dramatic effect of the level face of panels, the usual method of stiffening is to bond stiffening ribs to the back of the panel skin. These stiffening ribs are usually made of hollow sections of GRP that are overlaid with GRP as the laminate is built up. Figure 7.39 is an illustration of a ribbed panel stiffened with top hat and box section hollow ribs.

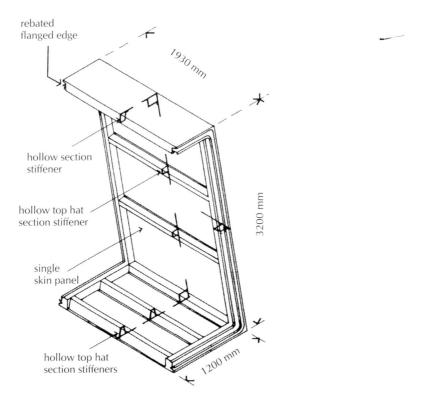

Figure 7.39   Ribbed GRP panel.

In common with the other thin skin panel material, metal, GRP can be formed with ease around a core of insulating material as a sandwich panel. The sandwich of GRP at once provides stiffening and insulation. These sandwich panels are made with two laminate skins of GRP moulded around the insulating core with the GRP skins joined around the edges of the panel to seal the sandwich and for fixing. Figure 7.40 is an illustration of a sandwich panel. The size of these panels should be limited to avoid too great a distortion of the finished panel face through differential expansion of core and skin material.

As an alternative to the use of a core of insulating material to one sandwich panel, panels may be formed of two separate sandwiches of GRP and insulation joined at the edges, as illustrated later in Figure 7.42. The purpose of using two separate skins is to distance the outer skin, which is most likely to suffer distortion due to solar heat, from the inner skin and so limit distortion.

Jointing

As with any other facing panel wall structure, the joints between the panels of GRP have to allow for differential structural, thermal and moisture movements between the supporting frame and the wall, and must serve as an effective

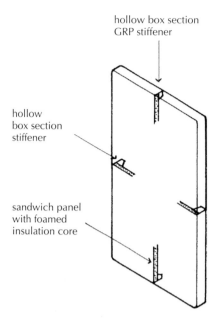

hollow box section
GRP stiffener

hollow
box section
stiffener

sandwich panel
with foamed
insulation core

Figure 7.40   GRP sandwich panel.

barrier to penetration of rain. The three types of joint that have been used are mastic sealant, gasket and open drained joint. There have been a number of failures of sealant joints due to over wide joints or poor workmanship or both, which resulted in unsightly mastic failure, and which have given this method of jointing a bad name. Provided the joint is reasonably tight and adequate to the anticipated movements, a skilfully applied sealant joint will give satisfactory performance for the life of the sealant material, which will need periodic attention.

Gasket jointing techniques, adopted from glazing, have been successfully used for GRP panels. Preformed gaskets of Neoprene or ethylene propylene diamine monomer (EPDM) fit around and seal the edges of adjacent panels, with the gasket compressed to the panels with adjustable clamps bolted to the carrier system. These gaskets have preformed cross over intersections at the junction of horizontal and vertical joints that are heat welded on site to straight lengths of gasket. Both Neoprene and EPDM gaskets oxidise, harden and lose resilience on exposure and may need replacement every 10 to 20 years. Open drained joints have the advantage that the jointing material is not visible and that the open drain serves as a check to driving rain. The open drained joint illustrated in Figure 7.41 has outer and inner drain channels and a baffle.

Support and fixing
Because GRP is a comparatively expensive material, it is used as a thin skin and in consequence GRP panels are lightweight and do not by themselves require

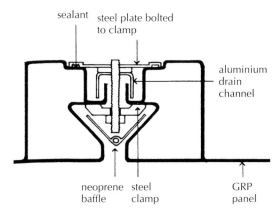

sealant   steel plate bolted
             to clamp

aluminium
drain
channel

neoprene   steel   GRP
baffle   clamp   panel

Figure 7.41   Open drained vertical joint.

substantial support fixings. Because of the considerable thermal movement of GRP and the thin skin form of its use, it is essential to support and restrain panels to or in fixings that will allow for thermal movement and restrain the lightweight panels in position. Usual practice is to clamp panels back to the structure or to the carrier system inside Neoprene or EPDM gaskets that hold the panels in place and act as weather seal. The double sandwich GRP panels shown in Figure 7.42 fit to an aluminium carrier into which the GRP panels are housed and secured with aluminium top hat sections that are screwed to the carrier and compress Neoprene gaskets to the panels. The aluminium carrier is bolted to a hollow rectangular section structural frame.

Another method of support and fixing is to incorporate timber battens or framing in either single skin or sandwich panels. The timber battens can be enclosed in the GRP laminate as stiffening ribs and used as a means of fixing the panel by screwing back to the carrier system or structure and as a means of fixing for windows.

GRP as a material for wall panels lost favour principally because of failure due to poor manufacturing techniques and lack of colour fastness to exposed surfaces. Since the introduction of this material for use as wall panels there have been considerable improvements in the mixing manufacture and use of thermosetting materials which, applied to this type of panel, could make it wholly acceptable as a wall panel material.

7.7 Sheet metal wall cladding

Sheet metal has for many years been used as a wall cladding material. Corrugated iron sheets were first produced about 1830 from wrought or puddled iron which was cold rolled to the traditional sinusoidal (corrugated) profile.

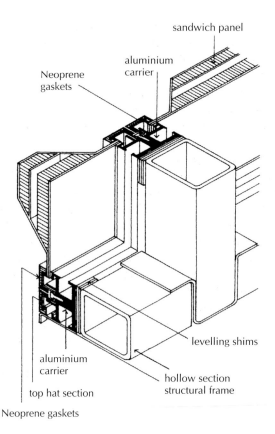

Figure 7.42   **Double sandwich GRP panels.**

The purity of the iron and the thickness of these original corrugated sheets gave them a useful life of up to 50 years. After steel making became a commercial proposition, in about 1860, thin strips of steel were cold rolled to form corrugated sheets from 1880. These sheets, which were considerably thinner and cheaper than the original iron sheets, had a useful life of about 12 to 15 years due to the destructive corrosion of the metal by rust. Coatings of tar or pitch improved the life of steel sheets to some extent. Galvanised, corrugated steel sheets were first used in the UK about 1883. The sheets were coated with a film of zinc, applied by dipping or spraying. The zinc coating did provide some protection against corrosive rusting for some years. The thicker the zinc coat, the longer the sheets lasted.

Improvements in the techniques of cold roll forming of steel strip, that were introduced in 1960, made it possible to form a range of trapezoidal section profiled sheets. The improved strength and rigidity of the deep profiles made it possible to use thinner strip, wider spacing of supporting rails and improvement in appearance. These thin trapezoidal section sheets were coated with zinc and later with zinc on to which coloured, organic plastic coatings were applied as protection and decoration. Cold roll forming of metal strip is limited to the production of sheets profiled in one direction only. Corrugated and trapezoidal profile sheets are best suited for use as roof and wall cladding to small and single-storey buildings where the side lap of profiles and end lap of sheets provide reasonable weather protection.

The use of rectilinear metal panels as a wall cladding began in the early years of the 20th century in France as metal faced infill panels were used to mask floors, walls and columns between fully glazed wall panels. These comparatively small panels were usually pressed into some decorative profile. This early system of infill panels was later developed into the form of larger metal panels fabricated as complete metal, storey height panels and panels incorporating windows. These metal panels were generally fabricated as a sandwich of two skins of metal enclosing a core of insulating material.

The two metals most used for cladding panels are mild steel and aluminium in the form of hot rolled strip. Mild steel, which has a favourable strength-to-weight ratio, suffers the considerable disadvantage of progressive, destructive rusting on exposure to atmosphere. To inhibit rusting, the metal is coated with zinc. Steel strip can readily be cold rolled or pressed to shape. Aluminium is a malleable metal, which can readily be cold rolled or pressed to shape and which on exposure to the atmosphere will form a dull, coarse textured oxide film that prevents further corrosion. Because of this dull, unattractive coating, aluminium is usually finished with an organic coating as decoration. Sheet metal cladding can be broadly grouped as:

❏ Laminated panels
❏ Single skin panels
❏ Box panels
❏ Rain screens

Laminated panels
Laminated, metal strip, wall panels are made from layers of metal formed around a central core of insulation with the long edges formed by welting or welding the inner and outer metal linings together. The combination of sheet metal and insulation provides a weather surface and insulation in one wall unit. The strip metal is usually of hot rolled strip. The outer strip is usually of cold rolled, corrugated or trapezoidal profile, and the inner of flat strip. The disadvantage of using a flat, hot rolled outer lining of strip metal is that the outer surface will tend to distort due to the considerable expansion of the metal caused by solar energy. The solar heating of the outer lining will not be transferred to the panel due to the insulation core, and the distortion of the outer surface will show obvious, unsightly rippling.

A profiled outer lining, which will also distort due to solar heating, will not show any signs of rippling due to the profiles which will take up and mask the effects of distortion. The principal advantage of profiled outer linings is that the depth of the profiles will give stiffness to the panel against bending between supports. The profiled laminated metal panel, illustrated in Figure 7.43, is made in lengths of up to 10 m for fixing with self-tapping screws to holes in horizontal sheeting rails. The vertical edges of sheets overlap as a weather seal.

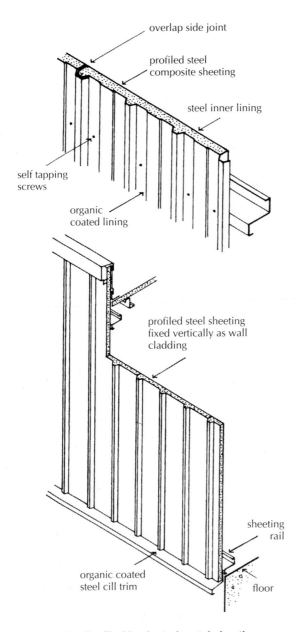

Figure 7.43   Profiled laminated metal sheeting.

The insulating core is of polyisocyanurate foam. Both inner and outer steel strip linings are galvanised, and both inner and outer linings of both steel and aluminium strip are usually coated with a coloured organic coating for protection and decoration. The difficulty of making a neat, weather tight and attractive finish to the ends of the one-direction profile panels at eaves, ground level and at corners and around openings limits their use to simple, shed forms of building.

Single skin panels
A single strip of steel or aluminium, by itself, has limited stiffness and the thin edges of the metal provide poor means of making a weather tight joint to surrounding framing or other panels. Single skin panels of metal strip have been most used as an outer face of the rain screen form of cladding. A panel of metal strip from 3 to 6 mm thick is hung as an outer screen to rain in front of the main cladding system, as illustrated later in Figure 7.48. To avoid obvious distortion of flat panels, due to differential thermal movement of the panel relative to the supporting frame, the flat sheet is hung on cleats fixed to the back of the panel, which hook over cleats fixed to the carrier frame, as illustrated in Figure 7.48. This allows sufficient movement of the thick, flat panel to avoid obvious rippling of the surface.

Single skin panels of hot rolled metal strip are usually stiffened by forming the strip into a shallow pan. The flanged edges of the pan provide a surface for joints to surrounding panels. The process of forming sheet metal into a shallow pan is a comparatively simple and inexpensive process. The corners of the metal are cut, the edges cold bent using a brake press and the corners welded. To provide additional stiffness to the shallow pan, and as a means of fixing to a carrier frame, it is common to weld a supporting frame of angles or channels to the back of the pan, as illustrated in Figure 7.44A. These single skin panels are fixed to the structural frame or to a separate carrier system fixed to the structural frame with insulation fixed to the back of the panels.

Single skin panels can be pressed to profiles, as illustrated in Figure 7.45, and fixed as a rain screen or formed as window panels, as illustrated in Figure 7.44B. The flanged, rounded edges for the panel and the window opening provide sufficient stiffness without the use of a supporting frame. The flanged edges and rounded corners are formed by drawing or pressing. This two-dimensional operation is considerably more expensive than one-way cold rolling.

Pressing of strip metal in one operation is performed by pressing around a shaped former or by deep drawing or a vacuum press where the strip is pressed or drawn to shape around a former. These one-off processes are generally limited to panels of 2 by 1 m for pressing and up to 5 by 2 m for drawing. The deep drawn, storey height aluminium panels, illustrated in Figure 7.45, are formed as window units that are fixed to an outer carrier frame. The outer carrier is connected to the inner carrier by plastic, thermal break fixings. A gasket provides a weather seal between outer and inner carriers at open joints.

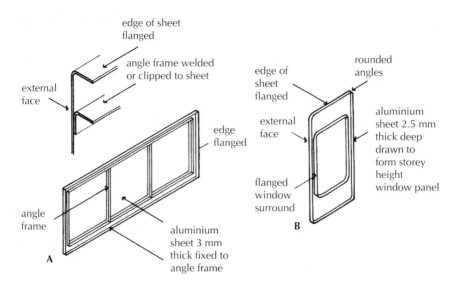

Figure 7.44   A, Single skin panel with frame. B, Profiled single skin panel.

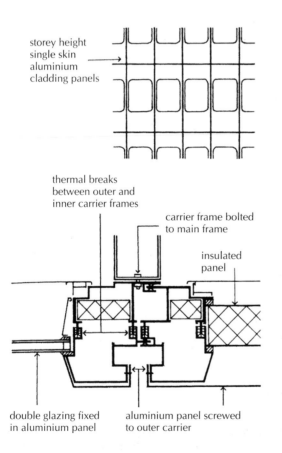

Figure 7.45   Single skin aluminium cladding panels.

The inner carrier, which is bolted to the structural frame, supports an insulated box panel system as inner lining and internal finish. The open joints and the separation of the outer carrier from the inner carrier allow some unrestrained thermal movement of the cladding panels to limit possible distortion due to solar heating. The complexities of the outer and inner carriers supported by a structural frame are necessary for the precision engineering skills required for this system of cladding.

Box panels

Box panels are made from two single panels with flanged edges formed around a core of insulation as a box, as illustrated in Figure 7.46. The flanged edges of the inner and outer panels are pop-riveted together around a Neoprene strip. The Neoprene strip acts as a thermal break, which allows a degree of movement of the outer panel to that of the inner panel.

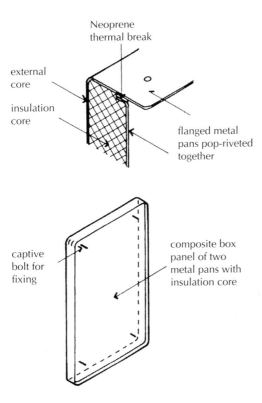

Figure 7.46   Metal box cladding panel.

As an alternative to riveting the flanged edges of metal panels to form a box, the flanged, rebated edge of one panel may be bonded to an edging piece of wood or plastic, which is bonded to the flat edge of the other panel. Both metal panels are bonded to an insulation core. The advantage of box panels is that an inner and outer lining can incorporate an insulation core in one prefabricated unit ready for site fixing, with the metal faces ready prepared as an inner and outer finish as required. Box panels are much used as an inner lining, to provide both insulation and an internal finish behind an outer sheet metal lining, and to provide the main weather and decorative finish. As an inner lining, the box panel is unlikely to suffer distortion of the faces of panels due to differential thermal movement. Where box panels are used as an external cladding there is a likelihood of the outer lining suffering surface distortion due to thermal expansion of the outer face of the panel. This likely distortion may be limited by the use of comparatively small panels and masked by profiling the outer metal skin. The aluminium strip box panels illustrated in Figure 7.47 were specifically designed for this building, which is a notable example of the integration of the components of a building.

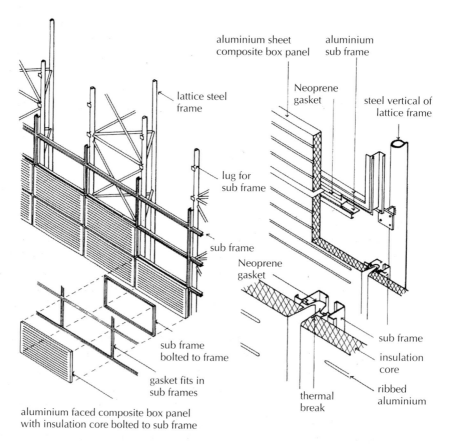

Figure 7.47   Composite box panel cladding.

The ribbed, anodised, aluminium panels were made by vacuum forming. The outer tray was filled with Phenelux foam, and the inner tray then fitted and pop-riveted to the outer tray through a thermal break.

Separate aluminium subframes for each panel are bolted to lugs on the structural frame. Continuous Neoprene gaskets seal the open drained joints between panels, which are screwed to subframes with stainless steel screws. The horizontally ribbed face of the outer skin of the box panels serves the purpose of masking any distortion that may occur, provides some stiffening to the aluminium strip and provides a decorative finish.

Rain screens

The term 'rain screen' has been used to describe the use of an outer panel as a screen to an inner system of insulation and lining so arranged that there is a space between the screen and the outer lining for ventilation and pressure equalisation. The open joints around the rain screen, illustrated in Figure 7.48, allow some equalisation of air pressure between the outer and inner surfaces of the rain screen, which provides relief of pressure of wind-driven rain on the joints of the outer lining behind the rain screen. Another advantage of the rain screen is that it will protect the outer lining system from excessive heating by solar radiation and protect gaskets from the hardening effect of direct sunlight.

All wall panel systems are vulnerable to penetration by rain, which is blown by the considerable force of wind on the face of the building. The joints between smooth faced panel materials are most vulnerable from the sheets of rainwater that are blown across impermeable surfaces. The concept of pressure equalisation is to provide some open joint or aperture that will allow wind pressure to act each side of the joint and so make it less vulnerable to wind-driven rain. Plainly it is not possible to ensure complete pressure equalisation because of the variability of gusting winds that will cause unpredictable, irregular, rapid changes in pressure. Provided there is an adequate open joint or aperture, there will be some appreciable degree of pressure equalisation, which will reduce the pressure of wind-driven rain on the outer lining behind the rain screen. A fundamental part of the rain screen is airtight seals to the joints of the panels of the outer lining system to prevent wind pressure penetrating the lining. Because of the unpredictable nature of wind-blown rain and the effect of the shape, size and groupings of buildings on wind, it is practice to provide limited air space compartments behind rain screens, with limited openings to control air movements. The profiled, single skin rain screen panels illustrated in Figure 7.49 serve as spandrel panels between window panels to provide some protection to the insulated inner panels. The proliferation of joints between window and solid panels provides more joints vulnerable to rain penetration than there are with solid storey height panels.

Rain screen panel systems have not been as extensively used as might be supposed. One reason is that a rain screen is most effective on walls devoid of

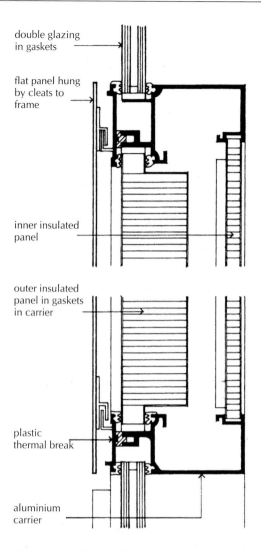

double glazing
in gaskets

flat panel hung
by cleats to
frame

inner insulated
panel

outer insulated
panel in gaskets
in carrier

plastic
thermal break

aluminium
carrier

Figure 7.48   Flat single skin aluminium panel rain screen.

windows and is of minimal advantage when used as spandrel screens between windows. Another reason is that the necessarily more complex and expensive carrier system required for rain screens, as compared to a straightforward panel system, may not seem to be justified by actual relief of pressure of wind-driven rain on joints behind the screen.

Jointing and fixing

Sheet metal panels have been in use as storey height and spandrel panels to wall cladding systems for many years and the jointing and fixing of these panels for use as external cladding has developed from their early use in curtain walling.

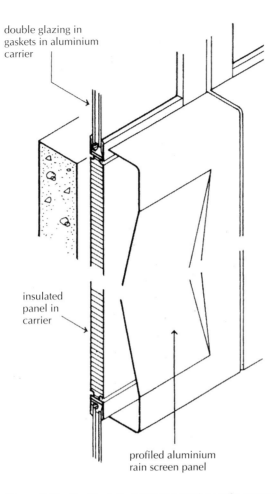

double glazing in
gaskets in aluminium
carrier

insulated
panel in
carrier

profiled aluminium
rain screen panel

Figure 7.49   Profiled aluminium panel as rain screen.

The use of preformed gasket seals in aluminium carrier systems has developed with changes in curtain wall techniques so that today the majority of composite panels are fixed and sealed in Neoprene gaskets fitted to aluminium carrier systems fixed to the structural frame, as illustrated in Figures 7.45, 7.47 and 7.49, in which the carrier system supports the panels and the gaskets serve as a weather seal and accommodate differential thermal movements between the panels and the carrier system. To reduce the effect of the thermal bridges made by the metal carrier at joints, systems of plastic thermal breaks and insulated cores to carrier frames are used.

Where open horizontal joints are used to emphasise the individual panels, there is a flat or sloping horizontal surface at the top edge of each panel from which rain will drain down the face of panels and in a short time cause irregular and unsightly dirt stains, particularly around the top corners of the panels.

The advantage of the single-storey-height box panel is in one prefabricated panel to serve an outer and inner surface around an insulating core that may be fixed either directly to the structural frame or a carrier system with the least number of complicated joints. To take advantage of this simplicity of fabrication and use, some composite box panels of aluminium strip are formed with interlocking joints formed in the edges of panels. The interlocking joints in panel edges, illustrated in Figure 7.50, comprise flanged edges of metal facing, formed to interlock as a male and female locking joint to all four edges of each panel. The edge of one panel is formed by a pressed edging piece, welted to the linings and formed around a plastic insert to minimise the cold bridge effect at the joint. This protruding section fits into the space between the wings of the linings of the adjacent panel with a Neoprene gasket to form a weather seal.

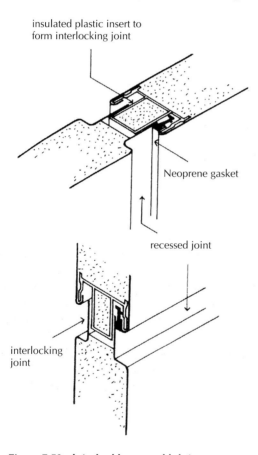

insulated plastic insert to
form interlocking joint

Neoprene gasket

recessed joint

interlocking
joint

Figure 7.50   Interlocking panel joint.

For this joint to be effective, a degree of precision in the fabrication and skill in assembly is required for the system to be reasonably wind and weather tight. As with all panel systems, the junction of horizontal and vertical joints is most vulnerable to rain penetration and requires precision manufacture and care in assembly. The interlocking joint system, illustrated in Figure 7.51, is used for flat, strip steel panels, which are used principally as undercill panels between windows or as flat panels without windows.

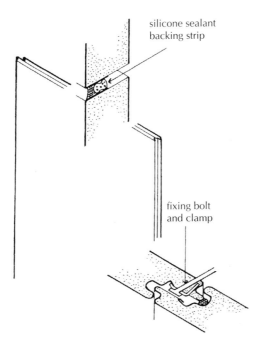

silicone sealant
backing strip

fixing bolt
and clamp

Figure 7.51   Interlocking panel joint

The vertical edges of the strip metal are cold roll formed to a somewhat complicated profile with the edges of the inner and outer linings either welted or pop-riveted together. The interlocking joint is designed to hide the fixing bolt and clamp, which connect to the edge of one panel and are bolted back to the carrier or structural frame. A protruding rib on the edge of one panel is compressed on to a Neoprene gasket on the edge of the adjacent panel. Horizontal joints between panels are made with a polyethylene backing strip and silicone sealant. These strip steel panels are made in widths of 900 mm, lengths of up to 10 m and thickness of 50 mm. The steel linings are galvanised and finished with coloured inorganic coatings externally and painted internally.

7.8 Glazed wall systems

Up to the beginning of the 20th century, glass was a comparatively expensive material. Window glass was made by hand in the spun, crown glass process and later the blown cylinder process. Window glass made by these processes was cut into comparatively small squares (panes) for use in the windows of traditional loadbearing walls. Plate glass was made by casting, rolling, grinding and polishing sheets of glass both sides. These laborious methods of production severely limited the use of glass in buildings. With the development of a continuous process of drawing window glass in 1914 and a process of continuously rolling, grinding and polishing plate glass in the 1920s and 1930s, there was a plentiful supply of cheap window glass, and rolled and polished plate glass. In the 1920s and 1930s window glass was extensively used in large areas of windows framed in slender steel sections as continuous horizontal features between undercill panels and as large metal-framed windows. During the same period, rolled plate glass was extensively used in rooflights to factories, the glass being supported by glazing bars fixed down the slope of roofs. Many of the sections of glazing bar that were developed for use in rooflights were covered by patents so that roof glazing came to be known as 'patent glazing' or 'patent roof glazing'.

Curtain walling
The early uses of glass as a wall facing and cladding material were developed from metal window glazing techniques or by the adaptation of patent roof glazing to vertical surfaces, so that the origins of what came to be known as 'curtain walling' were metal windows and patent roof glazing. The early window wall systems, based on steel window construction, lost favour principally because of the rapid and progressive rusting of the unprotected steel sections that in a few years made this system unserviceable. The considerable buckling and distortion of frames and fracture of glass that was due to rusting, rigid putty fixing of glass and rigid fixing of framing gave steel window wall systems a bad name.

With the introduction of zinc coated steel window sections and the use of aluminium window sections there was renewed interest in metal window glazing techniques. Cold formed and pressed metal box section subframes, which were used to provide a bold frame to the slender section of metal windows, were adapted for use as mullions to glazed wall systems based on metal window glazing techniques. These hollow box sections were used either as mullions for mastic and bead glazing of glass and metal windows or as clip on or screw on cover sections to the metal glazing. Hollow box section mullions were either formed in one section as a continuous vertical member, to which metal window sections and glass were fixed, or as split section mullions and transoms in the form of metal windows with hollow metal subframes that were connected on site to form split mullions and transoms. The complication of joining the many sections necessary for this form of window panel wall system and the attendant difficulties of making weather tight seals to the many joints has, by and large, led to the abandonment of window wall glazing systems. Glass for rooflights fixed in the slope of roofs is to a large extent held in place by its weight on the glazing bars and secured with end stops and clips, beads or cappings against wind uplift. The bearing of glass on the glazing bars and the overlap of bays down the slope act as an adequate weather seal.

To adapt patent roof glazing systems to vertical glazed walls it was necessary to provide a positive seal to the glass to keep it in place and against wind suction, to support the weight of the glass by means of end stops or horizontal transoms and cills, and to make a weather tight seal at horizontal joints. The traditional metal roof glazing bar generally took the form of an inverted T section, with the tail of the T vertical for strength in carrying loads between points of support with the two wings of the T supporting glass. For use in vertical wall glazing it was often practice to fix the glazing bars with the tail of the T, inside with a compression seal and on the outside holding the glass in place, as illustrated in Figure 7.52.

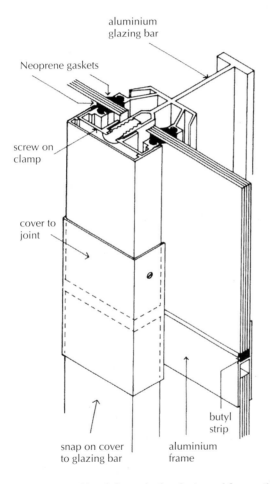

Figure 7.52   Aluminium glazing but used for vertical glazing.

The usual section of metal glazing bar, which is well suited to roof glazing, did not provide a simple, positive fixing for the horizontal transoms and cills necessary for vertical glazing systems. The solution was to use continuous horizontal flashings on to which the upper bays of glass bore and up to which the lower bays were fitted, as illustrated in Figure 7.52. Patent roof glazing techniques, adapted for use as vertical glazing, are still in use but have by and large been superseded by extruded hollow box section mullion systems.

Hollow box section mullions were designed specifically for glass curtain walling. These mullion sections provided the strong vertical emphasis to the framing of curtain walling that was in vogue in the 1950s and 1960s, and the hollow or open section transoms with a ready means of jointing and support for glass. The pattern of what came to be known as curtain walling was set by the United Nations Secretariat Building and Lever House in New York, in which the framing elements of slender vertical mullions supporting glass and smooth panels were emphasised by mullions as continuous verticals up the height of the building. Hollow box section mullions, transoms and cills were generally of extruded aluminium, with the section of the mullion exposed for appearance sake and the transom, cill and head joined to mullions with spigot and socket joints, as illustrated in Figure 7.53. A range of mullion sections was available to

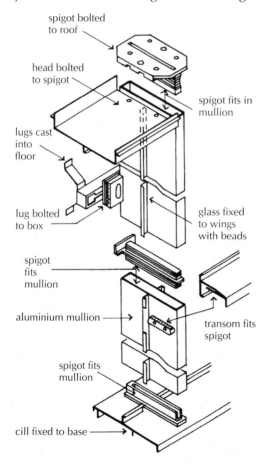

Figure 7.53   Aluminium curtain walling.

cater for various spans between supporting floors and various wind loads. The mullions, usually fixed at about 1 to 1.5 m centres, were secured to the structure at each floor level and mullion lengths joined with spigot joints, as illustrated in Figure 7.53. The spigot joints between mullions and mullions, and between mullions and transoms, head and cill, made allowance for thermal movement and the fixing of mullion to frame made allowance for differential structural, thermal and moisture movements. Screw on or clip on beads with mastic or gasket sealants held the glass in place and acted as a weather seal. This form of curtain walling with exposed mullions was the fashion during the 1950s, 1960s and early 1970s.

Stick system of curtain walling

This, the earliest and for many subsequent years the traditional form of glass curtain walling, has been typified as the 'stick system'. Typical of the stick system is the regular grid of continuous mullions, bolted to the structural frame, with short discontinuous transoms as a regular grid into which panes of glass are fitted and secured. The advantage of this system is the use of a small range of standard aluminium sections that can be cut to length and joined with spigot joints as a carrier frame for glass panes. The carrier system is secured to the floors of the structural frame with bolts and plates that allow for some small relative movement between structure and carrier frame. Initially the glass was secured in the carrier with sprung, clip on or screw on aluminium clips and later by gaskets that were compressed up to or both sides of the glass to secure it in place and act as a weather seal. While the very many joints between the members of the carrier frame and glass make allowance for structural and thermal movements, they are also potential points for penetration of wind and rain, particularly at corners where clips and gaskets are mitred to fit.

In the early forms of glass curtain walling, the section of the hollow, extruded aluminium mullions and transoms was exposed on the face of the walling. The squares of glass were held in position against wind pressure and suction by sprung metal clips that were fitted into wings on the mullions and transoms and bore against the face of the glass. These sprung clips were adequate to hold the glass in place but did not provide a wind and watertight seal, particularly at corners where the beads were mitre cut to fit. To show the least exposure of the aluminium carrier sections on the face of the glass walling, for appearance sake and as a means of fixing Neoprene gaskets to provide a more positive weather seal, the extruded hollow box section mullions and transoms, illustrated in Figure 7.54, were used.

The main body of the aluminium carrier was fixed internally behind the glazing with continuous head and cill sections, continuous mullions with transoms fitted to the side of mullions. The mullion and transom sections were fitted over cast aluminium location blocks screwed to the frame with a mastic seal or Neoprene gasket to the bottom of each mullion. The advantage of the joints between the vertical and horizontal carrier sections is that there is allowance for some

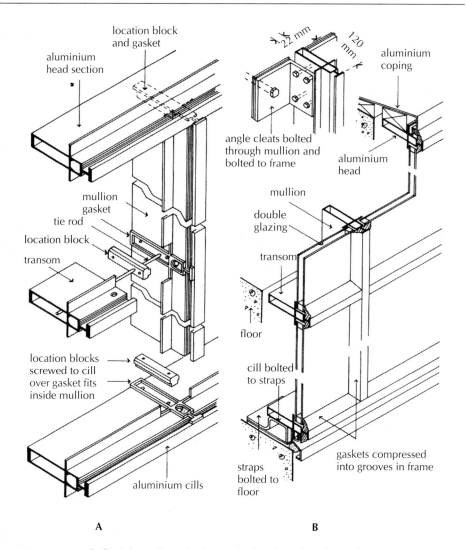

Figure 7.54   A, Curtain wall carrier frame. B, Double glazed panels.

structural and thermal movement. To provide the least section of carrier frame externally, the section outside the glass is the least necessary for bedding glass and for Neoprene gaskets that fit into the serrated faces of a groove. A slim section of aluminium and the gaskets are all that show externally. This system of extruded, hollow section aluminium carriers is designed to support the whole of the weight of the glazing and wind pressure and suction in the position of exposure in which the building is erected. The extruded aluminium carrier system is fixed to the structure by angle cleats, which are bolted together through the mullions and bolted to the structure at each floor level, as illustrated in Figure 7.54B.

The sealed double glazing units are fixed to the carrier frame with distance, setting and location blocks. Neoprene gaskets, mitre cut at corners, are compressed into the serrated edged groove in the carrier frame and up against the double glazed units. This gasket glazing system effectively secures the glass in position against wind pressure and suction, and acts as a weather seal. The slender section of the aluminium carrier frame, mullions and transoms that show on the external face provide the illusion of a glass wall.

With the traditional stick system of curtain walling, the aluminium members of the carrier frame are assembled and fixed to the structure on site, and the glass panes are glazed into rebates and weather sealed with gaskets which secure the glass in place and provide restraint against the considerable wind forces acting on the facade of multi-storey buildings. In this system of glazed wall cladding it is generally economic to use comparatively closely spaced mullions, to support panes of glass, from 600 mm to 1 m centres to use the least section of mullion and thickness of glass to support the weight of glass and the wind forces acting on the glass. The site glazing with mastic tape or gaskets used in this system may not provide long-term protection to the edge seals of double glazing units, which are vulnerable to decay due to the penetration of water, particularly at the bottom corners of glass.

Unitised or panel system of curtain walling

An alternative to the stick system is the unitised or panel system of curtain walling in which complete panels of an aluminium frame with glazing are fabricated as units ready for hoisting into position for fixing to the structural frame or to a carrier system. The advantage of this system is the facility of precision assembly of the components and glazing under cover in the conditions most favourable for successful glazing, particularly for insulating glass units, to provide the most effective protection to edge seals. By virtue of repetitive, precision assembly, large glazed panels may be fabricated at reasonable cost for fixing to the structural frame with narrow weather sealed joints between panels, the maximum area of glass and least exposed area of framing. In effect, this is a system of large, dead light window frames, ready glazed for fixing as a glazed wall system either between floors or as a curtain wall. The members of the frame are designed to show the least area on face necessary for satisfactory support, bedding and weather sealing of glass, and adequate metal section and depth to support the weight of glass and wind forces acting on the panel where it is fixed and restrained at its top and bottom edges to structural floors. The edges of the frames may be shaped for gasket glazing compressed into rebates by metal strips bolted from behind or may be shaped so that the long edges of panels interlock and the horizontal joints overlap over gasket seals.

Large panels may be fixed to a carrier frame, which is fixed to the structural frame. The carrier frame is designed to support the dead and live loads of the panels, provide a background for gasket or sealant joints and assist in the alignment of the panels across the face of the building. Figure 7.55 is an illustration of a glazed curtain wall in which panels of double glazed units are fixed to a system of aluminium mullions and transoms fixed to the structural frame at each floor level. The spacer bars at the edges of the double glazed units are shaped so that the two panes of glass can be secured to the wings of the spacer bar with adhesive silicone. The silicone acts as a powerful, long-term adhesive. Photograph 7.1 shows glass cladding on a building.

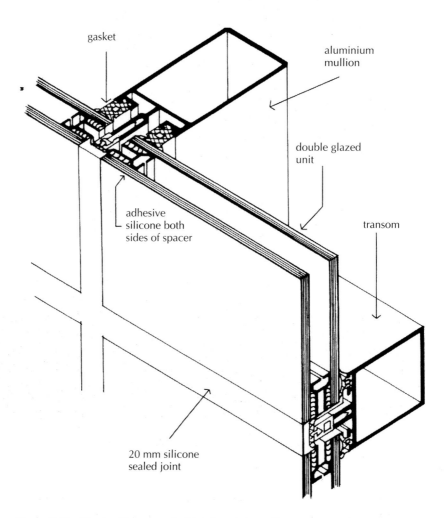

Figure 7.55   Flush silicone sealed joint curtain wall.

Photograph 7.1   Glass cladding: flush silicon sealed joint.

The glass panels are secured to the carrier frame with dumb-bells that fit to a bar that is screwed to the mullion, as illustrated in Figure 7.56. The dumb-bells fit into the wings of the spacer to secure the glazed panel at intervals on all four sides to hold the glass panel in position. A gasket fixed to the bar provides a backing on to which the silicone sealant jointing is run between glass panes.

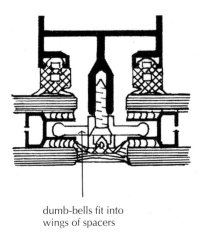

dumb-bells fit into
wings of spacers

Figure 7.56 Fixing glazed panels.

This system of curtain walling is used for the advantage of the flush silicone sealed joints that provide a flush glass external face. This system is much used with coloured glass for the dramatic effect of a large expanse of reflective material. For reasons of security the height of this cladding is limited to about 8 m, unless a system of mechanical retention of the outer panes is used. Mechanical retention takes the form of aluminium angles around the edges of each outer pane. This glazing system is used for small areas of cladding both internally and externally.

The traditional way of supporting and retaining glass has been inside a rebate in a wood, metal or plastic frame. To provide an edge clearance around the pane of glass, inside the rebate to allow for thermal movement, setting and location blocks, non-setting compounds or mastic tape are used. Putty, beads or gaskets around the outside hold the glass in place and serve as weatherseals. This process of glazing is used for both windows and curtain walling. Silicone has for some time been used as a gap and joint filling sealant for its strong, long-term adhesion, excellent weathering characteristics and high resistance to ultraviolet radiation, heat and humidity. The cured silicone, which has a firm, rubber-like consistency is sometimes referred to as silicone rubber as it may suffer some compression or extension and return to its former shape without loss of strength or adhesion. For some years one of these silicones has been used as an adhesive to bond sheet materials such as glass and metal.

Structural glazing sealant
The characteristics of making a powerful bond between glass and metal have been used in the system known as structural glazing. The advantage of this system is that glass may be bonded to the face of a metal frame through a narrow edge strip contact of silicone to the back of glass and the face of a metal frame. The glass is held firmly in place by the silicone, which will transfer a whole or

part of the weight of glass to the frame and the whole of wind forces acting on a glass panel in the face of a building. This comparatively straightforward system of glazing, which needs no spacing, bedding or weathering, has been exploited for the facility of a flush external glass face interrupted only by very narrow open joint or silicone sealant gap filling seals between panes of glass. Figure 7.57 illustrates diagrammatically the simplicity of this system of glazing with panes of glass bonded to a simple aluminium frame, with the joints between panes of glass silicone sealed. Single or double sheets of glass can be supported by an aluminium frame designed for fixing to a carrier system or curtain wall grid secured to the structural frame.

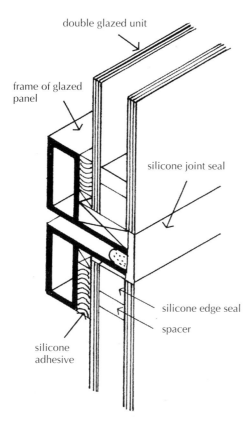

Figure 7.57   Four sided structural sealant.

Because of the adhesion of the silicone, both the weight of the glass and wind forces acting on the glass panel are supported by the metal frame in the diagrammatic illustration of four-sided structural glazing shown in Figure 7.57. Because of the natural resilience of the structural silicone bond, some small thermal movement of glass relative to the frame can be accommodated. Thermal breaks wedged into the frame and bearing on the back of the glass together with the silicone bond will serve to reduce the cold bridge effect at the junction of glass and metal. The gap filling silicone seal between the edges of glass can be run continuously around all joints. The seal is run on to a backer rod of polyethylene and tooled either flush with the glass or with a shallow concave finish. The joints between glass faces should ideally be no more than 20 mm wide.

Large single panes of clear, body or surface tinted, 20 mm heat strengthened or laminated glass may be four sided structural glazed to extruded aluminium frames for fixing to an aluminium carrier system of mullions, supported at each floor level and transoms as necessary to provide additional support for the glass. Structural glazing systems are designed to give a flush facade of glass with the supporting mullions and transoms of the carrier frame just visible through a transparent glass facade. The extent to which the carrier frame is visible will depend on the intensity of light reflected off the surface of glass. For structural silicone to be most effective in bonding glass to the supporting framework, it should be applied in conditions where the cleanliness of the surfaces to be joined, temperature and humidity can be controlled, particularly for safety where large sheets of glass act as the facade of tall buildings. The length, spread and depth of the silicone adhesive runs required depend on the size and weight of the glass pane to be supported and the anticipated wind forces in the position of exposure of the building.

Two-sided structural silicone glazing is used for the two vertical edges of single glass panes which are bonded to a metal frame fixed to mullions of a curtain wall system and supported at their base and restrained at their top edge by aluminium transoms. The wind forces acting on the glass are carried by the framing back to the carrier system, and the weight of the glass by the transoms of the curtain wall. Structural glazing to insulating glass (IG) units may be applied to the inside face of the inner sheet of glass in the form of four-sided glazing bonded to an aluminium panel frame, as illustrated diagrammatically in Figure 7.57. The edge seal to the spacer bars between the glass in the IG units is made with silicone sealant to provide the most effective seal against penetration of water. As there is no means of bonding the outer glass to the frame, the outer glass is held in place by the silicone edge seal. The bond of the edge seal will be insufficient to retain the outer glass firmly in place and the glass would sink under its weight. For this reason, silicone setting blocks and the fin of the panel frame are used in the glazing rebate to provide mechanical support. As a means of forming a narrow, open or silicone sealant joint between insulating glazed units, the edge of the units is formed as a stepped edge. The outer pane of glass projects beyond the inner pane in the form of a step, as illustrated in Figure 7.58, with the panel frame behind the glass edge.

The glazed curtain wall system illustrated in Figure 7.59 is designed to utilise the advantages of precision engineering fabrication to minimise site work and gain full advantage of the application of silicone adhesive, silicone sealant and gaskets in the most favourable conditions.

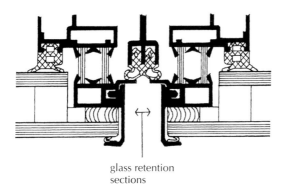

glass retention
sections

Figure 7.58   Glass retention sections.

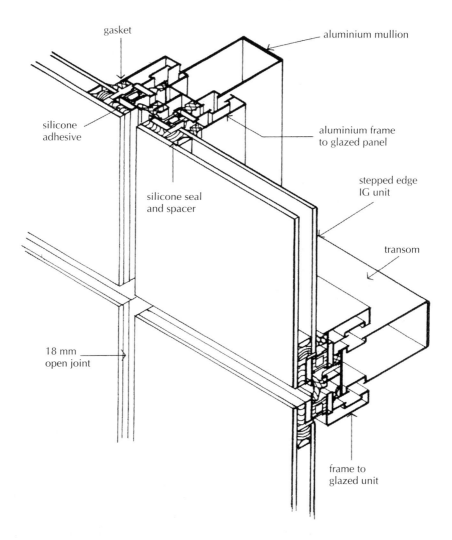

gasket

aluminium mullion

silicone
adhesive

aluminium frame
to glazed panel

stepped edge
IG unit

silicone seal
and spacer

transom

18 mm
open joint

frame to
glazed unit

Figure 7.59   Prefabricated flush face curtain wall.

Aluminium mullions are fixed to the structural frame with aluminium transoms cleat fixed to the mullions. Aluminium frame sections to each glazed panel are fixed to the edges of the outer pane with silicone adhesive, with the panel frame providing mechanical support to the bottom edge of the IG unit. Clips set into channels in the glazed panel frame and the carrier frame secure the glazed panel in position. EPDM gaskets set into grooves provide a seal between the glazed panel frame and the carrier mullions and transoms, and serve as thermal breaks. An 18 mm open joint is formed between the glazed panels, backed by a gasket which is set into fins on the mullion and bearing on the panel frames as a weather seal. As a precaution against the possibility of the outer panes of the IG units becoming dislodged by wind pressure, an aluminium glass retention section is fixed around the four edges of each pane and clipped back to the frame, as illustrated in Figure 7.58. These glass retention sections provide some mechanical restraint, particularly for the large panels of glass often used in multi-storey buildings. This sophisticated, precision system of glazed curtain walling is most used for large areas of glazed cladding where the considerable cost of prefabrication and fixing is justified by the durability and appearance of the finished result.

Suspended frameless glazing

Suspended glazing is a system of supporting large panes of glass by bolts secured to brackets fixed to glass fins or to an independent frame, without any framing whatever around the glass panes and with narrow joints between panes, gap filled with silicone sealant to give the appearance of a flush glass face. A more exact description would be suspended, frameless glazing. The principal use of this system is as a screen of glass as a weather envelope to large, enclosed spaces such as sports stadia, conference halls, exhibition centres and showrooms, where a clear view of activities inside or outside of the enclosure is of advantage. Frameless glass is also used in airports; whilst allowing passengers to watch the aeroplanes, the increased vision improves opportunity for surveillance and improves airport security. The prime function of this use of glass is not the admission of daylight or as an efficient weather shield.

Suspended glazing has also been used as a glass wall screen to a variety of buildings and entrances where little or nothing can be seen of outside or inside activities to justify the screen and the glass enclosure, and its visible supporting frame is used for the sake of appearance. Suspended glazing is usually vertical and limited to a height of about 20 m. Sloping, suspended glazing has been used as a wall screen and for roofs largely for effect rather than utilitarian purposes.

Suspended glazing depends on the use of stainless steel bolts that pass through holes drilled in the glass to connect to stainless steel plates that are fixed to glass fins or an independent framework. Holes are drilled near each corner of a pane far enough from edges to leave sufficient glass around holes to bear the weight of the glass and resist shear stresses. Plastic washers, fitted to the accurately drilled holes, provide bearing for the bolts, prevent damage to cut glass edges and make some little allowance for thermal movement. Bolts are screwed to stainless steel plates that are fixed to glass fins or a supporting metal frame. One, two or four back plates are used at the junction of four glass panes with fibre gaskets between the surfaces of plate and glass. Sleeves around bolts at connections to plates accommodate some small rotational movement. For appearance sake the external head of supporting bolts may be countersunk headed to fit to a washer in the counter sinking of the glass for a near flush or flush fit. Figure 7.60 is a diagram of the connection of a bolt to glass and plate.

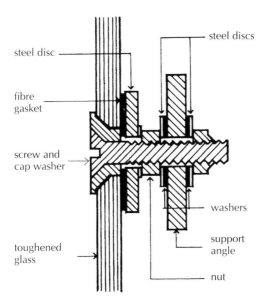

Figure 7.60   Bolt support for suspended glazing.

For the safety of those inside and outside a building either heat toughened or heat treated laminated glass is used for suspended glazing so that, in the event of a breakage, the glass shatters to small fragments least likely to cause injury. Toughened glass is made in maximum sizes of 1500 × 2600 mm and 1800 × 3600 mm and thicknesses of 10, 12 or 15 mm. Laminated glass is made from 4 or 6 mm glass for the thinner layer, and 10 or 12 mm for the thicker layer, in maximum sizes of 2000 × 3500 mm in glass thickness combinations of 4 or 6 mm with 10 or 12 mm with a 2 mm thick interlayer of polyvinylbutyral (PVB).

Insulating units of glass may be used for double glazed suspended glazing which is hung from bolts through both the inner and outer panes of glass with a clear plastic boss around the bolts to act as a spacer in the cavity. The usual combinations of glass are an inner pane of toughened glass 6 mm thick, a 16 mm air space and an outer pane of 10 or 12 mm thick glass. The edges of the insulating glass units are made with spacer bars and a silicone edge sealant. Holes for bolts are normally 60 mm from edges at corners. The maximum size of glass used is 2000 × 3500 mm for 6 mm glass. Both body and surface tinted glass may be used.

Large sheets of glass are suspended on bolts to produce the maximum area of unobstructed glazed screen for the benefit of a wide, clear view in and out of buildings. Where there is a conventional structural frame with floors at regular intervals of, say, 3 m it would seem logical to provide support for the glass panes at each floor level, with the glass close to the edge of the floors with internal guard rails at normal cill level for security. This practical solution would give little more sense of an unobstructed view than a between-floors glass window wall. It is practice, therefore, to hang the glazed screen some distance from the edges of floors and provide some separate, independent system of support and stiffening to the screen against wind forces to emphasise the effect of a flush, frameless screen of large panes of glass. Any independent system of support for suspended glazing has to provide support for each pane of glass and restraint against wind forces. Two systems of support are used. In the first, glass fins are fixed internally at right angles to the screen and bolted to angle plates bolted to the glass screen. In the second, lattice metal frames are fixed between structural floors and roofs, and bolted to the glazed screen at the vertical joints between glass panes with the frames projecting into the building. With both systems the glass fins and the frames are designed to provide support for each glass pane of the screen and resistance to wind forces that will tend to force the screen to bow out of the vertical plane.

Glass fin support

Of the two systems of support used for suspended glazing, glass fins fixed at right angles to the screen is the least obtrusive and provides the least visual barrier to a wide, clear view. The glass fins are the least width necessary to support the weight of the glass screen and anticipated wind loads in the position of exposure. Each fin is the same height as the glass panes it supports, and is bolted through stainless steel plates to the fin above and below and the corners of the four glass panes supported, as illustrated in Figure 7.61. For light loads separate small plates may be used to join fins and fins, to the glass screen. For heavier loads two plates are used to join fins, and two smaller plates to join the main panes of glass.

For small glazed screens the glass fins may be used to the top and bottom panes, or the top two panes of glass with the top and bottom panes bolted to the floor and roof. For larger and heavier screens the fins will usually extend the full height of the glazed screen, as illustrated in Figure 7.62. Whichever system is selected will be chosen as being the least visually intrusive compatible with adequate strength to give support.

Fins are usually cut from 19 mm thick toughened glass, which is holed for bolts and fixed with stainless steel plates over 1 mm thick fibre gaskets each side of the glass fin. Both single and double glazing may be used for the panes of glass to the screen for the fin system of support. The usually accepted maximum height for fin supported glazing is 10 m.

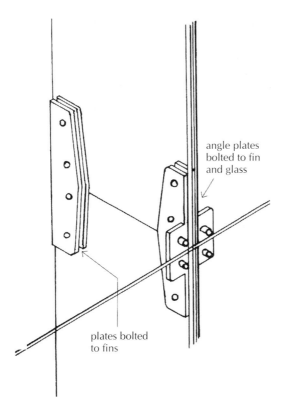

Figure 7.61   Glass fins.

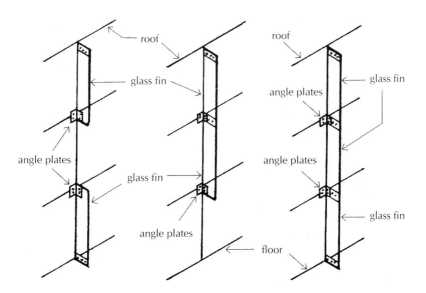

Figure 7.62   Glass fin support for suspended glazing.

Framed support

Glass fin support for frameless, suspended glazing, which is limited to a height of about 10 m, is used generally for sports stadia, and support to glazing hung as a screen to conventional framed structures where a clear, unobstructed view is critical. For large enclosures, suspended glazed screens are supported by systems of lattice steel frames and tensioned cable rigging fixed between floor and roof or to an independent steel frame used to support both wall and roof glazing to single-storey enclosures where a more sturdy system of support is necessary. Various single cell enclosures have been constructed with lattice framed supports and tensioned cable stays to support clear glass suspended glazing for both the wall and roof, to the extent that the glass acts more as a showcase for the complicated system of frames rather than another purpose. The frame and tension cable supports may be used separately or in combination. The most straightforward system of support is by lattice steel frames anchored between floor and roof level to a structural frame or to a separate frame at the junction of walls and roof. Each frame provides support to the glazed screen at the junction of four large panes of glass. Figure 7.63 and Photograph 7.2 show typical lattice frames.

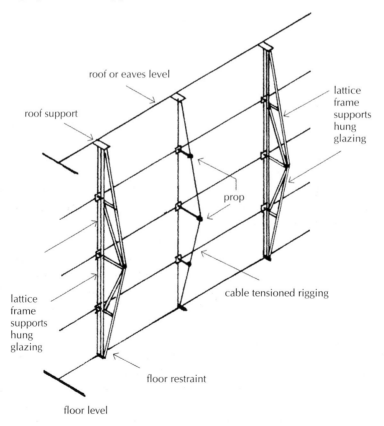

Figure 7.63   Lattice frame support for suspended glazing.

Fixing plate with four
bolts holds glazing in
position

Latice frame supports
hung glazing

Photograph 7.2   Suspended glazing systems with lattice frame.

The frames are fabricated from small steel sections welded together with stays and props to support rectangular panes of glass. The panes may be hung on one long edge to provide the maximum practical width between frames for appearance sake.

As an alternative a system of primary lattice steel frames and secondary tension cable rigging may be used in alternate vertical supports to suspended glass, as illustrated in Figure 7.63. In this secondary rigging the cable is tensioned between floor and roof across the solid props that are bolted to the glass at the junction of four panes of glass, which act with the rigging as a tensioned truss. This tensioned rigging is used as a less obvious and unobtrusive means of support, which provides some support for the glass and more particularly as resistance to wind pressure acting on the glass screen. Lattice frame and secondary rigging systems depend on anchorages to the floor and a structural roof for support, with the top panes of glass being supported by the roof and the bottom edge of lower panes by restraint from the floor.

Where suspended glass is used as an enclosure to walls and roof, a structural steel frame is used to support the whole of the weight of glass and wind loads. Various systems of light section vertical, horizontal and sloping lattice frames are used together with tensioned cable rigging systems for stability and effect. A variety of plates are used for bolting glass, the most straightforward of which is one plate shaped to take the bolts at the junction of four panes of glass. Figure 7.64 is an illustration of a comparatively simple system of suspended glass

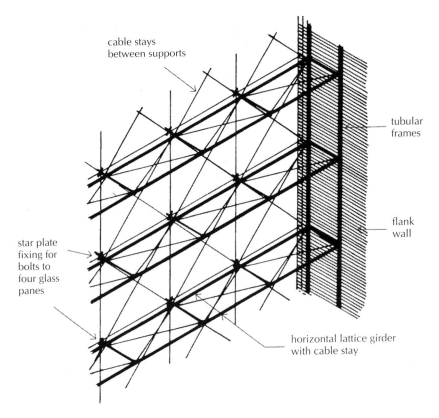

cable stays
between supports

tubular
frames

flank
wall

star plate
fixing for
bolts to
four glass
panes

horizontal lattice girder
with cable stay

Figure 7.64   Horizontal lattice girder and cable stay supports.

Star fixing plate secures the
edge of four sheets of glass
Tubular steel frame supports
suspended glazing

Photograph 7.3   Suspended glazing systems.

supported by horizontal lattice steel frames fixed between steel portal side frames to give support to the glass. The horizontal lattice frames are braced with tension cables and braced up the face of the glass with cables. Photograph 7.3 shows a suspended glazing system.

7.9 Internal walls and partitions

To limit the spread of fire inside a building it should be subdivided into compartments separated by walls and/or floors of fire resisting construction. The degree of subdivision depends on the use of, and fire load in, the building. The fire load is a measure of the inflammability of the contents. The construction of separating walls is also dependent on the height of the top storey above ground as an indication of the ease of escape from the building. See also Chapter 10.

Compartment walls

Compartment walls should serve, with floors, as a barrier to fire between compartments and be linked to the fire resisting construction of means of escape enclosures. Doors in compartment walls should be of fire resisting construction to match the resistance of the walls. Walls that are common to and separate buildings, including those separating semi-detached and terraced houses and flats, should be constructed as compartment walls. Compartment walls inside framed buildings are usually constructed on beams or as part of the reinforced concrete structure in cross wall and box frame forms of construction. Those walls supported by the frame, which will be non-loadbearing, may be of loadbearing thickness to provide the required fire resistance.

A wall of timber or metal studs at least 44 mm wide at not more than 600 mm centres covered with 12.5 mm plasterboard with all joints taped and filled and a 100 mm thick reinforced concrete wall with minimum 25 mm cover to reinforcement will give 30 minutes' fire protection. A wall of timber or metal studs at least 44 mm wide at not more than 600 mm centres covered with 25 mm plasterboard in two layers with joints staggered and taped and filled, a solid masonry wall, with or without plaster, at least 90 mm thick and a 120 mm thick reinforced concrete wall with a minimum of 25 mm cover to reinforcement will give 60 minutes' fire protection.

Sound transmission

A requirement of the Building Regulations is that a wall that separates dwellings, including houses and flats, from a dwelling from another building shall resist the transmission of airborne sound. Airborne sound is generated as cyclical disturbances of air from, for example, a radio that radiates sound with diminishing intensity with distance from the source. The vibrations of air caused by the sound source will set up vibrations in the enclosing walls and floors which will cause vibrations of air on the opposite side of walls and floors in the form of sound.

The most effective insulation against airborne sound is a dense barrier such as a solid wall, which will absorb much of the energy of the air vibrations before it vibrates significantly. The heavier and more dense the material of the wall, the more effective it is in reducing sound. Some common forms of construction that will provide adequate sound insulation for separating walls are: a 215 mm thick, bonded brick wall of density 1840 kg/m^3, a 215 mm thick, concrete block wall of density 1840 kg/m^3 and a 190 mm thick insitu concrete wall of density 2200 kg/m^3. The comparative requirements for the thickness of separating (compartment) walls for fire resistance and sound insulation are:

	Fire resistance	**Sound insulation**
Solid masonry	90 mm	215 mm
Concrete (insitu)	120 mm	190 mm

Partitions

The word partition is used to describe any non-loadbearing division used to provide visual and/or acoustic privacy in the enclosed space. The traditional partition used in houses and flats was built of thin concrete blocks or timber stud frames covered both sides with plaster or plasterboard. Concrete blocks are usually 60 or 75 mm thick and the stud frames formed with 75 × 50 or 100 × 50 mm timber studs.

Concrete block partitions provide poor acoustic privacy and are not particularly stable when built off timber floors. Because of the drying shrinkage of timber floor joists and the slamming of doors, these partitions are liable to develop cracks in hard plaster surfaces. Door linings should be storey height and fixed to both floors and ceilings to give some stability to the partition. These partitions provide poor support for heavy fittings, such as bookshelves, that are fixed to them.

Timber stud partitions which provide poor acoustic privacy are comparatively stable due to the secure fixing of the head and sole plate between floors. Secure fixing of heavy fittings can be provided by noggings fixed between studs.

For use as partitions, a system of cold formed, galvanised strip steel sections has been extensively used for all types of building. The floor and ceiling sections and noggings are fixed to floors, ceilings and walls, and stud sections fitted between the floor and ceiling sections as fixing and support for tapered edge plasterboards. The joints between the boards are finished with wet plaster finish applied over a reinforcing tape and finished level with the board faces ready for decoration.

Metal stud partitions

Two forms of metal stud partition are available. The first, which is specifically designed for houses and flats, is a standard 75 mm overall thickness to take up minimum floor area and for light duty where the occupants have an incentive to exercise care to avoid damage to the single skin plasterboard faces. The second is supplied in a range of five thicknesses for various floor heights, strength, robustness, and fire and acoustic requirements. The second form of partition comprises lipped channel sections 48, 60, 70 and 146 mm wide to friction fit into floor and ceiling channels 50, 62, 72 and 148 mm wide. Channels are screwed to the floor and ceiling with self-tapping screws and the studs are cut to length to friction fit inside the channels at 600 mm centres, with the studs facing the same way. End studs are screwed to abutting walls.

Door openings are framed with studs each side of the opening. The studs are screwed to the head and base channels. The head of the door opening is framed with a channel section. The door opening is lined with sawn timber framing, which is screwed to the studs as fixing for door linings.

Cut outs are formed in the studs for electrical cables with metal fixing channels screwed across studs for fixing switch and socket outlet boxes. Single and

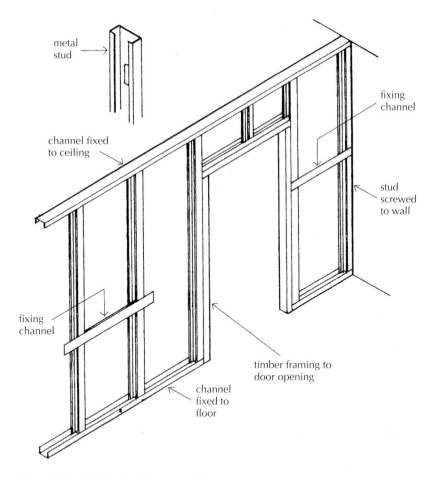

metal stud

channel fixed to ceiling

fixing channel

fixing channel

stud screwed to wall

timber framing to door opening

channel fixed to floor

Figure 7.65   Metal stud frame partition.

double layer plasterboards, usually 12.5 or 15 mm thick, are fixed to the stud framework with self-tapping screws. Tapered edge boards are used for a joint filling finish, and square edged boards for a plaster finish. Figure 7.65 is an illustration of a typical metal stud frame partition. Metal stud partitions may be used as separating walls by the use of two metal stud frames separated by a 100 mm cavity, inside which insulating blanket or bats are fixed and on each side of which single or double layers of plasterboard are fixed to provide fire resistance.

8 Prefabrication and Off-Site Production

Prefabrication is a term used to describe the construction of buildings or building components at a location, usually a factory, remote from the building site. Off-site production is another term widely used to describe the manufacture of a prefabricated building. The manufactured building or building parts are then delivered to the site and assembled in their final position. This method of construction enables a high degree of accuracy (precision) and quality control of the component parts, which are then transported to the site to a precise timetable and erected in position in a clearly defined sequence. To undertake this process effectively and efficiently requires clear design decisions and planning input early in the design process. Component parts need to be accurately designed, as do the joints between them, and attention must be given to fixing and positioning tolerances. On a large, and usually highly repetitive, scale, prefabrication may prove to be a more efficient alternative to more traditional site-based construction methods. For commercial applications the saving in time on the site is an important economic consideration, allowing a faster return on investment and earlier occupation of the building. Improvements in accuracy, quality and safety are other important considerations.

8.1 Terms and concepts

The majority of components that make up buildings are factory produced, e.g. doors, windows, staircases, sanitary ware etc., and are readily available from manufacturers' catalogues of standard products. Construction is essentially a process of assembly, fixing and fitting of manufactured components in a precise location, the building site. Putting these disparate components together in a location remote from the construction site is a logical development but by no means a recent phenomenon, since prefabricated buildings have been used for a long time. The early British settlers to America took prefabricated timber houses with them in the 1620s, and records show that prefabricated buildings of timber were exported from the UK for use in other countries. With the development of cast iron, and in particular the development of prefabricated cast iron components in the 1840s and 1850s, came the development of prefabricated iron buildings, with many houses being shipped to Africa, Australia and the Caribbean. Steel fabrication was developed in the 1930s in America and the UK, and aluminium fabrication followed after the Second World War. Concrete panels were developed during the 1900s and have proved to be popular in some countries (such as Denmark) but have had more limited application in the UK. Advances in the development of lightweight concrete panels and material technologies, such as carbon reinforcement, have helped to keep concrete an effective choice for some developers. The majority of systems currently in use in the UK are based on a framed construction of timber or lightweight steel (Photograph 8.1). Concrete systems are primarily based on loadbearing concrete panels. The main concepts relating to cut timber, lightweight metal and concrete, and the extent of off-site production associated with each technology, is discussed below.

A Timber frame house construction

B Steel frame house construction

Photograph 8.1   Timber and steel framed house construction.

Prefabricated units are usually produced at a location independent of the building site, and the term 'off-site' prefabrication is sometimes used. There are some situations where the prefabrication is undertaken at the construction site and the term 'on-site' prefabrication is used to describe this activity. The extent to which construction activities are moved to a factory (or workshop) setting will vary considerably on the type of prefabrication employed. Some buildings are built on site from factory produced elements while others are delivered to site as complete units, merely craned into position, bolted to the foundation and then 'plugged in' to the services supplies. The primary use of prefabricated units is for the new-build market, although the techniques and methods are equally suited to refurbishment and upgrading projects.

Volumetric construction and modular construction are terms that tend to be used quite interchangeably to describe the process of making large parts of buildings, or entire buildings, in a factory before transporting them to the site, where they are then placed in position. Fully complete modules, including wiring and plumbing, fixtures and fittings, and decoration are built in factories under controlled conditions, then transported to site and positioned by crane on a pre-prepared foundation. Rarely is there any need for scaffolding or on-site storage facilities. When using modular construction, consideration must be given to the sequence of lifting the prefabricated units into the building, and also to safely manoeuvring them into their final position. Clear access must be maintained; thus in situations where scaffolding is required, care should be taken to ensure that the scaffold does not block access.

Supermarkets, hospitals, schools, airports, hotel chains and volume house builders have successfully used modular construction techniques. Volumetric house construction is popular in North America, Scandinavia and Japan, although it has had a rather chequered history in the UK. Following the housing shortage after the Second World War, prefabricated housing was seen as a quick and effective solution to the UK's housing needs, although with the passage of time a whole raft of problems, both technical and social, led to a move away from volumetric modular construction. The Government's report *Rethinking Construction* (1998) led by Sir John Egan identified five drivers for change: a customer focus; a quality-driven agenda; committed leadership; integration of processes and teams around the product; and commitment to people. Off-site production, especially volumetric modular building, has been promoted heavily following the publication of the Egan report as it is seen to be 'compliant' with the aims and objectives in the report. Modular building is one means of helping to achieve efficiency, reduce wastage of materials and deliver improved quality of the finished product. Some specialist commercial applications, such as chains of hotels, supermarkets and fast food outlets, have exploited factors such as time and repetition of a particular style (associated with brand image) particularly well to make prefabrication and modularisation work for their business needs. For commercial applications the slight increase in initial build cost can be offset against savings in time and longer-term savings in the repetition of units.

With a large housing need in the UK, combined with a skills shortage in the building trades, attention has once again turned to prefabricated modular volume houses. Murray Grove in Hackney, developed by the Peabody Trust and designed by Cartwright Pickard as a prototype, represents an innovative example of prefabricated housing (Photograph 8.2). The project made use of Yorkon's standard modules, similar to those used for hotel bedrooms. Each

Murry Grove modular
housing – Peabody Trust

Foundations laid, first module
ready to be craned into
position

Modules lifted from the lorries
and craned into position

All modules are fully furnished
and serviced

Finished building

**Photograph 8.2   Modular housing (courtesy of Yorkon Ltd, www.yorkon.info and
Cartwright Pickard Architects.)**

module has a lightweight steel framing structure. Single bedroom flats comprised two 8 m × 3.2 m modules, and two bedroom flats were made from three modules. The bedrooms and living rooms have the same internal dimensions (5.15 m × 3 m), thus enabling living rooms to be used as an extra bedroom if required. The build cost of this modular development was more expensive than traditional methods; however, the cost benefits will increase when a similar approach is taken on other housing projects. The Peabody Trust has continued its commitment to volumetric construction with an affordable housing scheme at Raines Court. Major economical advantages are achieved with projects that have long runs of identical modules. For smaller developers the speed of construction may not be their primary concern; however, other factors such as more consistent quality and improved working conditions may be determining factors. For some architects and builders, off-site production provides an alternative approach to more traditional construction methods, especially given the increased choice and degree of customisation brought about by advances in IT and greater manufacturing flexibility.

Factory produced systems

Prefabricated foundations systems

Recently there has been an increasing trend to use prefabricated foundations systems. The use of driven piled foundations removes the need for mass excavation of the site: this can save time and expense on some sites (Figures 8.1 and 8.2 and Photograph 8.3). The development of brownfield sites is an example where it may be necessary to limit the amount of disturbance to the ground because of contamination within the ground. Where the risks of the

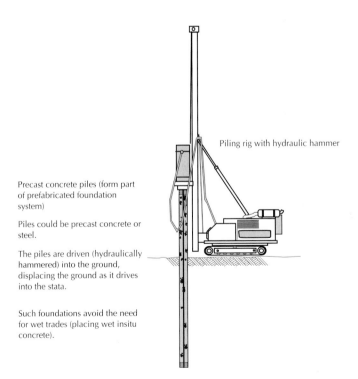

Piling rig with hydraulic hammer

Precast concrete piles (form part of prefabricated foundation system)

Piles could be precast concrete or steel.

The piles are driven (hydraulically hammered) into the ground, displacing the ground as it drives into the stata.

Such foundations avoid the need for wet trades (placing wet insitu concrete).

Figure 8.1   Piling rig with hydraulic hammer driving in concrete piles.

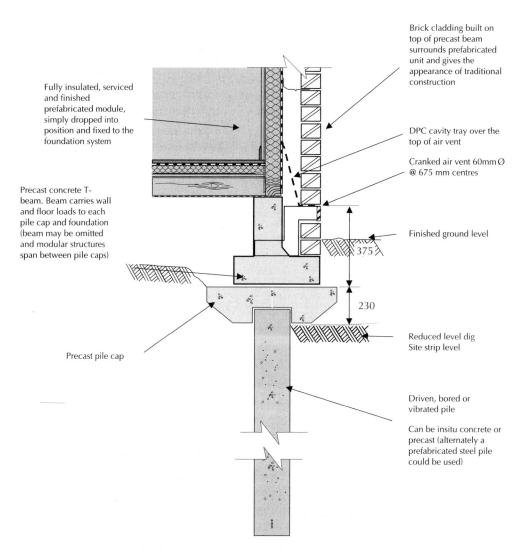

Fully insulated, serviced and finished prefabricated module, simply dropped into position and fixed to the foundation system

Precast concrete T-beam. Beam carries wall and floor loads to each pile cap and foundation (beam may be omitted and modular structures span between pile caps)

Precast pile cap

Brick cladding built on top of precast beam surrounds prefabricated unit and gives the appearance of traditional construction

DPC cavity tray over the top of air vent

Cranked air vent 60mm Ø @ 675 mm centres

Finished ground level

375

230

Reduced level dig Site strip level

Driven, bored or vibrated pile

Can be insitu concrete or precast (alternately a prefabricated steel pile could be used)

Figure 8.2 Prefabricated foundation systems and modular units (adapted from www.roger-bullivant.co.uk).

contamination are low then it may be feasible to leave the ground undisturbed and sealed. The use of driven piles and some bored displacement piles removes the need for soil disposal and excavation that would be needed for traditional foundation systems (Figure 8.1). Prefabricated foundations also reduce the problems associated with working around wet concrete foundations. As soon as the piles are driven into the required set, pile caps positioned and beams craned into position, the prefabricated units can then be delivered and positioned on the foundations (Figure 8.2 and Photograph 8.3). If the modular

Precast concrete or steel piles are driven into the ground. Once the ground has been levelled no further material needs to be excavated

Precast concrete beams are easily craned and placed in position

Extensive foundations can be installed quickly without the need for wet trades or removing excavated material

Photograph 8.3   Prefabricated foundation system (www.roger-bullivant.co.uk).

units are sufficiently strong in their construction, the foundation beams can be omitted, thus the modular units sit on, and span between, the pile caps.

Frames and panels

Prefabricated frames and panels (walls, floors and roof sections) can be fabricated, fitted and finished in the factory before being delivered to site. When designed as flat units they are relatively easy to transport and crane into position (Figure 8.3). The frames are designed so that they can be lifted and manoeuvred into position and can transfer the structural loads of the building. Because of

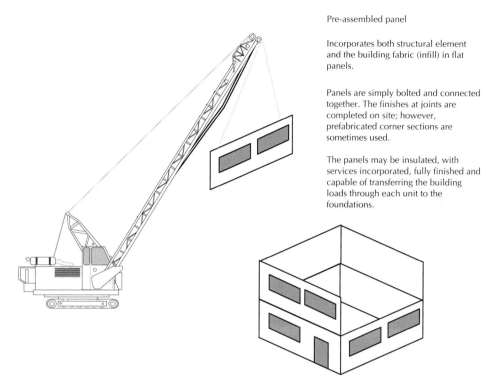

Pre-assembled panel

Incorporates both structural element and the building fabric (infill) in flat panels.

Panels are simply bolted and connected together. The finishes at joints are completed on site; however, prefabricated corner sections are sometimes used.

The panels may be insulated, with services incorporated, fully finished and capable of transferring the building loads through each unit to the foundations.

Figure 8.3   Panel and frame construction (flatpack construction).

this, the structural elements of the units are stronger than would be required for insitu construction. After the units are fitted together, the corner finishing pieces should be attached and the units sealed. Effective fitting of prefabricated units relies heavily on the use of sealants between panels. Prefabricated panels range from simple unfinished panels that make up the structural internal leaf of the external cavity wall to fully finished external wall units. Where the external wall panels are delivered as a finished unit, they will be loadbearing and come complete with services, finishes to the internal and external faces, and windows and doors.

The panels may be delivered with finishes or services already fitted or could comprise just the structural unit, as shown in Photograph 8.4.

Volumetric assemblies

Volumetric assemblies are three-dimensional units; their frame sizes are usually determined by transport, access and the size of the building site. The units can provide complete houses or rooms or can be connected together to make large offices, houses and rooms. As the vast majority of the work is carried out off site, the building can be erected extremely quickly on site. Photographs 8.5,

Standard panels and units are stored
ready for delivery

Panels can be assembled around a
frame or can form selfsupporting
structure

Photograph 8.4   Prefabricated concrete panels (www.roger-bullivant.co.uk).

8.6 and 8.7 show the assembly, from the factory to the finished product, of a £9 million hospital.

Common volumetric units include bedrooms, kitchens, bathrooms and toilet pods (Table 8.1). Volumetric construction is particularly suited where parts of the building are identical and the design relatively repetitive, for example, hotels, student accommodation, commercial offices, prisons, schools, food and retail outlets. Logistics is a major consideration when constructing buildings using large volumetric units. Once manufactured, the units need to be stored off site, labelled so that it is clear where they belong in the assembly and arranged so that they can be delivered in the correct sequence (Figure 8.4). If the finishes are highly sensitive to weather, units may need to be stored internally in controlled environments. To reduce demands on storage, 'just-in-time' manufacturing processes should be adopted. This means that the units needed first are fabricated off site and are completed just before they are required on site (Photographs 8.5, 8.6 and 8.7). Just in time reduces demands on storage; however, any delays at the factory, or during transportation, will result in delays on site. Units are designed with extra structural strength so that they can be transported safely without being damaged. Particularly large units may need to be escorted along roads by the traffic police. Any delays due to transport, e.g. blocked roads and inclement weather, may cause delays on site.

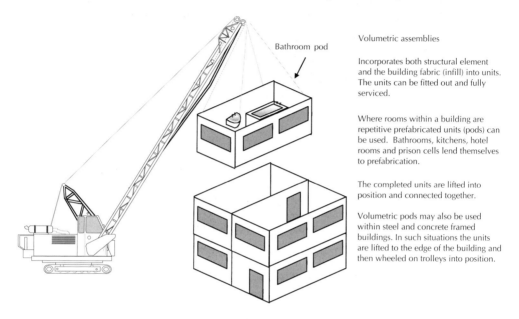

Bathroom pod

Volumetric assemblies

Incorporates both structural element and the building fabric (infill) into units. The units can be fitted out and fully serviced.

Where rooms within a building are repetitive prefabricated units (pods) can be used. Bathrooms, kitchens, hotel rooms and prison cells lend themselves to prefabrication.

The completed units are lifted into position and connected together.

Volumetric pods may also be used within steel and concrete framed buildings. In such situations the units are lifted to the edge of the building and then wheeled on trolleys into position.

Figure 8.4   Volumetric assemblies crane lifting volumetric module.

Modularised building services

Building services is one area of the construction industry that can benefit in a major way from prefabrication. A considerable part of building services is repetitive work. Many components can be pre-assembled and grouped together e.g. horizontal pipework, vertical risers, complex wiring systems, pre-wired and assembled electrical installations (light fittings, switches, heating units etc.). The findings of a study undertaken by BSRIA showed that there were cost saving benefits in excess of 10% to be gained, at every level of the supply chain, through the use of prefabricated and pre-assembled services (Wilson, Smith and Deal, 1999).

Buildings are becoming more reliant on technology and services. Clearly it is beneficial to do as much as possible of the assembly of services off site in clean and controlled environments. While it may not be possible to prefabricate long runs of cables and pipes that have to be fed around the building,

The Yorkon assembly line demonstrates the cleanliness and efficiency of off-site construction. Lifting gear, flat clean floors and the controlled factory environment provides much improved working conditions compared with that of a construction site

Volumetric assembly – fully fitted apartments being assembled in the factory

Roof assembly

The externals walls for the modular apartments are lifted into position

Photograph 8.5   Factory production of modular apartments (www.yorkon.info).

it is possible to assemble the fixtures, fittings and plant. This has led to the development of innovative jointing and fitting systems to ease assembly and future replacement, repair and disassembly work. It is also becoming common to break services down into units that can be delivered as discrete modules. Plant rooms with boilers, air handling units, power terminals and connecting cables and pipework can be made-up off site in structural frames, tested, then transported to site and lifted into place, where they are subsequently commissioned for service (Figure 8.5). Photograph 8.8 shows an example of a standard steel frame structure fitted out with prefabricated bathroom and toilet pods. Table 8.1 provides example of units suited to modular construction.

The site for the hospital has been cleared, the foundations and services installed ready for the delivery of the steel-framed modules.

The modules are delivered to site and craned into position in a controlled sequence. The modules are bolted in position and the services connected.

Within days, the final module is placed in position and the internal finishes are completed. Only the external cladding needs to be applied.

Photograph 8.6   On-site assembly: Bradford Hospital (www.yorkon.info).

The finished hospital looks no different to any other hospital.

The modular construction meant that the on-site construction time was more than halved.

The 4950sqm facility started treating patients just nine months after start on site.

The hospital accommodates three new 28-bed wards and six clean air operating theatres.

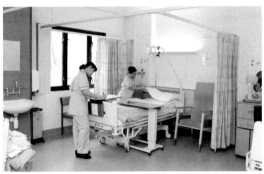

123 steel-framed modules up to 14m in length and weighing up to eight tonnes were craned into position. The hospital is clad in traditional York stone to complement the surrounding local architecture.

Photograph 8.7   Finished product: state-of-the-art hospital (www.yorkon.info).

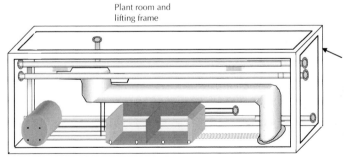

Plant room and lifting frame

Prefabricated services assembly

Structural frame is created so the service or plant unit can be craned and manoeuvred into position.

Often full plant rooms can be prefabricated and lifted into position.

Where possible the services are tested and commissioned off site. Reducing the level of commissioning necessary.

The level of on-site fitting, plumbing and testing is reduced.

Figure 8.5   Modular building services.

Hotel accommodation

The structural frame and floors have been erected.

Volumetric modules have been used for toilet, bathroom and service pods. Each pod is constructed within a steel structural frame, which allows the modules to be craned into position and wheeled into place.

Once in position the external cladding can be fitted.

The finishes, fittings and services in these modules are considerable and would have taken a long time to finish had they been assembled on site.

Photograph 8.8   Hotel accommodation: prefabricated bathroom and service pods.

Table 8.1   Buildings, fittings and assemblies suited to modular construction

Typical modular buildings	Pre-assembled units and modules
Housing	Roof, wall and floor panels
Hotels	Fully constructed roofs: domestic
Service stations	and commercial
Large retailers and food chains	Plant rooms
Prisons	Bathrooms and toilet pods
Commercial offices	Kitchens
Student accommodation	Lifts
Schools, colleges and universities	Electrical switch gear
	Computer server rooms
	Water tank rooms
Prefabricated horizontal and vertical distribution units	**Terminal unit that can be pre-assembled**
Cable management systems – ducting	Light units, luminaires, fittings and
Sprinklers	connecting cables
Pipework	Fan coil units
Plumbing and sanitary	VAV boxes
Rainwater pipework	Radiators
Data and telecommunications networks	Distribution boards
Modular wiring systems	

8.2 Functional requirements

Prefabricated (modular) buildings are no different to those constructed on site in that they must also comply with prevailing building control and associated legislation. Thus the functional requirements of prefabricated buildings are the same as those identified for elements of site-constructed buildings as described in *Barry's Introduction to Construction of Buildings*. The only exception to this is a requirement for increased strength (bracing) of the floor and wall panels to resist the loads imposed on the units during transportation and craning into position. Because prefabricated buildings are factory produced by one manufacturer, it is a little easier to determine the design life and service life of the entire unit. Prefabricated units produced for a commercial use, such as fast food units, are designed and built for a specific (often very short) design life, which is based on the predicted future market for a particular business use. Thus durability may be less of a concern than recycling of the redundant unit and rapid replacement with a new unit that better satisfies the business need. Some attention to routine cleaning and maintenance is still required, as is the ability to undertake repairs and minor alterations should the need arise.

Depending on the methods used, approximately 80–90% of work can be done in the factory. However, site-specific groundwork, construction of foundations and services connections, is still needed and should be completed before the module(s) is brought to site. Careful design of modular systems may help to reduce the size and hence initial cost of foundations, and advances in prefabricated foundation systems may also be used. Similarly, the careful grouping of services can save on pipework and connection costs. It is becoming increasingly common to assemble as much of the services (gas, electrical and water etc.) as possible. Preassembled lighting can reduce the amount of time spent working at height on scaffold and lifting platforms, helping to reduce the amount of overhead work and helping to increase safety and worker wellbeing. Off-site testing and commissioning can help to reduce potential problems, and hence delays, on site.

Skill on site is in managing the sequence of assembly, the craning, and joining the modules together safely. In the majority of cases scaffolding is not required, which is a considerable cost and time saving, while also helping to improve safety. Defects can be dealt with in the factory – the zero defect approach – so there should, in theory at least, be no problems at practical completion. However, defects tend to be associated with fitting units together and to the preprepared foundations and services. Damage to the units during lifting and positioning is also possible.

The recent return to prefabrication in the UK has seen an expansion of manufacturers, each offering bespoke systems, and so the performance of each system, for example for fire resistance, will be specific to one manufacturer.

The decision to use traditional or modular construction will be coloured by a number of factors, most of which are outside the scope of this book.

From a construction perspective it is important that the client sets out clear functional and performance requirements/specifications. Decision criteria (for set performance requirements) may include:

❑ Required quality standard
❑ Expected design life
❑ Time to manufacture the unit(s) off site
❑ Time to assemble the unit(s) on site
❑ Ease of assembly (tolerances)
❑ Extent of finishing required on site
❑ Economic value (whole life costing)
❑ Waste minimisation (both off and on site)
❑ Labour skills and cost
❑ Health and safety considerations
❑ Ease of adaptation once complete
❑ Ease of maintenance, repair and replacement
❑ Spatial flexibility
❑ Aesthetic considerations
❑ Materials choice
❑ Recyclability

All of these listed above have to be offset against the appropriate functional requirements, such as thermal performance, fire safety etc. Factors that should be considered when deciding whether to use prefabrication are identified in the Ishikawa (fishbone) diagram (Figure 8.6).

8.3 Off-site production

Off-site production is a process that incorporates prefabrication and pre-assembly to produce units and/or modules that are then transported to site and positioned to form a permanent work. A number of drivers are behind the desire for more off-site production; the key drivers are:

❑ Skills shortages
❑ Government and industry pressure
❑ Changes in Building Regulations
❑ Client pressure for better buildings

Although different manufacturers adopt different strategies, generically speaking the off-site production process has a number of advantages and disadvantages compared with more familiar approaches to the construction of buildings.

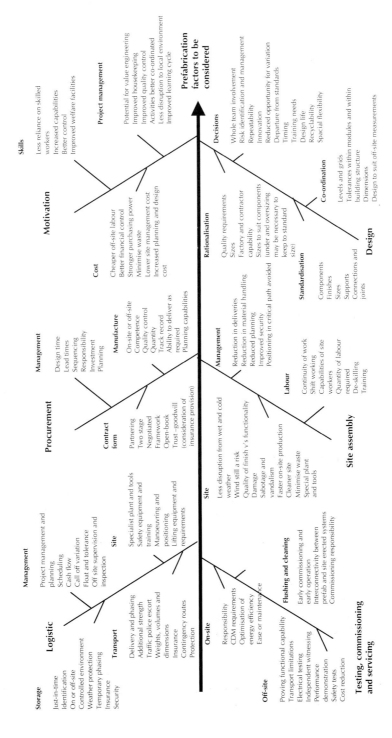

Figure 8.6   Factors to be considered when using prefabrication (adapted from Wilson, Smith and Deal, 1999).

Advantages

There are a large number of reasons why off-site production may be advantageous. Some of the most consistent arguments for moving construction process off site in to the factory are related to the age-old challenge of attaining and maintaining quality. The quality of buildings relies to a large extent on the weather at the time of production, the availability of appropriately skilled personnel to construct the building safely and the control of materials used in the construction of the building.

Control of the working conditions

Operations on-site will be influenced by the weather, with inclement weather leading to disrupted workflows and the possibility of inconsistent quality of work. With over 80% of the production process undertaken in a controlled indoor environment, the construction remains dry during assembly and the flow of work is consistent and easier to control to a specified quality standard. Skilled trades that are affected by weather conditions, such as painting, can be conducted under controlled and consistent conditions, thus helping to ensure better quality work. Control of dust and pollutants during production is easier and there is rarely any need for scaffolding, further helping to improve health and safety factors. Most units are delivered as complete modules and may be positioned on site in a day, an operation that can be done in most weather conditions. There is less reliance on scaffolding and working at height, thus helping to improve the safety of workers on the construction site by reducing their exposure to risk. In research that compared on-site work to prefabricated work, the factory-based workers were found to be more productive even though they worked fewer hours per day and took regular breaks. The factory workers had a better working environment, with more specialised plant and equipment, and were more efficient and reliable than the on-site construction workforce (Wilson, Smith and Deal, 1999).

Skills

There has been a shortage of skilled workers in the British construction sector for some time. Off-site production utilises large-scale equipment and robotic manufacturing, thus reducing the amount of labour required. Worker wellbeing is easier to control in the factory; for example, it is easier to identify and address problems. Poor posture position of workers and heavy lifting can be reduced and difficult tasks can be undertaken by specially designed machinery. Worker skills are applied in the factory, not on the site, which provides the opportunity for the development of skills specific to an allocated function on the production line. Thus development of worker skills may be easier in a factory setting, although this has to be offset against potential boredom of the worker engaged in highly repetitive tasks.

Control of the quality of materials

Given the high volume of production, manufacturers are able to purchase large quantities of materials and are able to demand high quality standards from their suppliers. Materials can be thoroughly inspected at the time of delivery to the factory, and all materials used in the assembly of a unit are known, recorded and traceable. Such demands may be difficult to achieve for small contractors and small developments. It is also easy to check that the customer is getting what they pay for. Such quality control is harder to implement with more traditional construction, quality checks are rarely as rigorous and it is difficult to prove that the contractor and sub-contractors have used the materials specified (indeed it is not uncommon for materials to be substituted for cheaper alternatives during construction, sometimes without the knowledge of those responsible for overseeing the quality of the constructed works). There is also less chance of theft of materials from the site and so site security can be reduced and is needed for a shorter period. Furthermore, the amount of material waste generated on the site should be reduced, if not eliminated. The use of lean production, or lean manufacturing, techniques will also help to eliminate waste during assembly in the factory.

Innovation

The use of prefabricated units should provide a wonderful opportunity to innovate, both in the technologies used, the management of the processes, and the architectural style of the building. While there are a few examples of innovative designs, the current situation appears to be geared to rather familiar and conservative designs, but there is no reason why this cannot be improved as architects, technologists and engineers work closely with manufacturers to explore the potential of off-site production. Other advantages include:

❑ Fast assembly time on site (time is required in the factory) – 'fast-track construction'
❑ Improved cost control and financial certainty
❑ Clear quality standards
❑ Lean production methods can be utilised (less waste)
❑ Warranties are available
❑ Improved safety and worker well-being
❑ Less disruption to neighbouring buildings

Cartwright Pickard Architects and Yorkon have developed some particularly innovative prefabricated structures that are starting to challenge ideas of what a prefabricated building looks like (www.yorkon.info).

Disadvantages

Off-site production may not be the right choice for all clients. There are a number of factors that appear to be disadvantageous.

Historical context and the perception of risk

As mentioned earlier in this chapter the perception of prefabricated buildings is not particularly positive with all members of society; the memory of the failures of the post-war prefabs are difficult to shift. There is still some concern about the long-term durability of prefabricated buildings. This is a perception that has yet to be tested with experience of buildings in use over the longer term. Manufacturers have started to address this with product warranties and guarantees. Specifiers need to be objective and base decisions on known and independently verified facts.

Town planning

Town planning restrictions may mean that prefabricated approaches are not particularly well suited to some sites. The issue of context needs careful consideration, and the appropriate town planning office should be consulted as early as possible to discuss any potential problems.

Choice

Manufacturers will use and market different systems, for example units built from timber or lightweight steel, and although all producers claim to offer considerable choice in design the reality is that there may be less freedom than is available with more conventional methods. This is because assembly lines are set up to produce a constant supply of identically sized modules, and a large amount of variety is difficult to accommodate unless there is a market to justify the high cost of tooling.

Factory-based production is based on bespoke systems that are patented and thus only available from one supplier. Purchasing a modular building has the effect of locking the customer into a relationship with one producer for repairs, maintenance and alteration works to the structure if the warranty is to be maintained. Indeed, some of the systems on the market may be difficult to repair or alter without input from professionals and specialist sub-contractors.

Costs

Additional material is required to brace the structure adequately to withstand the stresses and strains placed on the units during transportation to site and craning into position, thus increasing material cost and labour costs. This cost is built into the cost of the units and may be offset by the volume of units produced. For factory-based production to be economical the number of units or modules produced must be relatively large to cover the cost of tooling in the factory. The larger the scheme and the larger the amount of repetition, the greater the economic benefit to the customer. Similarly, the greater the repeat use of a design on other sites, the more economic the process of production.

Repair and maintenance

Depending on the construction used, the building owner may find that they have to use the original supplier of the modules for repair work and routine maintenance. This may be tied back to conditions of the warranty and/or simply be linked to the technology employed and the availability of expertise to carry out the required work. A similar concern may be expressed for future extension and adaptation of modular buildings.

Durability

Although manufacturers provide warranties for their modules (15 years is common), the long-term durability is less certain. Manufacturers tend to quote design lives of around 50 to 60 years, depending on the system. As more modules are produced and their durability monitored, we should be better informed about anticipated durability compared with more traditional forms of construction.

Access

Some construction sites have problems with clear and safe access, making the transportation and craning of large components very difficult or impossible. This is a constant challenge for those working on the extension of existing buildings, especially houses (e.g. Victorian terraces).

8.4 The production process

Given the repetitive nature of the manufacturing process, it is crucial from a business perspective that customer (market) needs are clearly identified and exploited. Thus research and development activities are concerned with market trends and technical (production) factors. Results of the research and development activities are applied to the design and specification of the production process to ensure a profitable manufacturing process. Some manufacturers may use lean manufacturing methods. The extent of robotic manufacturing processes will vary between manufacturers; however, most manufacturers will follow a production process similar to that described here, with rigorous quality control conducted by trained personnel at the end of each step in the production process.

A typical production process

The main steps in a typical production process are described here for a timber or lightweight steel framed unit:

❑ Discussion and confirmation of the customer's technical specification (in relation to production capacity and production constraints)

- ❏ Planning and scheduling of the manufacturing process, from ordering of materials through to site delivery and hand-over to the customer, is agreed prior to commencement of production
- ❏ Automated pick-up systems are used to coordinate production information and ensure that components are ordered from suppliers and delivered to the production line on time
- ❏ Components are allocated to a specific project and supplied to the production line (approximately 3000 components may be required for an average-sized house)
- ❏ Main floor, ceiling panels and external wall panels are assembled (e.g. automated nailing of plasterboard to joists)
- ❏ Frames are assembled in a box-shaped structure for rigidity (e.g. by automated spot-welding machines). Floor and ceiling panels are fixed to the frame, followed by the external wall panels. Fixing techniques vary but usually involve rivets, screws, nails, welds and glues. Joints between panels are filled using gaskets
- ❏ Partition walls and services are installed in accordance with the specific requirements of the customer
- ❏ Pre-assembled kitchen, bathroom and staircases are installed at the factory
- ❏ Painting and finishes are completed
- ❏ Final quality control check before the modules are protected with packaging (to avoid impact damage and protect from moisture and dust) prior to shipping
- ❏ Units are loaded on to trucks by large forklifts or cranes and transported to the construction site in accordance with the customer's delivery date
- ❏ Units are craned on to pre-prepared foundations and joined together using horizontal and vertical fixings for rigidity. Roofing units are delivered at the same time as the modules, craned into position and fixed
- ❏ Interior finishing work (if needed) is completed
- ❏ Final quality control check before the completed building is handed over to the client

Selecting a manufacturer

Before investing in modular construction, potential specifiers (purchasers) should:

- ❏ Visit the factory to see how the units are assembled, the quality control methods in place and the degree of flexibility available in the construction of the units (physical layout and choice of materials)
- ❏ Check the experience and financial stability of the manufacturer, ask for and take up references, check independent reports (if available); do not rely solely on the promotional material produced by the manufacturer
- ❏ Look for independent approvals. Check that the modular building system has been accredited by the BBA; ISO approval should apply to the whole

process; functional performance has been independently tested and endorsed (e.g. for quality, fire, acoustic insulation, thermal insulation and air leakage, and structural stability)
- ❏ Speak to fellow architects, engineers and contractors to get feedback on their experience with a particular manufacturer. What went well? What could have been done better?
- ❏ If applicable, investigate how the modular system will interface with traditional construction techniques and/or existing buildings
- ❏ Visit some of the schemes built based on that particular system. How are they weathering externally and standing up to use internally? What do the clients and users think?
- ❏ As with all other decisions about building components and products, try to consider at least three manufacturers and compare them to see who offers the best overall value

8.5 Pre-cut timber

Timber was the first material to be used for prefabrication, being readily available and easy to work in the factory with large machines and on-site with hand-held tools. Timber also has the advantage of being easy to work on if damaged in transit. England, North America and most of Scandinavia have all developed systems that encompass varying degrees of factory production, ranging from the fully built factory house, delivered to site and craned on to suitable foundations, to the 'kit-of-parts' which are assembled on site by hand and are popular with self-build (DIY) and self-help schemes.

Advantages

- ❏ Thermally good
- ❏ No waste material (all 'waste' timber can be recycled)
- ❏ Easy to work with hand-held tools
- ❏ Easy to repair

Disadvantages

- ❏ Some systems are difficult to extend and/or alter
- ❏ Design defects/assembly defects may lead to timber decay
- ❏ There will be some reduction in thermal insulation value at studs (although less compared with steel)
- ❏ Low thermal mass (high temperature fluctuation)
- ❏ Sound transmission may be a problem if not detailed well

The Segal self-build method

In the UK the architect Walter Segal developed a simple method of construction based on timber frame construction and modern materials specifically for self-help building projects. The Segal Method is based on a modular grid system that uses standard sizes of building materials as supplied by builders' merchants. The timber frame is built off simple pad foundations, which are dug at existing ground levels to avoid the need for expensive site levelling. Once the frame has been erected, the roof can be put on, services installed and walls added. The lightweight and simple design allows both men and women to build their home (individually or as part of a cooperative group) using simple tools and with limited knowledge or experience of building. This dry construction method eliminates what Segal called the 'tyranny of wet trades' (plastering, bricklaying etc.) and forms a lightweight, adaptable, ecologically sound building that is designed to suit the requirements of the users (and also the builders). Considerable cost savings are possible due to savings on labour and, to a lesser extent, materials due to the simplicity of the design.

Timber framed units

Volumetric production of timber units has been greatly assisted by developments in IT, allowing the production of a large variety of standard house types and providing the means to computer generate bespoke designs. As a general guide, the timber framed houses built in a factory use 20% to 30% more material in the framing than those framed on the site. This is to ensure a safe and secure journey from the factory to the site. The additional cost of the material is offset against time and labour savings. The majority of factories will glue and nail or screw the components together for a solid assembly. The main principles used are those outlined in *Barry's Introduction to Construction of Buildings* on timber framed construction. The main difference is that it is easier to control quality in the factory and the whole building assembly can be kept dry during manufacture, transportation and positioning, thus significantly reducing concerns about the moisture content of the timber.

8.6 Metal

Lightweight steel is the material most used for metal-framed units. Steel components and complete assemblies are constantly tested and some bespoke volumetric] house assemblies now have Agrément certification and product warranties.

Advantages

- ❏ No waste material
- ❏ Capacity for long clear span
- ❏ Quick construction times

Disadvantages

❑ Some reduction in thermal insulation, potential thermal bridging at the stud. High conductivity of steel is a concern for all steel framed units
❑ Low thermal mass (high temperature variation)
❑ Sound transmission may be a problem if not detailed well

Steel framed housing

Steel framed housing is becoming more common. Corus steel (previously British Steel Framing (BSF) fabricate a 75 mm deep galvanized steel frame system that provides the shell of a house, in a similar way to timber frame housing (Photograph 8.9). Corus offer a service whereby the steel frame for the house is designed, manufactured and erected. The assembly includes insulation and vapour barrier sheathing. Wall frames are delivered with integral bracing, which are easily manhandled into position. Floor joists are 150 mm deep 'Z' sections that are attached to the wall panels. The roof structure is assembled at ground level and lifted into place in one piece. Roof spaces can be designed to form extra habitable rooms. The windows and doors can be fixed to the steel frame before the brickwork is laid, giving the structure added weather protection, helping internal trades. On a 34 dwelling housing scheme in Southampton, Taywood Homes Ltd reported construction times of 8 weeks, with the steel frame only taking 3 days to erect for each dwelling.

Modular steel framing

Two of the best-known examples of modular steel framed construction are the schemes at Murray Grove and Raines Dairy in London for the Peabody Trust. Volumetric house construction comprises steel framed modular units (joined together for semi-detached and terraced units). The pitched roof is also prefabricated and craned into position. Average construction times are between six to eight weeks for a house, with the steel frame taking around three days to erect. The steel frame construction comprises cold-formed lightweight steel stud sections, commonly 75 mm deep galvanized steel framing members, which are sheathed in insulation on the external face and finished with fire resisting board on the inside face of the wall (thus creating a panel construction). The wall frames include integral steel diagonal cross bracing members and are designed to be easy to manoeuvre and fix on site. Floor construction typically comprises 150 mm deep steel joists fixed to a Z section element attached to the wall panels. Windows are installed on site and the external cladding (usually brickwork) is built on site once the frame is complete.

Construction costs are competitive with more conventional forms of house construction. Advantages include quick construction, dimensional accuracy, long life and long span capabilities (thus allowing for future adaptability).

The steel frame for the house is placed on the concrete floor slab ready for assembly.
The main frame and internal walls are made of light gauge structural steel.

Once erected the frame is insulated and vapour barriers envelope the building structure,
sealing in the structural steel and creating an airtight barrier. Finally the roof covering and
brick cladding can be applied. The whole construction process is very quick. Houses can
be erected and completed in just 12 weeks.

Photograph 8.9   Steel frame housing (courtesy of D. Johnston).

8.7 Concrete

Concrete panels have been in use since the early 1900s in the UK. Concrete is cast
in large moulds and the reinforced units transported to site before being craned
into position. Early pioneers would cast the concrete on site, but with concerns
over quality control and efficiency the casting of units has now moved to a few
specialist factories where quality can be carefully controlled and the casting
process made cost effective. Some of these units may be made from standard
mould shapes and are effectively available off the shelf; others are designed

and cast to a special order. The completed units are then delivered to site to suit the contractor's programme and lifted into position using a crane.

Considerable investment is required in making the moulds, and the units are heavy for transporting and positioning. More recent developments have been in the use of lightweight reinforced concrete units; however, there is still a large amount of work required on site to finish the concrete units; and this can add considerable time to the site phase.

Advantages

- ❏ Good sound reduction
- ❏ Good fire resistance
- ❏ Loadbearing capacity
- ❏ High thermal mass (less fluctuation of temperature)

Disadvantages

- ❏ A large amount of work is required on site compared with other materials
- ❏ Generates waste on site (e.g. drilling holes for services etc.)
- ❏ Changes are difficult to implement (both during and after construction)
- ❏ Units are heavy and awkward to manoeuvre
- ❏ High degree of finishing required on site

Reinforced concrete frames (rather than structural panels) can be used to provide the structural support for modules or pods, which are craned into position. It is, however, more usual for a steel frame to be used.

8.8 Joints and joining

Whatever system is chosen, the quality of the completed building will depend upon the way in which the unit is fixed to the foundations and also how individual modules are joined together. Tolerances and fixing methods are crucial to the final quality of the constructed work. There are three inter-related tolerances to consider:

- ❏ Manufacturing tolerances. Off-site production is capable of producing units to very precise dimensions that are consistent from unit to unit. Manufacturers will provide full details for their range of products.
- ❏ Positional tolerances. Maximum and minimum allowable tolerances are essential for safe and convenient assembly. The specified tolerance will depend upon the size of units being manoeuvred and the technologies employed to position the units.

❑ Joint tolerances. These will be determined by the materials used in the construction of the units (which determine the extent of thermal and structural movement), the size of units and their juxtaposition with other units.

Failure to work to the correct tolerance with prefabricated components can result in problems that are expensive and time consuming to rectify. Where on-site services are not positioned correctly it may be impossible to connect the modular unit. Problems with site plumbing and levels may mean that toilet and bathroom pods do not drain properly. Bathroom pods positioned out of level can result in water stand in the corners of shower units, rather than water flowing to the required drain. Adequate floor to ceiling heights must be maintained to ensure that the prefabricated unit and trolleys for manoeuvring the pod can be used. Occasionally contractors have forgotten to allow for plant and equipment necessary to position each unit.

Site work and connection to the foundation

The setting out and construction of the foundation must be carried out accurately since there is no (or very little) room for error. This also applies to the provision and accurate positioning of utilities such as electricity, gas, water and waste disposal. Scheduling of site activities must be complete before the modular units are delivered to site. Modules are usually bolted to a ground beam or directly to the foundation, after which the service connections are made and subsequently tested.

Connection of unit to unit

The connection of a modular house or small fast food unit to a foundation is a relatively simple operation. Larger schemes based on the assembly of several modular units require a greater degree of coordination to ensure structural integrity through vertical and horizontal fixing. Current practice is to fill the exposed joint with a flexible mastic joint, which serves both as a control joint and to keep the weather out.

9 Lifts and Escalators

Quick, reliable and safe vertical circulation is an essential feature of most commercial buildings and larger residential developments. Lifts (elevators) and escalators are the primary means of moving people, goods and equipment between different levels within buildings. Staircases are still required as an alternative means of escape in the event of a fire or when the lift or escalator is out of use (e.g. for routine maintenance). Lifts and escalators are prefabricated in factories by a small number of manufacturers, transported to site, installed and commissioned prior to use. Although the design and commissioning of lifts and escalators is the domain of engineers, there is a considerable amount of building work required to ensure that the mechanical equipment can be installed safely. This chapter provides a short description of mechanical transport systems.

9.1 Functional requirements

The functional requirements for staircases were set out in *Barry's Introduction to Construction of Buildings*. In buildings with a vertical change in floor level it is necessary to provide a means of transport from one floor to another, both to improve the movement of people within the building and to allow access to all parts of the building for everyone, regardless of disability. Lifts and escalators are the primary means of providing quick, reliable and safe movement between floors for people, goods and equipment. Stair lifts and platform lifts may also be used to allow movement of wheelchairs and pushchairs from one floor to another. Moving walkways are sometimes used to accommodate relatively small differences in floor level but are mainly used to transport people over long distances within large buildings, such as airport terminals. In all cases there is still a requirement for adequate provision of stairs to be used in the event of an emergency and to provide an alternative route should a lift be out of order, due to mechanical breakdown or routine maintenance.

Lifts and escalators

The primary functional requirements for lifts, escalators and moving walkways are:

❑ Safety
❑ Reliability and ease of maintenance

❑ Quiet and smooth operation
❑ Movement between floors (if moving walkway, speed along floors)
❑ Aesthetics
❑ Ease of use for all

Safety

Safety is paramount. Lifts, escalators and moving walkways must be designed and tested to ensure the highest safety standards. EN81-2 is the Lift Regulations standard for safety rules, construction and installation of lifts. Lifts should also comply with Disabled Access standard EN81-70. Well-maintained and serviced, these mechanical transport systems should provide relatively trouble free and safe transportation. Lifts must not be used in a fire. Fire fighting lifts, with independent electrical supply, may be built into high-rise buildings.

Reliability and maintenance

Given that lifts and escalators are crucial to the ease of circulation of people between different floor levels, reliability and the quality of the after-sales service is very important. Having mechanical transport systems out of order for a very short period of time is inconvenient, and a lengthy shutdown can be expensive in lost business and disruption to staff and customers. Many commercial buildings will require 'immediate' responses from the manufacturer's service department. Before a final choice of manufacturer is made, the proximity of the manufacturer's service branches to the building should be checked. The service level to be provided should also be checked and then clearly stated in the performance specification. It is also worth consulting with a number of previous clients to check the quality of service provided by the manufacturer and the reliability of the lift system.

Quiet and smooth operation

The finishes of the lift car, the lift speed and the smoothness of the ride will affect the experience of using the lift, escalator or walkway. Modern lifts are both quiet in operation and provide a smooth delivery to the required floor. Hydraulic lifts tend to provide a gentler but slower ride than traction lifts. Escalators tend to generate a small amount of noise during operation.

Speed

The comfort level of those using the apparatus limits the speed of escalators and walkways. Too fast and it tends to feel uncomfortable, especially when getting on and off. Lifts are sometimes marketed on their speed, especially in high rise buildings.

Aesthetics

The interior finish of the lift car and lighting, the ease of use of the call buttons, and finish of the landing doors all contribute to the overall experience of using the lift. With escalators and walkways the handrail/balustrade and surface finish contribute to the aesthetic of the apparatus.

Coordination and tolerances

An essential requirement when designing and planning buildings that include lifts and escalators is the coordination of dimensions. Lifts and escalators are manufactured to precise dimensions, and manufacturers set out specific requirements for the amount of tolerance in the built structure that supports and/or encloses their equipment. Failure to coordinate dimensional drawings prior to construction, and especially failure to liaise regarding any changes that occur during construction, may result in considerable rework on site. Critical dimensions are the finished floor to finished floor dimensions, the internal size of lift shafts (lift well) and the width and height of the structural opening to the lift shaft. The lift shaft must be constructed plumb and in accordance with the vertical tolerances provided by the lift manufacturer. The formwork and finished concrete shaft must be regularly checked as work proceeds to ensure the work is within the specified tolerance.

The widespread use of laser levels for setting out and checking work as it proceeds, together with the use of steel formwork, has helped to improve the accuracy of concrete poured insitu. Similarly the use of prefabricated units can assist in helping to achieve more accurate work. However, the quality of the work on site remains a determining factor. The work must be accurate, thus ensuring that there are no problems when the lift machinery is delivered to site. This requires high quality work, supervision and methodical checking of the work as it proceeds. Failure to do this will result in expensive and time-consuming rework. Critical areas/dimensions are discussed below.

Electrical supply

A suitable electrical supply will need to be made to the apparatus (lift car, escalator, walkway) as well as to the motor room and, in the case of lifts, the lift well. The motor room will require lighting and emergency lighting. The lift well will need to be lit at the top and bottom with intermediate lights spaced at a maximum of 7 m. 13 amp switched electrical outlets will be required in the motor room and the lift shaft for power tools. Heating, ventilation equipment and thermostats will also be required to maintain an ambient temperature (as specified by the lift manufacturer). The lift car will require lighting and emergency lighting.

9.2 Lifts (elevators)

The development of the skyscraper was dependent on developments in safety and speed of passenger lifts (elevators). Lifts are, however, not the sole domain of high-rise buildings, being a common feature in the vast majority of buildings with a change of level. There are a relatively small number of well-known manufacturers and installers of lifts. Thus choice of manufacturer is rather limited, although the choice of lift car size, its performance and internal finishes is quite extensive, with lift cars designed and manufactured to suit the requirements of a particular development. For many small to medium sized developments there are a standard range of lift sizes available from a standard range. The quality of the lift car will be determined by its function. For example a lift car that conveys people will be built to a different finish than one designed solely for transporting goods. All passenger lifts should comply with EN81-70 Accessibility to lifts for persons including persons with disability. There are two types of lift in manufacture: mechanical or traction lifts; and hydraulic lifts.

Traction lifts

The most common form of lift is the mechanical lift, sometimes described as a traction lift. The lift car is operated by a system of pulleys and steel wires, powered from an adjacent lift motor room. A safety system stops the lift from falling should the steel cables break due to prolonged wear (which is highly unlikely in a maintained lift). Traction lifts are the most common type of lift in use. They can be used in buildings with as little as two different ground levels (e.g. ground and first floor) to multi-storey buildings with numerous floors.

Lift motor room

The lift motor room may be positioned adjacent to the lift (Figure 9.1), or more typically at roof and/or basement level. The lift motor room must be large

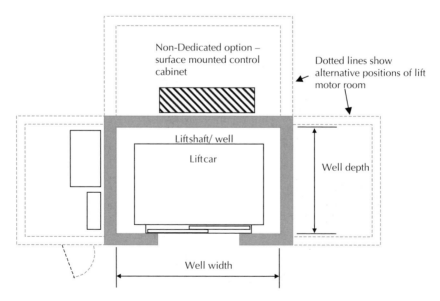

Figure 9.1   Plan of lift shaft.

enough to accommodate the necessary equipment and allow clear and safe access for routine maintenance and replacement activities.

Hydraulic lifts

For low to medium rise applications a hydraulic passenger lift may form a safe alternative to a traction lift. Loading on the hydraulic lift shaft is not excessive and some manufacturers provide hydraulic lifts with their own structure, which helps to keep building costs down. Hydraulic lifts require a smaller lift pit than traction lifts, since the pump can be located adjacent to the hydraulic ram, as illustrated in Figure 9.2. The hydraulic ramp can be side or rear mounted, providing typical speeds of 0.15–1.0 m/s and a smooth operation.

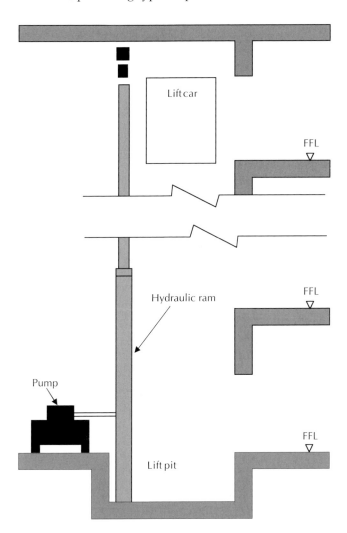

Figure 9.2   Hydraulic lift drive.

Hydraulic lifts offer a number of benefits over traction lifts. Life cycle maintenance costs may be lower, simply because there are less pulleys and lengths of wire ropes, hence less wear. A hydraulic lift uses power only in the ascent; it uses gravity to descend and a small amount of energy to operate the valves. Energy savings may be possible, compared with traction lifts. The hydraulic oil has a long lifespan and will usually last the life of the lift. If the oil needs to be replaced it can be recycled. In an emergency, for example a power cut, the hydraulic lift provides safe and convenient lowering for passengers and service engineers alike. An emergency button/switch will allow the lift to lower automatically under gravity.

Non-motor room option
An alternative to the dedicated motor room is a 'motor-roomless' option for some models. This space-saving design incorporates a cabinet with control equipment that can be mounted on, or recessed into, a wall adjacent to the lift shaft (Figure 9.1). In addition to saving space, the surface mounted cabinet allows maintenance work on drive and control equipment to be conducted safely (and outside the lift well).

Lift function

The function of the lift will determine the size, safe loading, speed and interior finish of the lift car. Lifts tend to be described as passenger lifts, goods lifts, trolley lifts, service lifts, stair lifts and vertical platform lifts.

Passenger lifts
Passenger lifts are usually specified by the maximum number of people carried per lift car. For example a six-person lift (450 kg) or eight-person lift (630 kg). The speed of the lift (both the response time to a call and the time to travel between floors) may also be a prime consideration for tall buildings. The quality of the interior finish is usually specified to match the quality of the building interior.

Goods lifts
Goods lifts are designed to be durable and functional. The lift car is usually constructed from mild steel sheeting with a baked enamel finish to the walls and a heavy-duty vinyl to the floor. They are usually specified by minimum size of the lift car and maximum loading. Speed of the lift car is not a prime consideration. Goods lifts may be built with a lift shaft or loadbearing wall for support. Alternatively, goods lifts with their own robust structure and motor assembly are available, which allows for greater flexibility in positioning. Self-contained lift assemblies alleviate the need for a separate motor room and are ideally suited to installation in existing buildings. Building work is required

to make the necessary openings, followed by installation and commissioning. Typical loadings range from 500 kg to 1500 kg. The size of the lift car will depend on the type of goods being transported between floors. Goods trolleys, palletised goods, warehouse stock, furniture and other bulky goods are typical loads. If the goods lift is to be used for passengers as well, then the lift car will need to be larger.

Trolley lifts

Some goods lifts are designed to accommodate goods trolleys only. Trolley lifts provide a quick, safe and efficient way of moving heavy loads on a trolley between different floor levels. Typical loadings are 250 kg or 300 kg. The size of the lift car is typically around 1000 mm wide and 1000 mm deep, with a height of approximately 1400 mm. Hinged or concertina landing entrance gates are common.

Service lifts

A service lift carries relatively small loads, for example food, drinks, beer crates, documents and laundry between floors of buildings. Originally termed a 'dumb waiter', the service lift allows quick, convenient and safe movement of goods between floors. Service lifts are widely used in restaurants, pubs and clubs for the transportation of bottled drinks and food, for example where the kitchen is located above or below the dining area. Typical maximum loadings are 50 kg or 100 kg. The size of the car will typically be around 500–650 mm wide and 350–650 mm deep with a service door (hatch) somewhere around 800 mm high. Service lifts are available from stock in a variety of standard sizes, although some variation in dimensions and finishes can be made on request to the manufacturers. It is usual to provide an intercom facility adjacent to the service lift to allow persons to communicate their requirements between floors.

Stair lifts

A stair lift, as the name implies, is installed on a flight of stairs to allow safe access between levels for disabled people. Stair lifts are usually installed in buildings where a lift is deemed not to be necessary, or cannot be installed economically in existing buildings. In public buildings it is common to provide stair lifts (or platform lifts) in addition to lifts. Stair lifts are designed to accommodate a wheelchair; the stair lift should be operational without assistance. Controls are located on the lift car. The lift car moves up and down a rail fitted to the wall adjacent to the stair flight. For new projects the stairs must be constructed to be wide enough to allow safe access of stair lift and people. For installation in existing buildings, for example houses, the stair lift will take up the whole width of the stair when in use. The stair lift may be designed to fold away when not in use.

Vertical platform lifts

A vertical platform lift comprises an open flat platform, with safety rails and door. It is sometimes positioned next to small flights of stairs to allow access for wheelchairs, pushchairs and goods trolleys.

Lift specification

The design and specification of lifts involves a large number of options; the main ones are listed here.

Capacity

The anticipated capacity of the lift car will determine its size and load carrying capacity. What is the lift to be used for? People only; people, equipment and goods; or goods only?

- ❑ Low volume (e.g. 6 person lift)
- ❑ High volume (e.g. 16 person lift)
- ❑ Wheelchair only or wheelchair plus attendant(s)?
- ❑ Goods only or goods and passengers
- ❑ How many lifts are required?
- ❑ How many floors need to be served (and the maximum travel distance)

Door type

Type of lift car door is specified by side or centre opening. Lifts usually have lift car doors and landing doors, and it is common practice to use the same door opening types:

- ❑ Single panel (side opening)
- ❑ Two panel side opening
- ❑ Two panel centre opening

Entrance/exit position

There are three entrance/exit configurations, shown in Figure 9.3:

- ❑ Front only. This is the smallest well and lowest cost
- ❑ Front and rear. Increases well size and cost
- ❑ Adjacent entrances. Available on some models. This will increase well size and cost

Clear entrance width (e.g. 900 mm) and height (e.g. 2000 mm) will need to be specified.

Mounting

There are two main options: structure supported or wall mounted lifts (Figure 9.4).

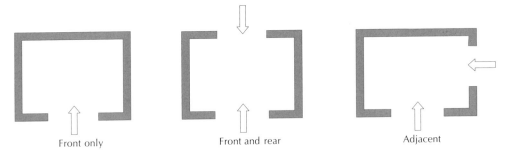

Front only   Front and rear   Adjacent

Figure 9.3   Entrance/exit positions.

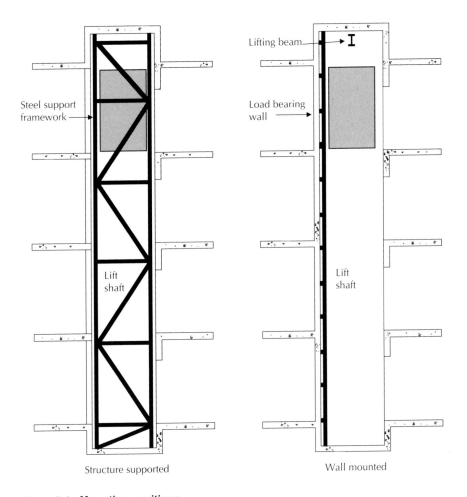

Figure 9.4   Mounting positions.

The structure supported lift tends to be cheaper to build because the load-bearing wall and associated lifting beam are not required. However, this needs to be offset against a higher lift cost (10–20% more than a wall mounted lift) and a slightly larger well size. In the majority of cases scaffolding is not required, which helps to save time and cost. Structure supported lift structures are typically used for goods lifts, trolley lifts and service lifts.

The wall mounted lift has a higher initial build cost, due to construction of a loadbearing wall and a lifting beam at the top. Scaffolding will be required to provide a safe working surface. The advantages are lower lift cost and smaller well size than the structure supported lift.

Well size

The interior of the lift shaft, the well size, will be determined by the size of the lift car and the space required around the car (Figure 9.5). Manufacturers provide details on minimum clear dimensions for each of their models. For example a lift car 1000 mm wide and 1250 mm deep would require a well size of approximately 1500 mm deep and 1600 mm wide.

Finishes

A wide range of finishes are available for lift cars to suit different budgets and design requirements. Consideration should be given to the ceiling finish, which may be flush or suspended to suit different lighting effects. Walls may be finished with decorative panels and/or mirrors. Round section handrails can also be added. Floors are typically finished with hardwearing vinyls, carpet or stone. A skirting helps to protect wall finishes from scuffing and damage.

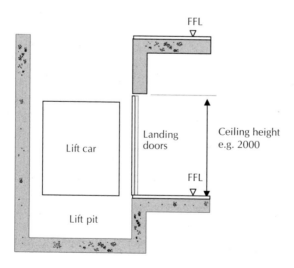

Figure 9.5 Vertical section through lift shaft showing critical dimensions.

Consoles

Push button consoles incorporate illuminated lights to indicate that a lift car has been called. Special pushes and raised (tactile) signage will assist people with a visual impairment. Braille may also be provided on the consoles. Recorded audio messages (voice annunciater) are usually fitted as standard. Consoles also incorporate an emergency voice communication system.

Construction of the lift shaft

The lift car must be supported on a loadbearing wall or frame. Alternatively the lift assembly may be self-supported, i.e. independent of the building structure, in a robust steel framework. The usual arrangement is to build a lift shaft with reinforced concrete or loadbearing masonry. Concrete is now regarded as the preferred option for cores, escape routes and lift shafts in Europe and the US. With increased sophistication of off-site manufacturing it is now possible to manufacture the lift shaft and associated assemblies as prefabricated modules. This can help to improve quality and also help with dimensional coordination. However, currently most lift shafts are constructed on site using slip form or climbing formwork systems (Figure 9.6 and Photograph 9.1).

Concrete lift shafts

Lift shafts are usually square or rectangular (with a corresponding square or rectangular lift car) and constructed of reinforced concrete. The most common construction method is to use insitu reinforced concrete to form the walls of the lift shaft. Circular lift shafts with corresponding circular cars are also available, although not very common.

   As noted earlier, the quality of the work carried out on the site is critical to ensure that the lift shaft is built to the specified tolerances. Tolerances are particularly important where the lift car and guide rails are to be fixed to the walls and frame. Critical dimensions are the finished floor to finished floor dimensions (Figure 9.5), the internal size of lift shafts (lift well), and the width and height of the structural opening to the lift shaft. Standard internal sizes of lift shaft are shown in Table 9.1

   In order to achieve the required running clearances, door alignment and reliability of lift installation and operation, lift shafts should be constructed to high standards of verticality (within vertical tolerances). While all shafts should be constructed to the tolerances set by the lift manufacturer, a standard guide is that the shaft should not deviate from the vertical by more than 1/600 of the height or by 5 mm per storey (whichever is the greatest). However, the total deviation throughout the full height of the building must not be more than 50 mm (Figures 9.7 and 9.8 and Table 9.1).

   The solid core of a concrete lift shaft is often used to add lateral stability to the building. The steel or concrete frame is tied into the concrete lift shaft, and the solid walls of the shaft act as bracing. When lift and stair shafts are

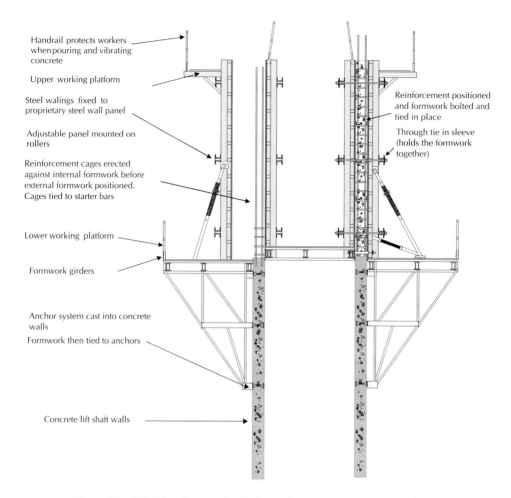

Handrail protects workers when pouring and vibrating concrete

Upper working platform

Steel walings fixed to proprietary steel wall panel

Adjustable panel mounted on rollers

Reinforcement cages erected against internal formwork before external formwork positioned. Cages tied to starter bars

Lower working platform

Formwork girders

Anchor system cast into concrete walls
Formwork then tied to anchors

Concrete lift shaft walls

Reinforcement positioned and formwork bolted and tied in place

Through tie in sleeve (holds the formwork together)

Figure 9.6   Climbing formwork platforms (adapted from www.peri.ltd.uk).

combined, the concrete enclosure can be designed to form the protected shaft for the purpose of escape during a fire. If designed with the correct concrete cover to the reinforcement and the concrete uses aggregates that don't expand and spall when exposed to heat, the concrete structure will have good heat resisting properties. If designed and placed properly, concrete can be used to form a compartment wall capable of resisting fire for up to 4 hours.

9.3 Escalators and moving walkways

Similar to lifts, the manufacture and installation of escalators and moving walkways is limited to a small number of manufacturers. Because of the problem of keeping the moving parts rust free, escalators and walkways should not be used outside unless they are adequately protected from rain and snow.

Climbing formwork.

Upper working platforms and formwork enclosed by protective sheeting

Legs of scaffolding securely anchored into lift shaft allows the formwork to be cantilevered out

Fixing bracket. Part of the fixing is cast into the concrete. Rebates are also cast into the internal face of the shaft to allow fixings for the climbing formwork

In this formwork the lower scaffolding lifts are suspended from the upper platform, this allows the concrete shaft to be cleaned as it rises and formwork fixings to be easily accessed.

Photograph 9.1   Lift shaft: climbing formwork (www.doka.com).

Table 9.1   Well dimensions for general purpose or intensive traffic lift installations (adapted from Ogden, 1994)

Persons carried	Internal car sizes			Recommended well size	
	Width	**Depth**	**Height**	**Width**	**Depth**
8	1100	1400	2200	1800+K	2100+K
10	1350	1400	2200	1900+K	2300+K
13	1600	1400	2300	2400+K	2300+K
16	1950	1400	2300	2600+K	2300+K
21	1950	1750	2300	2600+K	2600+K
24	2300	1600	2300	2900+K	2400+K

K = tolerance:
    25 mm for shafts less than 30 m height
    35 mm for shafts exceeding 30 m height, but less than 60 m
    50 mm for shafts greater than 60 m, but less than 90 m

K = Wall tolerance
25mm for shafts not exceeding 30mm
35mm for shafts not exceeding 30m, but not exceeding 60m
50mm for wells over 60m, but not exceeding 90m

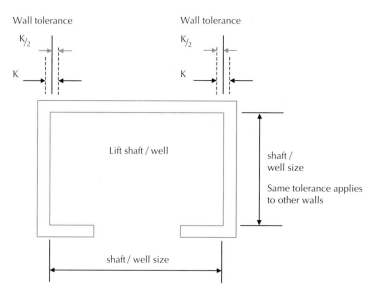

Figure 9.7   Lift well sizes and horizontal tolerances.

Escalators

Escalators move people vertically from one level to another, and are a popular feature of commercial shopping centres and large retail stores, travel inter-changes and other commercial buildings. Usually installed as an architectural feature, they help to guide people from one level to another without the need to use staircases or lifts. The escalator comprises a moving steel mat that moulds itself to the profile of the transport system underneath. This forms a series of steps on which one stands until reaching the top or bottom of the escalator. Pitches tend to be steeper than staircases, typically around 35–45° to the horizontal.

Critical dimensions are primarily the finished floor to finished floor level. A pit is required at the base of the escalator to house motors and associated equipment. This is usually covered with a steel plate, which can be removed for routine maintenance and repair.

Moving walkways

Moving walkways provide a flat moving surface to move people horizontally. They are common in large buildings, such as airports, where people would

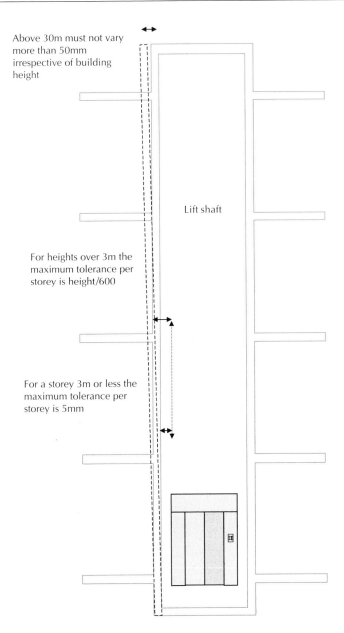

Above 30m must not vary more than 50mm irrespective of building height

Lift shaft

For heights over 3m the maximum tolerance per storey is height/600

For a storey 3m or less the maximum tolerance per storey is 5mm

Figure 9.8   Lift shaft construction tolerances (steel or concrete frame).

otherwise have to walk long distances. The idea is that people stand on the continually moving surface of the walkway and are transported from one end to the other without the need to walk. Moving walkways tend to be slightly slower than normal walking speed (remember people have to step on and off each end), and it is common for people to continue to walk along the moving surface, thus making it faster than walking on a static surface. The walkways are usually designed and constructed so that people have the option of walking alongside the walkway. A safety 'stop' button is positioned at the end of the walkway so that it can be stopped in an emergency. Walkways are also manufactured to be installed at a low pitch to the horizontal to accommodate small changes in floor level.

The walkway is positioned within a shallow pit in the floor slab. The depth of the pit will, to a certain extent, be determined by the manufacturer to be used. Walkways are protected with solid balustrades along their length. These are usually constructed of glass to improve visibility and hence safety.

10 Fit Out and Second Fix

Many commercial buildings are built to a 'shell' finish, i.e. the main structure of the building is completed, but the internal fittings and decoration is carried out as part of a separate contract to suit client/user requirements. Electrical trunking and light fittings, suspended floors, suspended ceilings, partition walls, shelving, display units, internal fittings, painting and decorating, and signage all form part of the fit out and second fix operations.

10.1 Commercial fit out

The term 'commercial fit out' is normally used to cover the fit out and second fix to offices, retail units and industrial buildings. The main components of a fit out, the raised floor, suspended ceiling and internal partition walls are described below, together with a brief description of painting and decorating. The vast majority of raised floor, suspended ceiling and internal partition wall systems are manufactured to a patented design. Quality of the system and adaptability varies between manufacturers, from budget systems to expensive systems for the most prestigious of developments. Careful research is required to choose the most appropriate system to suit the requirements of the building users.

Offices and commercial buildings

Speculative office developments are a common feature in our towns and cities. These are usually built to a relatively standard design and completed to a shell finish under the main contract. At the most basic level the shell finish includes the structure and external fabric of the building, staircases, lifts and service core. A more common approach is to include the raised floor and suspended ceilings, sanitary fittings and electrical outlets. Internal partition walls are usually installed as part of the fit out to suit the needs of the particular user. The office space is usually leased and it is not uncommon for users to move to another address, or lease more or less floor space at the end of the rental period. Thus flexibility of office space is a prime consideration.

Retail

The term 'shop fitting' is used to describe the process of installing the furniture (display shelving and units, points of sale etc.) and associated equipment (chiller cabinets, freezer cabinets etc.) to a retail unit. The vast majority of retail units are designed to be relatively flexible in terms of use. This is because the units will normally be leased out to a particular business for a certain period of time. Thus it is common practice to build the shell of the unit, with the fitting out carried out by the retailer to suit a particular house style (corporate image). This applies to shops located on the high street and also to units located in large shopping centres.

Retailers tend to change their display arrangements on a regular basis. Minor changes can usually be accommodated in adjustable shelving units, but more major changes associated with an update in corporate image usually necessitate a complete refit of the retail unit, often resulting in a lot of wasted materials. Similarly, with a change of retailer there is usually the need for new shop fitting.

Food retail units will require additional drainage points for the condensate drain to freezer cabinets and chiller units. The store layout tends to be changed less frequently because the condensate drains determine the position of freezer and chiller units. Changes in position usually necessitate changes to drain positions, which can be disruptive to the sales area.

Shop signs

Provision for shop signage is usually provided at strategic places on the exterior face of the shop unit and this too will usually be installed (subject to town planning consent) by the organisation leasing the shop. The shop signs are printed on to relatively thin backgrounds (e.g. acrylic sheet) or on to a translucent material so that the sign can be illuminated from behind. The shop sign is then fixed to the face of the wall, usually with screws. From a design perspective it is important to provide a structure suitable for supporting the signage and any electrical connects required for the lighting.

Industrial

Industrial production facilities have special requirements to suit a particular manufacturing process. These may include one or a number of the following features: clean rooms, security (of staff and materials/processes), wash down facilities, special materials handling areas and secure storage, specialist fire protection systems etc. These are outside the scope of this book but readers should be aware that these special requirements have a bearing on the choice of construction methods used and how services are integrated, e.g. sealing services as they pass through compartment walls in clean room construction. Often some compromises need to be made. For example, in fast track projects it would be sensible to use prefabricated, framed construction to save time on the building site; however, some production processes may require solid masonry walls.

On fast track projects, such as new pharmaceutical production facilities, it has become common practice to manufacture and test industrial plant prior to installation in the building. The equipment is then transported to site, moved into its final position and 'plugged-in'. It is then put through a final series of conformity and safety tests, i.e. it is commissioned for use. This is usually done while the main build contract is still underway. Large access doors facilitate the delivery and subsequent maintenance and replacement of large pieces of equipment.

10.2 Raised floors

A raised floor (access floor) is used to conceal services, usually electrical cables and air conditioning ducts, in the cavity between the raised floor and the structural floor. Raised floors are particularly useful in large open plan spaces, such as offices, where it is not practical to house services in partition walls. Air-conditioning grilles and services outlets, such as electrical and telephone sockets, can be provided within the floor to provide a flat and even surface finish. The cables and air conditioning ducts are hidden from view in the cavity between the raised floor and the structural floor finish, with strategically placed access panels to allow convenient access for repair, maintenance and upgrading. Hence the term 'access floor'. Concealing the cables not only improves the visual appearance of the floor area but also helps to keep trip hazards to a minimum. Most manufacturers of raised floor systems also supply ancillary items such as electrical floor boxes, cable ports and cable management systems.

Raised floors are used in new building developments as well as in refurbishment work where there is sufficient vertical height between structural floors to add a raised floor. By adjusting the depth of the supports to suit uneven floors and changes in floor levels, it is possible to provide a flat level surface and eliminate the need for ramps and short flights of stairs.

Functional requirements

The functional requirements are to:

- Accommodate and conceal services:
  - Provide ease of access to services
  - Allow changes to services below the floor
  - Create a clear cavity for services
- Be flexible:
  - Provide a level, attractive and durable floor finish
  - Accommodate changes in surfaces
  - Remove changes in structural floor
  - Facilitate changes in level
  - Provide required surface level
- Sustain and resist imposed loads:
  - Rigid and stable
- Have good appearance and aesthetics:
  - Hide unsightly structural floors
  - Present feature in its own right e.g. lighting

❑ Sound control:
   ○ Resist passage of impact and airborne sound and
   ○ Provide acoustical control (absorption and reflection)
❑ Provide required thermal resistance and prevent formation of condensation
❑ Provide protection against fire:
   ○ Control spread of fire and maintain structural stability in a fire
❑ Durability:
   ○ Resist wear and tear
❑ Ease of maintenance:
   ○ Safety of installation and maintenance – CDM Regulations
   ○ Provide a comfortable, easy to clean finish

Floor assembly
There are two primary components to a raised floor, the floor panel and the support pedestal. Combined, the components provide a rigid and stable floor surface for a variety of uses. The design life of the floor panels tends to be anything up to 25 years, that of the supports around 50 years. Loading capacity will vary with different systems, usually expressed as a point load over 25×25 mm, e.g. 3 kN, and uniformly distributed load, e.g. 8 kN/m^2. In situations where heavy furniture or equipment is to be installed, the loading of the system can be improved by the installation of additional support pedestals. Manufacturers will also provide details of the air leakage (at 25 Pa), the fire rating (e.g. Class 0), the combined weight of the system (e.g. 40 kg/m^2), which will vary with height, and a statement on electrical continuity of the system (e.g. that it complies with IEE Regulations). Acoustic performance of the system (which depends on the type of floor finish) should also be provided. Floor to wall junctions are usually sealed with an air seal.

Floor panels
Floor panels are manufactured to a nominal size of 600×600 mm. Depending on the construction, the depth of the panel will be from around 25 mm to 40 mm. The majority of panels are constructed of high-density particleboard encapsulated in a galvanised steel finish, approximately 0.5 mm thick. Non-metallic systems comprise high-density particleboard or specially treated moisture resistant chipboard, usually applied with a protective finish. It is common in UK properties to install carpet tiles on a trackifier but timber and marble may also be used. Metallic panels are gravity fitted, laid into the structural support grid using a hand-held suction device. Indentations in the underside of the panel allow positive location and subsequent retention within the support system. Rapid access is available through the use of a suction device to lift the panels out. The non-metallic systems tend to be screwed to the structural supports through a countersunk hole at each corner. Unscrewing the panels from the supports provides access.

Support pedestals

The pedestal is usually constructed of steel. Concrete and timber can also be used. Steel pedestals are adjustable to suit variations in the structural floor (Figures 10.1 and 10.2 and Photographs 10.1 and 10.2). They are fixed on a 600 mm^2 grid and a laser level is used to ensure a level finish. Electrical earthing is required when metallic components are used. Angled supports are also manufactured to provide a ramped floor finish. Pedestals are manufactured in a range of heights, providing clear cavity spaces from as little as 30 mm up to 1000 mm.

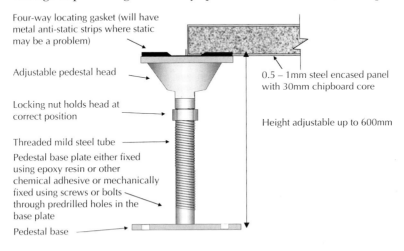

Four-way locating gasket (will have metal anti-static strips where static may be a problem)

Adjustable pedestal head

0.5 – 1mm steel encased panel with 30mm chipboard core

Locking nut holds head at correct position

Height adjustable up to 600mm

Threaded mild steel tube

Pedestal base plate either fixed using epoxy resin or other chemical adhesive or mechanically fixed using screws or bolts through predrilled holes in the base plate

Pedestal base

Figure 10.1   Raised floor pedestal.

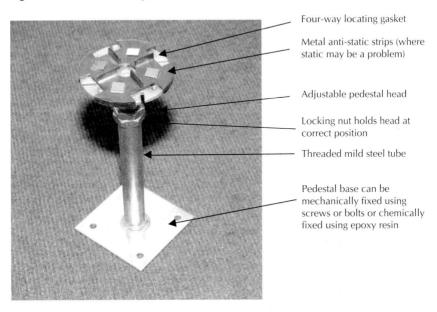

Four-way locating gasket

Metal anti-static strips (where static may be a problem)

Adjustable pedestal head

Locking nut holds head at correct position

Threaded mild steel tube

Pedestal base can be mechanically fixed using screws or bolts or chemically fixed using epoxy resin

Photograph 10.1   Raised floor pedestal.

Example of pedestal and panel floor

❑ 31 or 32 mm overall galvanised steel/chipboard sandwich panel 600 × 600 mm, which is placed on polypropylene cruciform locating lugs and sits at the top of the pedestal. The panel can be rested on or fixed to the pedestal.

❑ The pedestal is made of mild steel tube and base plate, with threaded components to allow for adjustment. Locking nuts are also used to hold the pedestals securely at the correct level.

❑ Pedestals are fixed with polyurethane or epoxy adhesive to sub-floor at 600 mm centres or mechanically fixed (plug and screw or gun nailed) if the pedestals are over 600 mm high (deep void). As panels are square, pedestals are fixed on a grid based on the module sizes.

❑ Panels are often constructed from a 0.5 mm steel top and bottom with 30 mm chipboard sandwiched between, or 1 mm steel top and bottom for heavy loading. Other types of panel construction are also available.

❑ A typical floor depth is about 600 mm, but can be increased to over 2 m (using considerable support, framework).

❑ Fire barriers should be used under compartment walls and where necessary to prevent the cavity acting as a conduit for fire.

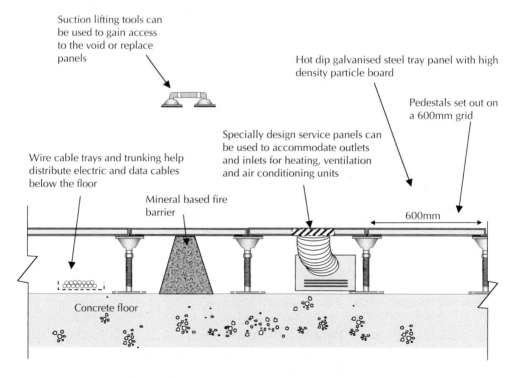

Suction lifting tools can be used to gain access to the void or replace panels

Hot dip galvanised steel tray panel with high density particle board

Pedestals set out on a 600mm grid

Specially design service panels can be used to accommodate outlets and inlets for heating, ventilation and air conditioning units

Wire cable trays and trunking help distribute electric and data cables below the floor

Mineral based fire barrier

600mm

Concrete floor

Figure 10.2   Raised floor.

Cable trays and service ducts are easily positioned under the raised floor

Fire barriers are installed to prevent the passage of fire through the raised floor void

Photograph 10.2   Raised floors.

Maintenance

Raised floor cavities need to be regularly cleaned to remove dust and debris. The cavity should be vacuumed using static-free tools and special air filters at least twice per year. Suction tools can be used to gain easy access into the cavity (Figure 10.2).

10.3 Suspended ceilings

Suspended ceilings are primarily used to provide an attractive finish to the ceiling while at the same time concealing services. A suspended ceiling comprises a lightweight structural grid, which is supported by wire hangers attached to the underside of the structural floor, and into which lightweight tiles are positioned. The structural grid is based on a 600×600 mm module. Suspended ceilings may also be used in refurbishment projects to provide a level ceiling finish to areas of the building that differ in height.

Suspended ceilings not only provide an attractive, level ceiling finish, with convenient access to services when required (Photographs 10.3 and 10.4), but are also used to house a wide variety of equipment, which may include ventilation grilles, light fittings, fire sprinklers, detectors (e.g. smoke, heat), security cameras, movement sensors and alarms.

Functional requirements

The functional requirements (adapted from BS 8290: Part 1 1991 Suspended Ceilings Code of Practice for Design) include:

❑ Concealment of structure – To hide changes in the structure, beams, floors and ties, and to provide a level and attractive ceiling finish.
❑ Concealment of services – To create a clear cavity for services, hide services such as ducts, equipment and cables, and allow easy access for maintenance.
❑ Decorative appearance – The aesthetics of finishes are always important.
❑ Thermal insulation – Thermal insulation may be introduced if the ceiling is adjacent to the roof. It must be designed and positioned to prevent interstitial condensation.
❑ Acoustic control: sound insulation and absorption – The two main aspects of acoustic control are absorption of a sound within an enclosed space, and the reduction of sound passing through a material or structure. The measures necessary to control these two are quite different. Sound absorption is achieved by incorporating porous material of a low density in the ceiling. Sound insulation requires an impervious material of a very high density. Mineral wool, with its high density, is often placed on top of the ceiling grid.
❑ Fire control: control the development and spread of a fire, containment of a fully developed fire and protection of the structure against damage or collapse – A fire resisting ceiling should not be confused with a fire protecting ceiling; different test methods and criteria apply. A fire protecting ceiling offers protection to the structural beams and floors. The term 'fire protection' implies that the ceiling can satisfy the stability, integrity and insulation requirements for a stated period. Sprinkler heads and systems may form part of the fire protection system.
❑ Control of condensation – Placing thermal insulation over a suspended ceiling can increase the risk of interstitial condensation. For guidance on controlling condensation reference should be made to BS 5250 (Code of practice for the control of condensation in buildings) and BRE 143 (Thermal insulation avoiding risks).

- ❑ Hygiene control – Smooth cleanable finishes may be required to facilitate a clean room.
- ❑ Ventilation – The services may form an integral part of the suspended ceiling unit. Where services are concealed within the ceiling they may need additional support. Each piece of plant can be individually tied to the building structure.
- ❑ Heating, air conditioning, illumination – The luminaires may be surface mounted or independently supported from the structure. The ceiling may also be used to control the amount of light reflected and diffused into the room.
- ❑ Electrical earthing and bonding – The ceiling grid must be earthed. In the event of any electrical fault, the parts of the ceiling capable of carrying an electrical current are earthed and protected.

Ceiling assembly

A wide variety of ceiling systems are available (Figures 10.3 and 10.4 and Photographs 10.3 and 10.4). There are two primary components to a suspended ceiling, the ceiling tile and the support grid. Combined, the components provide a lightweight and level ceiling finish. The design life of the ceiling tiles tends to be anything up to 25 years, that of the support system around 50 years.

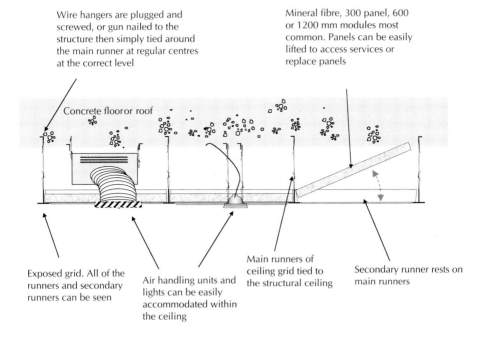

Wire hangers are plugged and screwed, or gun nailed to the structure then simply tied around the main runner at regular centres at the correct level

Mineral fibre, 300 panel, 600 or 1200 mm modules most common. Panels can be easily lifted to access services or replace panels

Concrete floor or roof

Exposed grid. All of the runners and secondary runners can be seen

Air handling units and lights can be easily accommodated within the ceiling

Main runners of ceiling grid tied to the structural ceiling

Secondary runner rests on main runners

Figure 10.3   Suspended ceiling.

Services housed above
the suspended ceiling
grid.

Suspended ceiling grid
and partition wall frame
installed.

Ceiling tiles installed and
glazed panels inserted
into the frame.

Photograph 10.3   Suspended ceiling (courtesy of D. Highfield).

Loading capacity of the system is sufficient to carry the weight of the system and associated equipment. Heavy fittings will require independent support from the underside of the structural floor. Manufacturers should also provide details of the air leakage (at 25 Pa), the fire rating (e.g. Class 0), and the combined weight of the system, which will vary slightly with the depth of the hangers. Acoustic performance of the system should also be provided.

Ceiling tiles

Ceiling tiles are manufactured to a nominal size of 600×600 mm. Tiles are made from pressed lightweight steel or a lightweight material, such as particleboard, and usually have a paint finish. The majority of systems rely on a gravity fit, with the tile placed into the grid. In areas where wind uplift may be a problem, for example in entrance lobby areas, it is common to provide a more rigid fixing to avoid accidental lifting and damage to the tiles. Most manufacturers produce plain tiles as well as tiles with a variety of pre-formed apertures into which lights and sensors are placed. Installing perforated tiles can provide additional sound insulation.

Support grid

The support grid is made from lightweight wire hangers, fixed to the underside of the structural floor. These wire hangers support the lightweight modular grid into which the tiles are placed. Special fittings (edge details) are manufactured for intersections at structural columns and the walls. The support grid is set out and levelled using a laser level. The grid can be exposed, semi-concealed and fully concealed (Figure 10.4 and Photograph 10.4).

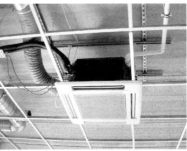

Services are suspended on their own brackets prior to the fixing of the ceiling grid

Double railed fixing bracket allows the position of the air handling unit to be accurately positioned once the ceiling grid is in place

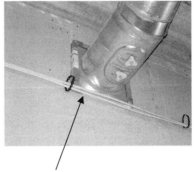

Where the ceilings passes over compartment walls fire blankets are needed to prevent the passage of fire

Services that pass through compartment walls also need to be fitted with special fire barriers, which seal the duct in the event of a fire

Photograph 10.4 Suspended ceilings: accommodation of services (courtesy of G. Throup).

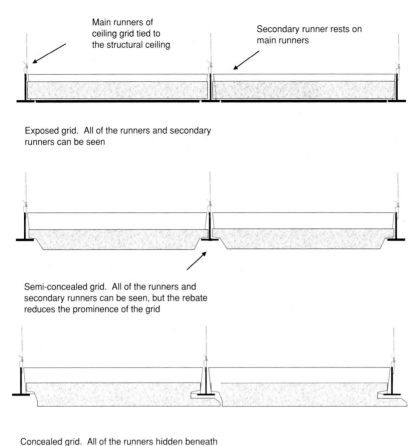

Exposed grid. All of the runners and secondary runners can be seen

Semi-concealed grid. All of the runners and secondary runners can be seen, but the rebate reduces the prominence of the grid

Concealed grid. All of the runners hidden beneath the ceiling panels

Figure 10.4   Suspended ceiling grids.

Maintenance

Careless handling and repositioning within the grid may damage tiles. To maintain an attractive finish, damaged tiles will need to be replaced. Tiles may also become dusty and marked through careless handling, thus some cleaning may be required.

10.4 Internal partition walls

Internal partition walls are used to create discrete areas within large interior spaces. A wide variety of proprietary systems are available, which are designed for different uses and for different quality levels. Alternatively partition walls may be constructed as a framed structure (in timber or mild steel) or as a masonry wall (brick or block). In office developments the emphasis tends to be on flexibility and future adaptability of the workspace, while in industrial units emphasis tends to be more on durability and ability to withstand minor impacts.

The term 'partition wall' is used rather loosely in the construction sector to cover both loadbearing and non-loadbearing internal walls. These have already been described in *Barry's Introduction to Construction of Buildings*. In the context of this chapter we have limited the description of partition walls to some of the more flexible and adaptable systems that are used in commercial developments.

Functional requirements

Partition walls may be freestanding units positioned on the floor; be tied to the structural floor; or tied to the structural floor and the ceiling structure. In buildings with a high roof, for example conversion schemes, it is not uncommon for the partition wall to be supported off the floor only and a suspended ceiling installed to create a sensible ceiling height in relation to the room proportions. Alternatively partition walls may be used to define space only, supported off the floor and without the need for ceilings. All partition wall systems need to be structurally stable and, especially in office space, easy to reposition without causing damage to floor or ceiling finishes.

Functional requirements are:

❑ Flexibility
❑ Adaptability
❑ Structural stability

Other performance requirements, which the partition wall may need to address, include:

❑ Fire resistance
❑ Resist the spread of fire
❑ Accommodate services
❑ Provide required acoustic control
❑ Transfer own weight and any fixtures and fittings (non-load bearing)
❑ Transfer building loads (load bearing)
❑ Allow the passage of light
❑ Allow cross ventilation
❑ Demountability

Fire resistance
The partition may be required to prevent the passage of fire acting as a compartment wall. Blockwork with a plastered finish provides a good resistance to fire, for example a 100 mm block wall finished with plaster easily offers 2 hours' fire

resistance, and a double skin plastered blockwork wall will achieve a 4 hours' fire resistance. Timber stud and proprietary walls only offer half an hours' fire resistance unless they are specifically designed as a fire resisting structure. There are many fire resisting plasterboards that can be easily applied to stud walls in single, double and triple thicknesses. All joints must be effectively sealed with fire stops. Any gaps, services ducts, ventilation units, doors and windows provide weaknesses in fire resisting structures and should be addressed in the detailing. It may be possible for the fire to pass around the wall under raised floors or suspended ceilings; effective fire barriers should be provided under and above the wall, if it forms a compartment wall.

Fire resisting walls should be fire stopped at their perimeter, at junctions with other fire resisting walls, floors and ceilings, openings around doors, pipes and cables. Fire stopping materials include:

❑ Mineral wool
❑ Cement mortar
❑ Gypsum plaster
❑ Intumescent mastic or tape (intumescent strip)
❑ Proprietary sealing systems

Figure 10.5 provides an example of a fire-resisting partition.

Resist the spread of fire
The surface material should not allow flames to pass across it and should not fuel the fire. In public buildings, walls should be designed so that the risk of flame spreading across the surface is minimal. Compartment walls must have a low risk of spread of flame (Class 0).

Accommodate services
Allow for maintenance and repositioning of services. Some proprietary partitions are prefabricated with conduits, pipework or cables already positioned. Skirting boards and dado rails are often a good place to provide access for services. Conduits and channels run behind plastic or wooden boards and entry boxes are positioned to allow services to be installed. Alternatively the services would need to be surface mounted.

With timber stud walls, timber grounds can be easily positioned to accommodate electrical plug sockets, pipework and other fittings. Steel and timber stud walls have become increasingly popular in flats and offices. However, when installing services within these walls care should be taken to ensure that the acoustic and fire resisting properties of the wall are not compromised by the penetration of services through the plasterboard. Ideally, services should be surface mounted to avoid the problem. Alternatively sound resisting material, such as acoustic resisting plaster board, should be positioned in the middle of the wall with timber or metal studs either side of the panel, thus allowing services to be accommodated without affecting the sound resisting material (Figure 10.5).

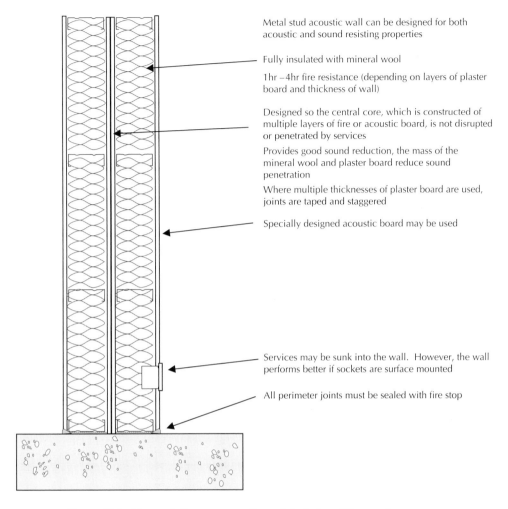

Metal stud acoustic wall can be designed for both acoustic and sound resisting properties

Fully insulated with mineral wool

1hr – 4hr fire resistance (depending on layers of plaster board and thickness of wall)

Designed so the central core, which is constructed of multiple layers of fire or acoustic board, is not disrupted or penetrated by services

Provides good sound reduction, the mass of the mineral wool and plaster board reduce sound penetration

Where multiple thicknesses of plaster board are used, joints are taped and staggered

Specially designed acoustic board may be used

Services may be sunk into the wall. However, the wall performs better if sockets are surface mounted

All perimeter joints must be sealed with fire stop

Figure 10.5   Fire resisting or acoustic metal stud partition wall.

With masonry walls the services can be surface mounted or chased into the wall. Chasing should not exceed one-third of the wall's thickness vertically and one-sixth horizontally. Chases in the wall will reduce the strength, acoustic properties and fire resistance. Care should be taken to ensure that these do not compromise the specified performance.

Provide required acoustic control

The degree of sound absorption and reflection in a room is particularly important in auditoriums, lecture theatres, concert halls etc. Lightweight porous materials will help to absorb the sound and reduce reflection whereas heavy, hard, smooth surfaces will reflect the sound. In an auditorium or lecture theatre, reflective surfaces should be used close to the source of the sound (e.g. position of the speaker) and absorbent materials should be used towards the back of the room, where noises reflecting would cause an echo.

Transfer own weight and any fixtures mounted on the wall

The ability of a partition wall to accept fixtures and fittings is often overlooked. Overloading a wall with shelves or other fittings may cause the wall itself to break or topple over, or the load of the fittings may simply pull the fittings out of the wall fabric. The wall materials must be capable of restraining the loads applied. Studs and other reinforcing materials may need to be positioned so that the wall can accommodate the load. The head of the wall will also need to be restrained if loads are anything other than minimal.

Transfer building loads (load bearing)

While it is common to have large, flexible open spaces divided by non-loadbearing, potentially demountable walls, it is often economical to have intermediate loadbearing supports, such as internal walls. The walls can be load bearing, carrying loads from floors, beams and components above the wall down to the building's foundations. By using intermediate loadbearing walls, the floor beams do not need to span as far, and the section and depth of each beam can be reduced. Smaller, shallower beams are less expensive per unit length than long, deep section beams.

Allow the passage of light

Light may be allowed to pass with or without vision through the material (usually glass). Windows and vision panels are easily introduced into stud and masonry walls. It is important that the windows are carefully selected and fitted so that they comply with the other performance criteria of the wall e.g. fire resistance, acoustic properties etc.

Allow cross ventilation

Mechanical and natural ventilation ducts may be installed through the wall. If the wall is a compartment wall, these service ducts will need to be fitted with a fire stop that is capable of sealing the duct in the event of a fire.

Demountability of partition walls

To accommodate changes of use, large spaces are often divided using de-mountable partitions. In such situations the ease with which a wall can be dismantled, reassembled and repositioned is of considerable importance. Any non-loadbearing wall is demountable but it may not be possible to re-erect the structure using the same components.

Some patent walls have been designed so that they can be easily repositioned, e.g. folding concertina doors, sliding doors or walls, without the need for tools, whilst others are bolted, clipped or fixed into place but can be relatively easily repositioned with minimum disruption.

Partition wall assembly

Partition walls are manufactured to a modular size. Systems are usually based on a lightweight steel stud system. Alternatively timber studs may be used (Figure 10.6). Using multiple layers of acoustic or gypsum fire resisting plaster board and filling the studs with mineral wall, the fire resisting and sound reducing properties of the wall can be significantly increased.

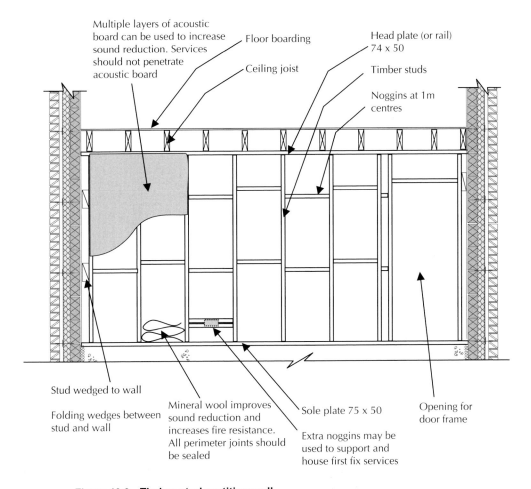

Figure 10.6   Timber stud partition wall.

The arrangements and systems of partition walls are constructed to provide the required properties. Often selected purely on economic grounds, different arrangements of brick, timber, concrete, steel and patent systems can be selected to provide the required finish, flexibility, acoustics and fire resisting properties (Figure 10.7).

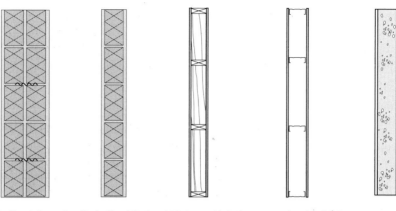

Double skin of plastered blockwork reinforced and tied with brick mesh. 4hr fire resistance, high mass, good sound insulation

Single skin of blockwork 2hr fire resistance, high mass, reasonable sound insulation (as long as all gaps filled)

Timber stud $\frac{1}{2}$ hr fire resistance, low mass, limited sound reduction

Metal stud wall. $\frac{1}{2}$ hr fire resistance, low mass, limited sound reduction

Insitu concrete, plastered both sides, 200mm thick = 4hr fire resistance, high mass, good sound insulation, plaster seals gaps and improves sound reduction

 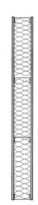  

Timber stud, fully insulated with mineral wool. $\frac{1}{2}$ – 1hr fire resistance, reasonable sound reduction, all joints must be sealed with fire stop. Acoustic board may be used

Metal stud, fully insulated with mineral wool. $\frac{1}{2}$ – 1hr fire resistance, reasonable sound reduction, all joints must be sealed with fire stop. Acoustic board may be used

Metal stud acoustic wall, fully insulated with mineral wool. 1hr – 4hr fire resistance (depending on layers of plaster board), designed so the central core is not disrupted by services, good sound reduction, all joints must be sealed with fire stop. Acoustic board may be used

Demountable proprietary partition. Sound reduction, and fire resisting properties vary with system. Important that all joints are sealed and fire stopped if sound or fire resisting properties required

Figure 10.7   Types of internal partition.

10.5 Painting and decorating

The term 'painting and decorating' is used to encompass a wide range of paints, wood stains, varnishes and oils for internal and external use, as well as the covering of internal walls with decorative wallpaper. The function of paints, stains, varnishes and oils is twofold: to provide a protective surface to materials, for example timber; and to provide a visually attractive finish to the material. Colour and surface texture does affect our mood and can be used to create a strong or subtle visual effect. The function of wallpaper is primarily to provide an applied decorative finish to internal (plastered) walls.

Colour

An important factor in the description and specification of paints, stains and varnishes is colour. Colour for building purposes is usually defined by British Standard (described below). Other systems such as Colour Dimension, Munsell, Pantone and RAL are also used in the construction sector to specify colours. This allows the contractor to purchase paint from any of the paint manufacturers, safe in the knowledge that the colour will be as specified. Paint manufactured for the domestic market (purchased and applied by householders) tends to be described by a wide range of exotic names, some of which relate directly to British Standard or RAL, others are specific to a paint producer. Paint manufacturers produce colour cards and small sample pots of paint (which cover approximately $2\,m^2$ of wall) in an attempt to help people make a choice from the vast range of colours available. Colour appears to change in different sized rooms, and against other colours. Natural light also varies in its quality through the day and this too will influence the appearance of a particular paint colour.

Patterned and textured wallpaper, by its very nature, has to be specified by manufacturer, pattern name and reference number. Rolls carry a pattern reference number and a batch number. Rolls of wallpaper with the same batch number should provide the best match for colour and texture.

British Standard
Under the British Standard system a specific colour is defined by a three-part code. For example Magnolia is defined as 08 B 15, Black as 00 E 53. The first part of the code defines the hue (ranging from 00 to 24), the second part the greyness (A to E) and the third part the weight (a subjective scale from 01 to 58, which incorporates reflectivity to incident light and greyness). To specify a paint colour it is only necessary to specify the relevant reference, e.g. 08 B 15, although it is common practice to give the paint name (in this case Magnolia) as well.

Required finish
The type of paint finish required will also influence the visual appearance of paint colour. These range from flat, full gloss, gloss, low sheen, matt and semi-gloss and must be specified along with the appropriate reference number. Manufacturers may produce some colours in one finish only, while other paints are manufactured in several different finishes. For example, a paint finish to an old property would most likely be in a flat finish to echo the type of paint

finish available at the time. Paints may also be formulated for particular uses, e.g. bathroom and kitchen paint that can be wiped down without affecting the finish, and paints that do not need to be so robust, e.g. ceiling paints. A number of 'effects' paints are also available, for example water-based paints for interior use that give a metallic finish, and acrylic-based paints for exterior effects.

Paint systems

Paint is composed of a number of ingredients, each with a specific purpose. Typically paint will include a binder, driers, solvents, pigments and a base material. The medium, or binder, solidifies after application to create the protective paint film. Alkyd resins and vinyl or acrylic resins have replaced linseed oil as the medium for the majority of paints. Driers are added to induce the polymerisation of the binder to ensure rapid drying. The solvent, either water or an organic material, helps to create fluidity to the paint to facilitate painting with a brush and/or roller and spraying. Water-based paints have better environmental and health and safety credentials than those made from organic solvents. Colour is added with organic and inorganic dyes and pigments. Opacity is achieved through the addition of a base material, usually titanium dioxide, and other inert extenders such as silica.

The finished paint system is made up of a series of 'coats' or layers, each coat of paint performing a specific task. For internal emulsion paint it is common practice to apply a 'mist coat' to the plastered wall, the substrate. This is emulsion thinned with water to act as an undercoat, followed by two finishing coats of the emulsion. For external and internal gloss paint, the base material, the substrate, requires a coat of primer, followed by an undercoat and one or two finishing coats. Paint manufacturers provide extensive guidance on the different types of paint that they manufacture and typical specifications for application to a wide range of materials. Higher specification paints are manufactured for external use in exposed conditions and marine/coastal climates.

Substrate

The surface of the material to be painted is known as the substrate. The type of material and its condition will determine the type of paint to be used. Preparation of the substrate is an important factor in achieving a durable paint finish, and preparation can account for as much as 50% of the cost of painting a surface.

Primers

The purpose of the primer is to adhere to the substrate, to provide protection from corrosion or deterioration and to offer a good base for the undercoat. The substrate material must be free of all loose material, be clean and, in the case of timber, have low moisture content at the time of application. Different primers are available for different materials and circumstances.

Undercoats

Undercoats provide an opaque cover and a good base for the finishing coat. It is usual practice to paint the undercoat in a different colour to the top, finishing coats to help the decorator monitor progress more easily. Undercoats are normally based on acrylic emulsions or alkyd resins.

Finishing coats

The finishing coat provides the protective and decorative surface finish. The exact colour and type of finish must be specified. The finishing coat should be inspected regularly for signs of damage and promptly repaired by removing loose paint and repainting to ensure protection to the substrate.

Application of paints

Paints and stains can be applied by brush, roller or application pads. For special effects, cotton rags and sponges may be used. All paints should be applied in accordance with manufacturers' instructions and in accordance with prevailing health and safety legislation and guidance. Low-odour paints should be specific where possible. Internal rooms should be well ventilated to avoid a build-up of fumes. Where paints are sprayed, protective breathing apparatus and protective clothing must be worn. The temperature of the surface to be painted is also a determining factor in achieving a durable finish. Most paint manufacturers recommend application at a surface temperature of 10°C or above and recommend against application of paint in extremely hot or cold weather.

Storage and disposal

Storage of paint presents a fire safety hazard and it must be stored in accordance with current health and safety regulations. Disposal of paints should also be undertaken in accordance with the manufacturer's instructions and due regard for environmental laws.

Special paints

Specially formulated paints have a wide variety of uses; the more common ones are described here.

Masonry paints

External walls of fair face brick, block, stone, concrete and render may be painted with masonry paint. These are predominantly water-based acrylic resin-based products, although solvent-based systems and mineral silicate paints are also produced. The paint usually contains fungicides to prevent, or at least delay, the growth of moulds and algae on the paint surface. Masonry paint is available in a range of colours and is produced to provide a smooth finish or a sand textured finish. The sand textured finish is more suited to covering fine cracks in render than the smooth finish. Larger cracks should be repaired with flexible filler prior to decoration.

Fungicide paints

In internal areas where mould growth is a problem (for example kitchens and bathrooms) it is common to paint walls with fungicide paint (sometimes marketed as mould protection paint). This acrylic paint contains a mix of fungicides to resist mould growth and is both washable and durable. The paint is formulated to release the fungicide gradually over a long period. While a cheap and effective solution, efforts should also be made to improve ventilation to the problem area, to determine the cause of the mould growth and to fix the problem (if possible) prior to decoration.

Multicolour (fleck) paints

Multicolour paints, applied by roller or spray system, incorporate coloured flecks to help disguise unsightly surfaces (and cover graffiti) and provide some deterrent to new graffiti. Some of these systems are designed to make the removal of graffiti easier compared to other paint systems.

Water-repellent paints

Water repellent paints may be applied to porous surfaces to prevent water penetrating the wall, yet still allowing the evaporation of water from the masonry. These silicone based paints can be applied to brickwork, concrete, stone and render to provide a moderate degree of protection to the building fabric.

Waterproofing paints

Epoxy waterproofing paints provide an impervious surface finish to surfaces and tend to be used where high humidity or wet operations would cause damage to normal paint finishes. Epoxy ester paint systems are also highly resistant to spillages of oil and some chemicals. In industrial buildings these paints may be used to provide a surface finish to a 'wet room' for washing equipment etc. In existing buildings, epoxy waterproofing systems may be used to improve the wall finish to damp basement walls, provided that a good bond can be achieved between the wall and the epoxy system. Bituminous paints serve a similar function, forming a waterproof finish to masonry and metal.

Heat resisting paints

Heat resisting paints are formulated to resist high temperatures. For example, aluminium paint is resistant to temperatures of up to around 250°C.

Flame-retardant paints

A flame-retardant paint will give off non-combustible gases when subject to fire, thus retarding the surface spread of flame.

Intumescent coatings

Fire protection to structural steel is achieved through a thin coat (1 to 2 mm) of intumescent paint, usually applied by spray guns in the steel fabricator's paint shop to control paint thickness. When subjected to fire, the paint expands to form a thick layer of insulating foam. Intumescent coatings are applied to give fire protection times of 30, 60 and 120 minutes. Intumescent emulsion paints and clear varnishes are also available for use on timber.

Wood stains

Wood stains are applied by brush and penetrate the surface of the timber, creating a permeable and water repellent sheen finish. Wood stains for exterior use are water-based or solvent-based and include a preservative basecoat, which helps to control rot and mould growth. Stains for interior use are also available. Wood stains are commonly specified as low, medium or high-build systems. For example, timber cladding and rough sawn timber would be applied with a low-build stain system, while external timber joinery, such as a smooth window cill, would benefit from a medium or high-build system. The stain is usually applied in two coats. The first coat will penetrate the surface of the timber and also adhere to the surface. The second coat creates a microporous surface finish that is water repellent. The finished stain is permeable to moisture movement, thus allowing the timber to 'breath' (unlike a gloss paint finish), which may help to reduce moisture-induced movement of the timber. The finished stain will fade on exposure to weather, and application of new top coats will be required to maintain an attractive finish and adequate weather protection to the substrate.

Varnishes for timber

Varnishes are applied by brush to create a protective surface film to timber. Polyurethane varnishes are either water-based or solvent-based systems and are available in gloss, satin or matt finishes. The solvent-based varnishes produce the harder and more durable coatings, e.g. 'yacht varnish'. Varnishes are available in a clear finish, to maintain the natural wood colour, are manufactured to enhance the natural colour of the wood (usually by darkening a light timber) and can be used to add some colour to the timber. For most applications two coats of varnish provide adequate protection.

Oils and waxes for timber

Oils and waxes are mostly used for internal applications to provide a protective finish to timber furniture. Natural oils, such as linseed oil, are liberally applied to timber with a soft cloth and any excess wiped off. Waxes, such as beeswax based polishes, are applied in a similar manner but the timber is polished with a soft cloth to bring out the natural grain of the timber and also to provide a gloss finish. Oils and waxes need to be applied on a regular basis to maintain appearance. Oils, such as teak oil, are also used externally to provide weather protection to, for example, timber garden tables and chairs. The oils for external use are formulated to produce an ultraviolet resistant and microporous finish. Regular application will be required to maintain an attractive appearance and weather protection in situations where the furniture is exposed to sunlight and rain. A wide variety of finishes is available, ranging from a transparent ('natural') finish to an opaque one.

Wallpaper

Wallpaper is the term used to cover a wide variety of sheet papers produced for use on internal walls. The papers are mainly used in domestic premises, applied by professional decorators or by householders. Where papers are applied in commercial premises it is important to check that the application of a paper does not compromise fire safety (surface spread of flame).

Papers adhere to the wall by means of a water-based adhesive paste. The paste may be mixed with water and applied to the wall or more commonly to the back of the paper. The sheet of pasted paper is then 'hung' on the wall. The paper is carefully positioned and then the surface is brushed over with a soft brush to apply enough pressure to remove any air bubbles that have formed under the paper. Once positioned, the sheet of wallpaper can be trimmed at the ceiling and skirting with a sharp knife. Some papers are manufactured with an adhesive back which is activated by soaking the paper in water for a few minutes or by brushing water to the back of the paper before it is applied to the wall. Some skill is required in keeping the sheets of paper plumb and patterns lined up at the join (especially if the wall is slightly out of true).

Lining paper

Lining paper is a relatively dense plain paper with a smooth finish that provides a good key for emulsion paints. The paper is applied to walls that are slightly uneven to provide a relatively smooth surface on which to apply emulsion paint.

Plain textured papers

In situations where internal walls are in poor condition, i.e. cracked and uneven, then a heavily textured paper may be used to disguise the poor wall finish. The texture is created when the paper is rolled or created by adding texture to the paper, for example very small wood chips. These textured papers provide a good key for the application of emulsion paint. This is a cheaper solution than applying a skim coat of finishing plaster (which provides a smooth and level finish) but the textured finish may not suit all tastes. Woodchip paper tends to be used on walls where the plaster is in very poor condition and/or the walls are badly cracked, something worth remembering when inspecting a property.

Patterned papers

Patterned papers are applied to walls to create an attractive internal finish. The vast majority of wallpaper is machine printed and mass produced, the popular ranges being the cheapest. Hand-painted wallpaper may occasionally be commissioned for prestigious refurbishment projects. These papers are expensive and require special skills to apply them correctly. The choice of wallpaper is a personal one.

Thin strips of paper are also produced to provide a border, for example at the junction of ceiling and wall, in a range to complement the chosen pattern.

Patterned textured papers

Wide ranges of papers are also available that have both a pattern and a textured surface.

Washable papers

Ranges of washable vinyl papers are manufactured for use in rooms where they may need to be wiped down. Examples may be in a kitchen, bathroom or a child's playroom.

11 Heating, Cooling and Services Provision

Innovation in building services provision since the Industrial Revolution has primarily focused on the control of the internal environment through mechanical and electrical systems that consume vast amounts of energy. Large internal spaces such as office buildings, retail units and leisure buildings are largely isolated from changes in climate, with a constant temperature and humidity maintained by sophisticated services equipment. With greater attention being given to the amount of energy consumed by buildings over their life, attention has started to turn to the specification of more energy efficient mechanical systems and also to passive systems that consume little or no energy. More thought also needs to be given to the way in which we incorporate services into the building fabric to make them more accessible and hence easier to maintain, replace and upgrade.

11.1 Functional requirements

Services design must form an integral part of the design philosophy for the building. Input from appropriate specialist engineers and suppliers must be made early in the design process to ensure compatibility and ease of constructability on site. The manner in which the building is to be constructed and the way in which the various services are to be included deserves much more attention than it is often given. Heating, cooling and services equipment will require regular maintenance and will have a much shorter service life than the main building structure, thus ease of access for repair, maintenance and upgrading is an important consideration for all building types. Encasing pipes and wires in concrete (for example) is not uncommon, yet clearly is not a very sensible approach.

Functional requirements will vary depending on the type of building and the anticipated use of space within the building. Generic functional requirements are:

- ❑ Easy and safe access for installation, maintenance and replacement
- ❑ Low (preferably zero) energy consumption during operation
- ❑ Low (preferably zero) emissions
- ❑ Ease of control by building users (manual and/or by smart technologies)

Buildings and energy useage

The majority of the energy produced in the UK comes from the combustion of fossil fuels. This process produces carbon dioxide (CO_2). Buildings use energy, and our building stock accounts for approximately 50% of all CO_2 emissions. Approximately two thirds of these building-related emissions are from the domestic sector and the remaining third from the commercial sector. The biggest energy use is for heating internal space (typically somewhere around 50%), followed by lighting (around 15%). Building design and the care with which buildings are constructed will influence the energy consumption of buildings and hence the CO_2 emissions for the building as a whole. For example, poor work may result in unnecessary air leakage. Reducing energy use can be addressed relatively easily with new buildings but is more problematic for the majority of the existing building stock.

Emphasis should be on reducing the amount of energy consumed by the building in use and on the provision of clean and sustainable energy sources. Increasing the thermal insulation of the building fabric and eliminating unwanted air leakage are obvious areas to address. Similarly, installing or upgrading to more energy efficient space heating systems will help to reduce energy usage, and the associated pollution. There are a growing number of zero energy demonstration buildings around the UK and some good examples in Europe from which to take inspiration. Some of these buildings also incorporate passive ventilation principles.

11.2 Indoor climate control

The control of the indoor climate to create a healthy environment has been a cause for concern for a long time. When we fail to create a healthy and enjoyable indoor environment, we can experience what is known as sick building syndrome (SBS). This is where the occupants of a building feel unwell when they are in the building, the discomfort caused by a combination of physical and psychological factors.

Technological innovations in the materials used for the building fabric (the building envelope) can play a significant role in the creation of a comfortable internal environment that is also energy efficient. The manner in which the building envelope responds to solar radiation, how heat loss (and gain) is controlled, how ventilation is provided and how noise is excluded are key factors.

Comfort

A large number of variables, both physical and psychological, influence our feeling of comfort within a building. These include:

❑ Air quality
❑ Air movement

❑ Presence of odours
❑ Ambient water vapour pressure (humidity)
❑ Radiant temperature
❑ Air temperature (and temperature gradients within the space)
❑ Activity level of users
❑ Type of clothing worn by users
❑ Natural and artificial lighting levels
❑ Background noise levels
❑ Interaction with other users
❑ Degree of individual control over the internal environment

Our comfort levels are subjective and individual. Discomfort will be felt if there is a large variation in one or more of the conditions, for example the air movement being too high (perceived as a draught and hence uncomfortable). The degree of control an individual has over his or her internal environment is an important consideration in the design of appropriate systems for shared spaces, e.g. office buildings. This too may influence our perception of comfort.

Heating

Space heating may be provided by a number of energy sources, the most common of which are described briefly here. Efficiency of the system is dependent upon the heat source and the design of pipe or duct runs.

Gas-fired boilers

Conventional boilers have the lowest initial cost due to simplicity of design and relatively limited insulation levels to the boiler. High-efficiency boilers have a higher initial cost and have more efficient heat exchangers and higher levels of insulation to the boiler casing. Condensing boilers have an enlarged, or additional, heat exchanger and are the most efficient of the three boilers. Under all conditions the condensing boiler will recover heat from the flue gases and will also recover heat from the condensation of water vapour in the flue gases under suitable conditions. High efficiency and condensing boilers should be specified since they are the most efficient and the reduced energy usage is accompanied by a reduction in CO_2 emissions. Manufacturers of gas boilers have made significant advances in their product range in recent years, thus specifiers are urged to consult manufacturers for technical advice and guidance on initial cost, cost in use, maintenance requirements and emissions of CO_2, nitric oxide and nitrogen dioxide (collectively known as NOx), and sulphur dioxide and sulphur trioxide (collectively known as SOx).

Electric systems

Electricity may be used for wall mounted convection heaters, storage radiators and underfloor heating systems. The benefits of electricity over gas are mainly associated with ease of installation and adaptability, as well as reduced installation costs and maintenance costs. Running costs may be lower than gas if off-peak electricity is used; however, we would urge readers to carry out a comprehensive life cycle economic evaluation of all systems prior to installation.

Active solar heating

Collectors convert solar radiation into heat for heating hot water and for space heating. Collectors comprise a black surface to absorb incoming radiation and are glazed and insulated to help reduce heat losses. Collectors are positioned (often on the roof) to optimise the amount of energy falling on them, usually on south-facing roof slopes. The heat generated in the collector is circulated to a storage device (hot water cylinder) by a water circuit. A back-up system (e.g. electrical immersion heater) provides heat during periods when solar heating is insufficient. Recent technological advances have combined photovoltaic technology with active solar heating systems.

Earth to air heat exchangers

Earth to air heat exchanges work on the principle that the temperature below the surface of the ground is more constant than the air temperature; also at considerable depth, the temperature below ground increases. By taking advantage of the warmer or cooler temperatures below ground a heat exchange system can be used to keep dwellings warmer in the summer and cooler in the winter. The heat exchange pipes can be run horizontally at shallow depths or vertically, sunk in bored holes, at much greater depths (Figure 11.1). Systems

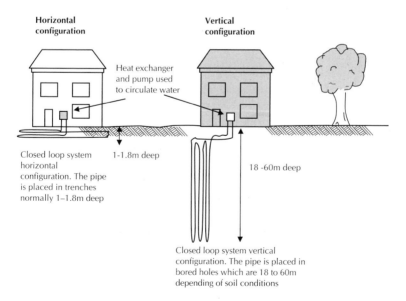

Figure 11.1   Earth to air heat exchanger.

are normally closed, meaning that water, mixed with an antifreeze liquid, is contained in the pipes and runs around a complete circuit, being continually pumped around. Alternatively, if there are deep wells or other water sources nearby, the system can simply draw off the water, feed it around the building, through the heat exchanger, and back to the original water source. Open systems are usually used for cooling purposes.

Heat distribution systems

Choice of heat emitter must be made to suit the construction of the building and the needs of the users. Consideration should be given to:

❑ Thermal response of building fabric
❑ Pattern of use of internal building spaces
❑ Appearance (available space)
❑ Individual control
❑ Installation and maintenance costs

The type of thermal response is usually defined as rapid, intermediate or slow. If the building is used sporadically it may be beneficial to have a rapid thermal response. Conversely if the building were used continually throughout the heating period, a slow thermal response would normally be a better choice. The common heat emitters are:

❑ Rapid: electric fan convectors, forced warm air systems. Quick to respond to changes in external air temperature. Good for lightweight building fabric.
❑ Intermediate: panel radiators, electric convectors, perimeter heating. Common in many buildings. Wide applicability.
❑ Slow: underfloor heating, electric storage radiators. Slow to respond to changes in external air temperature. Good for buildings with a high mass.

Warm air systems (rapid)

Gas fired warm air systems can be designed to include a mechanical ventilation heat recovery unit. Ductwork to carry the warm air is typically run in floor and roof voids. Outlets tend to be positioned in the floor and/or at a low level in the internal walls. Heating of the room is rapid but so too is cooling.

Water filled radiators (intermediate)

Hydronic heating systems with a gas-fired energy source are the most common systems installed in the UK. Pipework is run to wall-mounted panel radiators, which come in a wide range of sizes and designs. The boiler will have a thermostatic device for general control of output. A thermostatic valve connected to each radiator controls individual room temperature locally. The thermostatic valve is a relatively simple valve that is adjusted by hand to control the amount of hot water flowing through the radiator, and hence the temperature of the radiator and surrounding space.

Underfloor heating (slow)
Underfloor heating is more energy efficient than radiators because the heat is more evenly distributed around the internal space. The absence of radiators also allows greater flexibility in the use of wall space. Underfloor heating is particularly efficient when used with ceramic and stone tiles as a floor finish but can equally be used with timber and carpeted floor finishes. There are two systems, either 'wet' or 'dry'.

The wet system comprises hot water running through plastic pipes, usually fed from a gas-fired boiler. The water temperature in the pipes is less than that required for radiators, hence making the system suitable for use with solar collectors. The dry system comprises electric cables placed on a reflective sheet placed on top of the floor insulation. Both systems warm the room from the floor up, creating a decreasing heat gradient from the floor upward. Although slow to warm the air temperature compared with other systems, the heat retained in the floor structure will ensure a relatively stable internal air temperature throughout the heating season. Systems can be zoned (by room or area) to provide individual control to minimise energy consumption. Access for repair and/or replacement can be disruptive since the floor finish will need to be lifted and replaced.

Ventilation

Naturally ventilated buildings are a feature of vernacular architecture, with carefully positioned opening windows and louvered vents providing sufficient air flow through the building. Vernacular buildings were designed and built to respect their local climate and use it to maximise internal comfort. Mechanical ventilation systems are a feature of more recent, highly serviced buildings, which are sealed to the outside environment. In many modern buildings the local climate is largely ignored and the internal environment controlled solely by mechanical and electrical equipment.

Naturally ventilated spaces
The use of passive ventilation is gaining widespread acceptance as an environmentally responsible way in which to ventilate buildings. The art is still being rediscovered and is set to evolve further over the coming years as mechanical systems lose favour. Natural ventilation is, in principle, rather simple. The reality is that there is a lot more work required before consistent and reliable design guidance can be given. Naturally ventilated buildings are usually

ventilated from one side, or cross-ventilated, and also rely on the stack effect for effective ventilation (illustrated in Figure 11.2). Because it relies on natural forces, natural ventilation is less predictable than mechanical systems but, when implemented successfully, it is possible to make extremely large savings on capital and running costs.

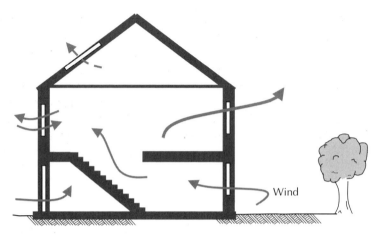

Section: Showing stack effect and cross ventilation

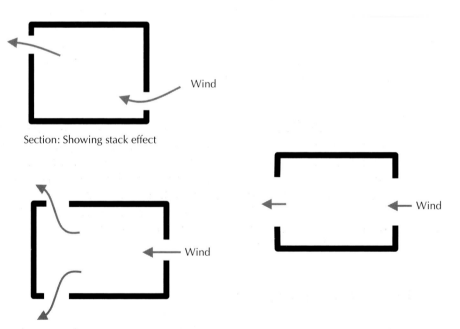

Section: Showing stack effect

Plans showing simplified cross-ventilation paths

Figure 11.2   Stack effect and cross ventilation.

Assisted natural ventilation systems

Assisted natural ventilation systems offer the benefit of knowing that the system will be able to cope with peak demands. Simple ceiling-mounted fans may be sufficient for the purpose. Alternatively systems are available that incorporate fans and heat recovery devices.

Mechanical ventilation systems

Mechanical ventilation with ductwork systems is widely used in commercial buildings. Mechanical air handling plant takes up a lot of space (usually positioned within the basement or/and on the roof) and consumes a lot of energy. Ductwork tends to be quite large in cross-section, and necessitates the use of raised floors and suspended ceilings to conceal the ducts. Thus the design of the building must take into account the quite considerable space requirements of mechanical plant. This is reflected in the initial cost, and complexity, of the building. Although advances have been made recently to reduce the energy requirements of mechanical ventilation systems (and hence the associated pollution), the majority of systems are expensive to run and maintain.

Cooling

With rising concern over global warming there has been increased interest in mechanical cooling systems for residential and commercial premises. We have only provided a very small amount of space here because a well-designed and constructed building anywhere in the UK, using natural ventilation and appropriate shading devices, should have no need for mechanical cooling systems. Unfortunately many buildings are not well designed and so some provision is, and increasingly will be, necessary.

Shading devices (with pv cells)

Shading to windows can be provided by curtains and blinds and/or by physical shading devices fixed or built into the external face of the building. Traditional designs featured deep window reveals that provided a degree of protection from unwanted solar glare and gain during the summer months. The trend for flat facades to buildings means that the shading has to be incorporated into the facade design, with blinds and solar control glazing. Vertical fins can be bolted to the facade to provide shading and also to provide some relief to an otherwise 'flat' building.

11.3 Hot water supply systems

There are two hot water supply systems, the central and the local, illustrated in Figure 11.3. The central system is suited, for example, to houses, hotels, offices and flats where a central boiler fired by solid fuel, oil, gas or electricity heats water in bulk for distribution through a straightforward vertical distributing pipe system with short draw-off branches leading to taps to sanitary appliances on each floor. In large buildings, one heat source may serve two or more hot water storage cylinders to avoid excessively long distribution pipe runs. The local system is used for local washing facilities where gas or electricity is run to the local heater either to avoid extensive and therefore uneconomic supply or distributing pipe runs, or because local control is an advantage. In some buildings it may be economic to use a combination of central and local hot-water systems. The difference between these systems is that with the central system hot water is run to the site of the sanitary appliances from a central heat source, and with the local system the heat source, gas or electricity, is run to the local heater, which is adjacent to the sanitary appliances.

Hot water supply

Central hot water supply

From Figure 11.3 it will be seen that water is heated and stored in a central cylinder from which a pump circulates it around a distributing pipe system from which hot water is drawn. In the two-storey house used to illustrate hot water distribution or supply pipe systems, the hot water was drawn directly from single branches. In a small building, such as a house, where the sanitary fittings are compactly sited close to the cylinder, the slight inconvenience of

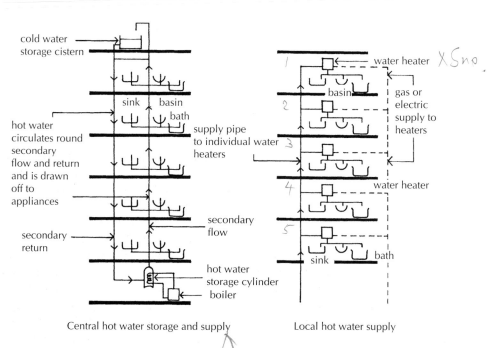

Central hot water storage and supply   Local hot water supply

Figure 11.3   Hot water supply systems.

hot water is pumped

running off the cooled water in the single branches before hot water is discharged is acceptable. In larger buildings the inconvenience of running off cooled water from long single branches (known as 'dead legs') is an unacceptable waste of water and energy. The length of these dead legs is limited to 20 m for 12 mm pipes ranging to 3 m for pipes more than 28 mm, unless the pipes are adequately insulated against loss of heat.

The storage cylinder contains hot water sufficient for both anticipated peak demand and demands during the recharge period. The system is therefore designed to supply hot water on demand at all times. The one disadvantage of the system is that there is some loss of heat from the distributing pipes no matter how adequately they are insulated. This is outweighed by the economy and convenience of one central heat source that can be fired by the cheapest fuel available, and one hot water source to install, supply and maintain hot water being at hand constantly by simply turning a tap. Where a mains pressure supply system is used, the supply pipe connects to the unvented hot water storage cylinder from which supply pipes connect to the fittings, and there is no roof level storage system.

Local hot water supply

A water heater, adjacent to the fittings to be supplied, is fired by gas or electricity run to the site of the heater. The water is either heated and stored locally or heated instantaneously as it flows through the heater. The advantages of this system are that there is a minimum of distributing pipework, initial outlay is comparatively low, and the control and payment for fuel can be local, an advantage, for example, to the landlord of residential flats. The disadvantage is that local heaters are more expensive to run and maintain than one central system.

Hot water storage heaters

The local hot water storage heater consists of a heat source and a storage cylinder or tank, and the instantaneous heater of a heat source through or around which cold water runs and is heated instantaneously as it is run off. The larger water storage heaters are used to supply hot water to ranges of fittings such as basins, showers and baths used in communal changing rooms of sports pavilions and washrooms of students' hostels.

The large, gas-fired, water storage heater illustrated in Figure 11.4 consists of a water storage cylinder through which a heat exchanger rises to a flue from a combustion chamber. A thermostat in the water storage chamber controls the operation of the gas burners in the combustion chamber, cutting in to fire when the temperature of the water falls. A cold water supply pipe is connected to the base of the storage cylinder. Hot water is drawn off either through a dead-leg draw-off pipe where pipe runs are short, or by a circulating secondary pipe system where runs are lengthy. The storage heater is heavily insulated to

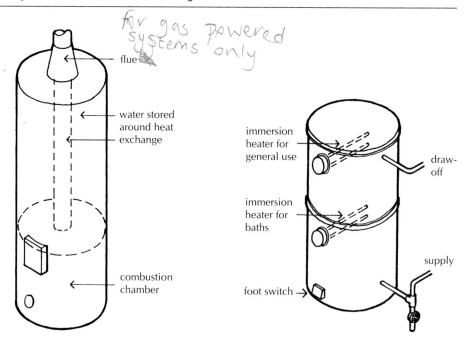

for gas powered systems only

Figure 11.4   Gas water storage heater.   Figure 11.5   Electric water storage heater.

conserve energy. The size of the heater is determined by the anticipated use of hot water at times of peak use.

The electric water storage heater, illustrated in Figure 11.5, is for use in communal washrooms of students' hostels and residential schools where peak demand for hot water in bulk is generally confined to mornings and evenings, between which times the heater automatically reheats the water. These heaters, which are heavily insulated to conserve energy, are housed in a separate enclosure away from the wet activities they serve, for safety reasons. Hot water for basins is drawn from the top of the cylinder, which is heated by an upper immersion heater. Hot water in bulk for baths is boosted by the operation of both the upper and lower immersion heaters. A thermostat and/or a timed switch operate the lower immersion heater. An advantage of the electric storage heater is that it does not have to be fixed close to an outside wall, which the gas heater does because of its flue.

The small electric, single point water heater, illustrated in Figure 11.6, is designed to heat and store a small volume of water for the supply to single basins. The hot water storage cylinder and electric immersion heater, which are heavily insulated to conserve energy, are housed in a glazed enamel metal casing for appearance sake. A thermostat controls the electric supply to the immersion heater, cutting-in to reheat the water as it is drawn off. These heaters are used for basins in single toilets where it is convenient to run water and an electrical supply for the occasional use of hot water.

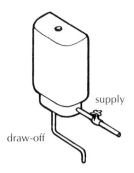

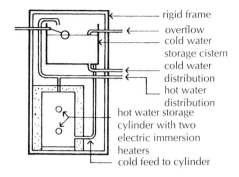

rigid frame
overflow
cold water
storage cistern
cold water
distribution
hot water
distribution
hot water storage
cylinder with two
electric immersion
heaters
cold feed to cylinder

**Figure 11.6 Electric storage single
point water heater.**

**Figure 11.7   Combined cold and hot water
storage unit.**

A combined cold cistern and hot water storage heater that is designed to fit into a confined space is illustrated in Figure 11.7. Inside a rigid frame, a cold water storage cistern and an insulated hot water storage cylinder are combined with connections for water and electric supplies, and hot and cold water draw-off connections. These units are designed specifically for use in small flats where space is limited, and they can be fitted close to the bathroom and kitchen. A disadvantage of these units is that there is poor discharge of water from outlets because of the small head pressure of water from the cold water storage cistern.

Instantaneous water heaters

These water heaters operate by running cold water around a heat exchanger so that water is heated as it flows. The heat exchanger only operates when water is flowing, hence the name instantaneous water heater. Because the temperature of the water at the outlet is dependent on the rate of flow of water, there is a limitation on the rate of flow from the outlet if the water is to be hot. Consequently the rate of flow from these heaters is limited.

Most instantaneous water heaters are fired by gas, which is ignited by a pilot light immediately water flows, to provide hot water instantaneously. Cold water running through a coil of pipework, wrapped around a combustion chamber and heat exchanger over a gas burner, is heated by the time it reaches the outlet. The cold water supply valve controls these heaters. When the valve is opened, the flow of water opens a gas valve to ignite the burners to heat water. A single point, gas, instantaneous water heater is designed to supply hot water to single fittings such as a basin or sink. These heaters are usually fixed above the fitting to be supplied, and the hot water is delivered through a swivel outlet. A typical single point heater is illustrated in Figure 11.8. Because of their small output and limited use, the air intake from the room and the exhaust outlet to the room is acceptable. Where effective draught seals are fitted to windows to rooms in which these heaters are fired, there should be permanent ventilation to the open air.

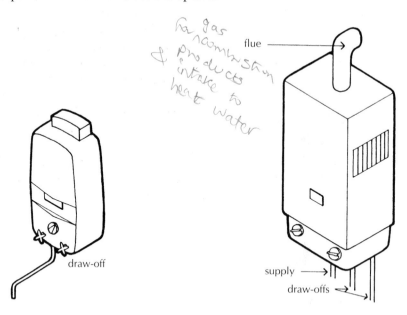

for gas combustion & products intake to heat water

Figure 11.8   Gas instantaneous single point water heater.

Figure 11.9   Gas instantaneous multi-point water heater.

to basin, sink & bath

Multi-point instantaneous gas water heater

These heaters, illustrated in Figure 11.9, can supply hot water to a sink, basin and bath through dead-leg draw-off pipes to fittings, hence the name multi-point. When a tap over one of the fittings is opened, the flow of cold water through the coils of pipe around the heat exchanger is heated to deliver hot water. The initial flow of cold water opens a gas valve and the pilot light ignites the gas burners to provide heat. When all the taps to fittings are shut and there is no water flow, the gas valve shuts.

The rate of flow of hot water from these heaters is limited by the need for sufficient time to allow an adequate exchange of heat to the water coils in the heat exchanger. When more than one tap is opened there will be a restricted rate of flow of hot water, and filling a bath can be a somewhat lengthy process. For these reasons these heaters are much less used than they used to be. The comparatively large output from these heaters necessitates a flue to open air to exhaust combustion gases, and also an adequate intake of air for efficient and safe combustion of gases. A flue and permanent air vents or a balanced flue are necessary. The gas valves in these heaters will only operate when there is comparatively high water pressure, such as that from a main supply, or a good head of water from a cistern.

flue

Water head

Instantaneous electric water heater

Water is heated as it flows through coiled heating elements immersed in a compact, sealed tank. A flow switch in the cold water supply inlet operates the electric supply. The control valve on the cold water supply pipe is used to adjust the temperature of the hot water outlet. These heaters, illustrated in Figure 11.10, are fixed over the single appliance to be supplied, with hot water delivered through a swivel outlet discharging over the appliance. The output of hot water from these heaters is limited by the rate of exchange of heat from the heating elements to cold water. In hard water areas limescale will coat the heating element and appreciably reduce the efficiency of these heaters. The heat exchange tank, which is heavily insulated, is housed in a glazed enamel casing for appearance sake. The advantage of these heaters is that they are compact, require only one visible supply pipe and may be fixed in internal, unventilated toilets as they have no need for air intake or a flue.

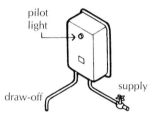

Figure 11.10   Electric instantaneous single point water heater.

11.4 Water services to multi-storey buildings

Mains water is supplied under pressure from the head of water from a reservoir, or a pumped head of water, or a combination of both. The level to which mains water will rise in a building depends on the level of the building relative to that of the reservoir from which the mains water is drawn, or relative to the artificial head of water created by pumps. In built-up areas there will, at times such as early morning, be a peak demand on the mains supply, resulting in reduced pressure available from the water main. It is the pressure available at peak demand times that will determine whether or not mains supply pressure is sufficient to feed cold water to upper water outlets. The pressure available varies from place to place depending on natural or artificial water pressure, intensity of demand on the main at peak demand time and the relative level of the building to the available supply pressure.

Mains pressure supply

Figure 11.11 is a diagram of an eight-storey building where the mains pressure at peak demand times is sufficient to supply all cold and hot water outlets to all floors. Two supply pipe risers branch to provide cold water to ranges of sanitary fittings in male and female toilets on each floor, and another riser branches to feed hot water storage cylinders on each floor for the toilets. This is the most economic arrangement of pipework where sanitary fittings are grouped on each floor, one above the other. To provide reasonable equality of flow from outlets on each floor, it is usual to reduce pipe sizes. In the arrangement shown in Figure 11.11, the bore of the risers will be gradually reduced down the height of the building to provide a reasonable flow from all outlets to compensate for reduced flows as pumped head pressure increases down the height of the building. Where the mains pressure is sufficient to provide a supply to a multi-storey building, it is not always possible to provide a reasonable equality of flow to fittings on each floor by varying pipe sizes alone. An increase in pipe size will provide a little reduction in pressure loss from the frictional resistance to flow of larger bore pipes and fittings. There is a limit to the resistance to flow that can be effected, because of the limited range of pipe sizes, without using uneconomic pipe sizes. Another method of providing equalisation of rate of flow from outlets floor by floor in multi-storey buildings is by the use of pressure-reducing valves at each flow level.

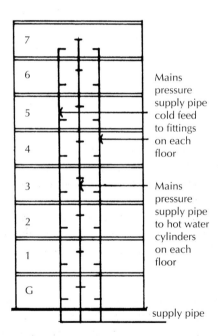

Figure 11.11   Mains pressure supply.

Multiple risers

Another method of equalising flow, being used experimentally, is the use of multiple rising pipes as illustrated in Figure 11.12. One rising supply pipe branches to supply the three lower floors and then rises to supply the three

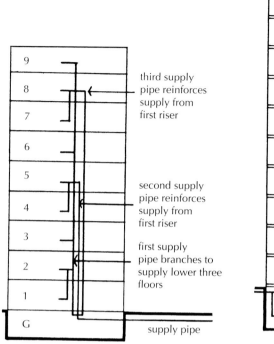

third supply
pipe reinforces
supply from
first riser

second supply
pipe reinforces
supply from
first riser

first supply
pipe branches to
supply lower three
floors

supply pipe

Figure 11.12 Multiple risers.

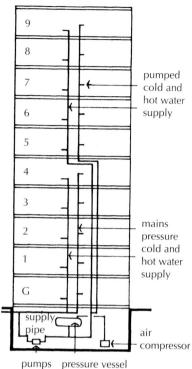

pumped
cold and
hot water
supply

mains
pressure
cold and
hot water
supply

supply
pipe

air
compressor

pumps   pressure vessel

**Figure 11.13 Pumped and mains
pressure supply.**

floors above, where it is joined and its supply reinforced by the second rising supply and then up to the top three floors where it is joined and reinforced by the third rising supply pipe. The logic in the use of this arrangement is that in multi-storey buildings in multiple occupation, such as a hotel, there will be unpredictable short periods of peak use. During this period there may be heavy call on water use on one floor, which may cause unacceptable starvation of water supply. By providing alternative reinforcing sources of supply and judicious arrangements of pipe sizes, a reduction of flow rate may be avoided or at least smoothed out.

It is good practice in the design of pipework layout to make an assumption of frequency of use of draw-off water to sanitary appliances. The assumption is based on an estimate of peak period use, which does not allow for unpredictable heavy use. To provide for possible maximum use would involve uneconomic pipework. Where mains pressure is insufficient at peak demand times, it is necessary to install pumps in the building to raise water to the higher water outlets. In this situation it is usual to supply from the supply pipe those outlets that the mains pressure will reach, and those above by the pumped supply, to limit the load on the pumps as illustrated in Figure 11.13.

Two mains pressure supply pipes rise to supply the lower five floors, one with branches to each floor level to cold water outlets and the other with branches at each floor level to hot water storage cylinders. The five upper floors are supplied by two rising supply pipes under the pressure of the pump in the basement. At each floor level, branches supply cold water to sanitary fittings and there are branches to hot water cylinders. There are two pumps, one operating and the other as standby in case of failure and to operate during maintenance. The pumps are supplied by the mains through a double check valve assembly to prevent contamination of the supply by backflow should the pumps fail.

Auto-pneumatic pressure vessel

The auto-pneumatic pressure vessel indicated in Figures 11.13 and 11.14 is a sealed cylinder in which air in the upper part of the cylinder is under pressure from the water pumped into the lower part of the cylinder. The cushion of air under pressure serves to force water up the supply pipe to feed upper-level outlets as illustrated. Water is drawn from the auto-pneumatic pressure vessels as water is drawn from the upper-level outlets so that when the water level in the pressure vessel falls to a predetermined level, the float switch operates the pump to recharge the pressure vessel with water. Thus the cushion of air in the pressure vessel and its float switch control and limit the number of pump operations. In time, air inside the pressure vessel becomes mixed with water and is replaced automatically by the air compressor.

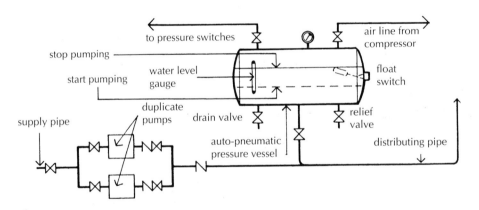

Figure 11.14   Auto-pneumatic pressure vessel.

Gravity feed cold water supply

In multi-storey buildings, where mains pressure is insufficient to raise water to roof level, cistern feed to cold water outlets may be used with a covered drinking water storage vessel or cistern. Figure 11.15 is a diagram of a ten-storey building in which the drinking water outlets to the lower five floors are fed by mains supply. Because the mains pressure is insufficient to raise water to roof level, the upper five floors are supplied with drinking water from a covered drinking water storage vessel or cistern. The roof level cold water storage cistern, the drinking water vessel or cistern and the drinking water outlets to the top five floors are fed by a pumped supply. The cold water storage cistern supplies cold water to all cold water outlets, other than the sink, and to hot water cylinders on each floor.

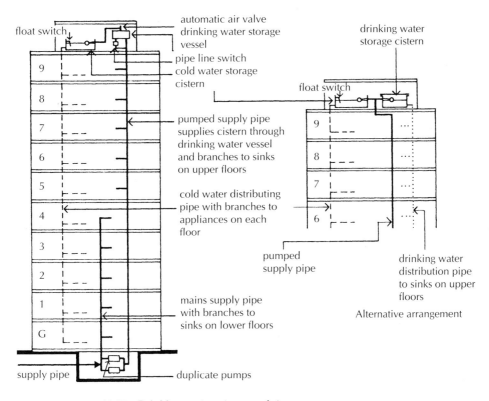

Figure 11.15   Drinking water storage cistern.

There is a duplication of rising pipework to the five lower floors, which is considered a worthwhile outlay in reducing the load and wear on the pumps. The pumped supply feeds both the higher drinking water outlets and cold water storage cistern through a drinking water storage vessel. The sealed drinking water storage vessel and the cold water storage cistern are filled through the pumped supply, in which a pipeline switch is fitted. A pipeline switch, illustrated in Figure 11.16, is used to limit pump operations. As water is drawn from the drinking water vessel, the water level falls until the float in the pipeline switch falls and starts the pump. The cold water storage cistern is supplied through the drinking water vessel, from which it can draw water to limit pump operations. When the water level in the cold water storage cistern falls to a pre-determined level, a float switch starts the pump to refill the cistern through the drinking water vessel.

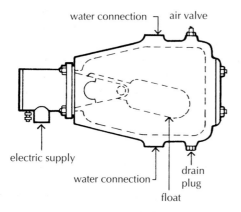

Figure 11.16   Pipeline switch.

Drinking water storage cistern

As an alternative to a sealed drinking water storage vessel, a drinking water storage cistern may be used, as illustrated in Figure 11.15. The cistern has a sealed cover and filtered air vent and overflow to exclude dirt and dust, as illustrated in Figure 11.17. A pump cut out, or float switch, controls the pump operations at a predetermined level. A float switch controls pump operations to fill the cold water storage cistern. The pumped service pipe feeds both cisterns

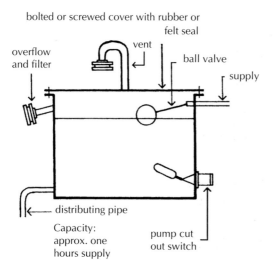

Figure 11.17   Drinking water storage cistern.

so that whichever switch operates, the pump operates to fill both cisterns. The advantage of the sealed drinking-water storage vessel is that it requires less maintenance than the drinking water cistern whose ball valve and filters require periodic maintenance. But the roof level switches have to be wired through a control box down to the basement level pumps, and the switches will require regular maintenance in a position difficult to access. As a check to the possibility of backflow from a pumped supply into the main, and as a reservoir against interruption of the mains supply, it has been practice to use a low level cistern as feed to the pumped supply to a roof level cistern, so that the air gap between the inlet and the water level in the cistern acts as a check to possible contamination of the mains supply.

Low level cistern

Where the mains pressure is insufficient to supply cold water outlets on upper floors of multi-storey buildings, and a separate drinking water supply is required, a cistern feed supply to cold water outlets may be used in combination with a pumped supply to upper level drinking water outlets, as illustrated in Figure 11.18. Here a low level cistern is fitted as supply to the pumps. A low level cistern is used as a form of check against contamination of the supply

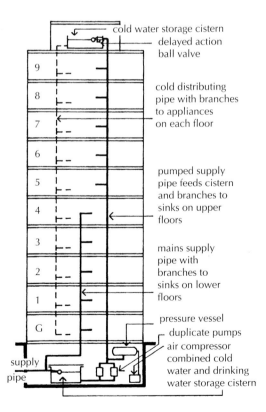

Figure 11.18 Low level storage cistern.

and as a standby against interruption of the mains supply. The covered low level cistern serves to supply the upper level drinking water outlets and the roof level storage cistern. The operation of the pumps is controlled at low level through a pressure vessel similar to that illustrated in Figure 11.14.

Drinking water outlets to the lower floors are connected to the mains supply pipe, which in turn feeds a low level storage cistern from which a supply is pumped to the upper level drinking water outlets and the roof level storage cistern. An auto-pneumatic pressure vessel, illustrated in Figure 11.14, and a delayed action ball valve to the roof level cistern, limits pump operations. As the low level cistern supplies both drinking and cold water outlets, it has to be sealed to maintain the purity of the drinking supply. A screened air inlet maintains the cistern at atmospheric pressure. The pumped supply pipe shown in Figure 11.18 feeds the roof-level cold water cistern, which is fitted with a delayed action ball valve.

Delayed action ball valve

A delayed action ball valve, illustrated in Figure 11.19, is fitted to the roof level cistern to control and reduce pump operations. (1) The delayed action ball valve consists of a metal cylinder (A) that fills with water when the cistern is full and in which a ball (B) floats to operate the valve (C) to shut off the supply. (2) Water is drawn from the cistern and, as the water level falls, float (F) falls and opens valve (D) to discharge the water from cylinder (A). The ball (B) falls (3) and opens the valve (C) to refill the cistern through the pump, so limiting the number of pump operations.

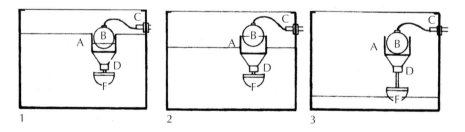

Figure 11.19   Delayed action ball valve.

Intermediate water storage cistern

Current practice in multi-storey buildings of more than about ten storeys is to use a roof level and one or more intermediate water storage cisterns to supply outlets other than drinking water taps. The intermediate cisterns spread the very considerable load of water storage and also serve to reduce the pressure in distributing pipes, for which reason they are sometimes termed 'break pressure cisterns'. Figure 11.20 shows a ground-level storage cistern supplying a pumped supply to an intermediate and roof-level cistern from which distributing pipes supply sanitary appliances on the lower and upper floors respectively, thus spreading the weight of water storage and limiting pressure in distributing pipes. The pumped supply also feeds drinking water outlets to upper floors. A float switch in the pressure vessel and delayed-action ball valves in the cisterns limit pump operations. Intermediate cisterns are used at about every tenth floor.

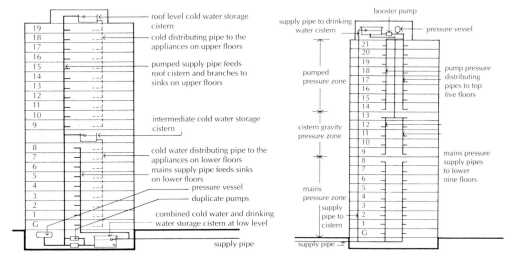

Figure 11.20   **Intermediate level storage cistern.**   Figure 11.21   **Zoned supply system.**

Zoned supply system

The supply to the 22-storey building illustrated in Figure 11.21 is divided into three zones in order to help equalise pressure and uniformity of flow from outlets on all floors. The supply to the lower nine floors is through two rising supply pipes taken directly from the mains supply. The supply to the eight top floors is from a pumped supply through a roof level pump and pressure vessel, and the supply to the intermediate five floors is by gravity from a sealed drinking water cistern at roof level. In this way the loss of residual head to the lower and upper floors is limited, and the loss of head from the cistern is limited by feeding intermediate floors only. By dividing the building into three zones of supply, loss of head is limited in the main supply and distribution pipes, and by reduction in pipe diameter in each zone, reasonable equalisation of flow from taps on each floor is possible.

11.5 Estimation of pipe sizes   22/12/15

To provide a reasonable rate of flow from outlets, the required size (bore) of pipe will depend on the static or pumped head of water pressure, the resistance to flow of the pipes, fittings and bends, and the assumed frequency of use of outlets. For small pipe installations such as the average dwelling where there are five to ten outlets, pipes of sufficient bore are used to allow simultaneous use of all outlets at peak use times. As only small bore pipes are required for this maximum rate of flow, there is no point in making an estimate of pipe size. With larger installations, such as the pipe system for a block of flats, it would be unrealistic and uneconomic in pipe size and cost to assume that all outlets will be in use simultaneously. It is usual, therefore, to make an assumption of the frequency of use of outlets, to estimate required pipe sizes that will give a reasonable rate of flow from outlets that it is assumed will be in use simultaneously at peak use times. If the actual simultaneous use is greater than the estimate then there will be reduced rate of flow from outlets. This 'failure' of the pipe system to meet actual in-use flow rates has to be accepted in any estimate of frequency of use of outlets.

Design considerations

Rate of flow
The rate of water flow at taps and outlets depends on the diameter of the outlet and the pressure of water at the tap or outlet. The water pressure depends on the source water pressure from a cold water storage cistern or a pumped supply, and the loss of pressure to the frictional resistance of the pipework and its fittings such as elbows, tees, valves and taps. The design of pipework installation is concerned, therefore, with estimating the resistance to flow and the selection of pipes of sufficient size to allow a reasonable rate of flow at taps, where the source water pressure is known.

Water pressure hydraulic or static head
In the design of pipework for buildings, it is convenient to express water pressure as hydraulic or static head, which is proportional to pressure. The static head is the vertical distance in metres between the source, the cold water storage cistern and the tap or outlet. This head represents the pressure or energy available to provide a flow of water from outlets against the frictional resistance of the pipework and its fittings. The frictional resistance to flow of pipes is expressed as loss of head (pressure) for unit length of pipe.

Loss of head
Loss of head values are tabulated against the various pipe diameters available and the pipe material. The frictional resistance to flow of fittings such as elbows, tees, valves and taps is large in comparison to their length in the pipe run. To simplify calculation, it is usual to express the frictional resistance of fittings as a length of pipe whose resistance to flow is equivalent to that of the fitting. Thus the resistance to flow of a tee is given as an equivalent pipe length. The equivalent length of pipe is given in Table 11.1. From this the frictional resistance of a pipe and its fittings can be expressed as an equivalent length

of pipe, i.e. the actual length plus an equivalent length for the resistance of the fittings. The head (pressure) in that pipe can then be distributed along the equivalent length of pipe to give a permissible rate of loss of head per metre run of equivalent pipe length, and the head remaining at any point along the pipe can be determined. From this, the pipe diameter required for a given rate of flow in pipework and at outlets can be calculated.

Table 11.1   Equivalent pipe lengths

Pipe fittings	Equivalent length of pipe in pipe diameters
90° elbows	30
Tees	40
Gate valves	20
Globe valves and taps	300

To select the required pipe sizes in an installation, it may be useful to prepare an orthographic or isometric diagram of the pipe runs from the scale drawings of the building. This diagram need not be to scale as the pipe lengths and head available will, in any event, be scaled off the drawings of the building. The purpose of the diagram is for clarity in selecting pipe sizes and tabulating these calculations. Figure 11.22 is an isometric diagram, not to scale, of a cold distribution pipe installation for a small building. The head from the base of the cistern is measured and all pipe runs to sanitary appliances are indicated. Each pipe run is numbered between tees and tees, and tees and taps.

A change of pipe diameter is most likely to be required at tees and it is convenient, therefore, to number pipes between these points and taps. There are various methods of numbering pipe runs. The method adopted in Figure 11.22 is a box, one corner of which points to the pipe or one side of which is along the pipe run. The box contains the pipe number on the left-hand side, the actual pipe length top right and the rate of flow in litres per second bottom right. The rate of flow in a pipe is the rate of flow of the single sanitary appliance it serves, or the accumulation of all the rates of flow of all the sanitary appliances it serves. In Figure 11.22 the head is measured from the base of the cistern to the taps or pipe runs. Some engineers measure from midway between the water line and the base of the cistern, and others from some short distance below the cistern to allow a safety margin and to allow for furring of pipes. It is usual to take head measurements from the base of the cistern for most building installations.

Frequency of use

In the following calculations to determine pipe sizes it is assumed that it is possible that all the taps served may be open simultaneously, and the pipes are sized accordingly. In large installations such as those of multi-storey blocks of

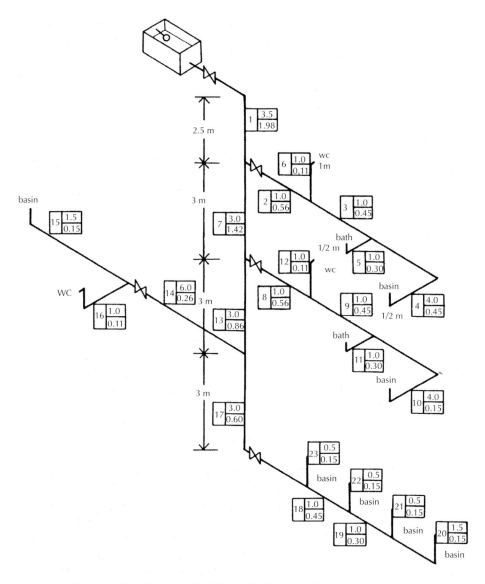

Figure 11.22 Diagram of cold distribution pipe layout.

flats, it is unlikely that all taps will be open at the same time, and a frequency of use assumption, described later, is made to avoid the expense of over-large pipes.

Critical run of pipes

The procedure for selecting pipe sizes is first to determine the first index or critical run of pipes, those along which taps are most likely to be starved of water when all the taps are open. This starvation or loss of rate of flow is most likely to occur in the branch closest to the cistern, where head pressure is least. The first index run in Figure 11.22 is pipe run 1, 2, 3 and 4. If pipes of adequate size for this run are selected to provide the required rates of flow at taps, then pipe run 1 will be large enough to supply all the other taps to the rest of the installation. The head of water available in the first index run 1, 2, 3 and 4 is 2 m, the vertical distance from the base of the cistern to the tap of the wash basin. The rate of flow from taps determines the cumulative rate of flow in the pipe runs. The unknown factors are the frictional resistance of the pipework and fittings, and the size of pipes required for the given rates of flow. It is necessary, therefore, to make an initial assumption of one of these unknowns: the pipe sizes.

Assumption of pipe size

To make this initial assumption it is necessary to calculate a rate of loss of head in the index pipe run. The actual length of pipes is known and it is necessary to make an estimate of the likely length of pipe whose resistance to flow is equivalent to the resistance of the pipe fittings. This is usually taken as a percentage of the actual pipe length, and may vary from 25 to over 100. With experience in pipe sizing this assumed percentage will approach a fair degree of accuracy. In general, the greater the number of fittings to each unit length of pipe, the higher the percentage. In Figure 11.22, pipe run 1, 2, 3 and 4 has an actual length of 9.5 m. Assume an equivalent length of about 50% or, say, 5.5 m. Thus the total equivalent length of pipe is $9.5 + 5.5 = 15$ m. The head is 2 m. The permissible rate of loss of head is therefore $^2/_{15} = 0.13$ per metre of equivalent length of pipe, and this rate of loss of head should not be exceeded at any point along the pipe run.

The graph in Figure 11.23 is used to select pipe sizes where the rate of flow is known and rate of loss of head has been assumed. In our example the rate of loss of head is 0.13. From that point on the base line, read up to 1.98 litres per second, the flow required in pipe 1. These two intersect roughly midway between the heavy oblique lines indicating 35 and 42 mm pipe sizes. If the 35 mm pipe were selected then the rate of loss of head would be greater than the permissible loss of head figure of 0.13, so the next larger size, 42 mm, is selected for pipe run 1. Similarly, for pipe runs 2, 3 and 4, from the 0.13 rate of loss of head on the graph read up to 0.56, 0.45 and 0.15 rates of flow to select pipe sizes 28, 28 and 18 mm respectively, choosing as before the pipe size above the intersection of points. These assumed pipe sizes may now be used to make a more exact calculation of friction losses in pipes and fittings, and a more exact selection of pipe sizes. For this purpose it is usual to tabulate the calculations, as set out in Table 11.2. The pipe numbers, design flow rates and actual pipe lengths of pipe runs 1, 2, 3 and 4 – the first index run – are entered in columns 1, 2 and 3, and the assumed diameters in column 7. The friction losses for each pipe run due to elbows, tees, valves and taps are now calculated. These may be determined by a multiplier of the assumed pipe diameters to give an equivalent length of pipe from Table 11.1, or more accurately from tables published by the Chartered Institution of Building Services Engineers.

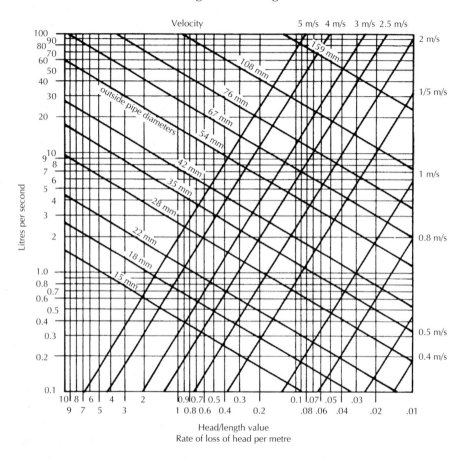

Figure 11.23 Graph for selection of copper pipe sizes. Pipe sizes are outside diameters.

In pipe run 1 there is a gate valve, an elbow and a tee. The equivalent pipe sizes for these, from Table 11.1, are 20, 30 and 40 respectively. The sum of these, 90, multiplied by the assumed pipe diameter of 42 gives an equivalent pipe length of 90 × 42 or 3.8 m. Thus the total actual length of pipe 1, and equivalent length for fittings, is 3.5 + 3.8 = 7.3 m, and this figure is entered in column 5 of Table 11.2. Similarly, for pipe run 2 where there is a valve and a tee, 20 + 40 = 60, which multiplied by the assumed pipe diameter of 28 gives an equivalent length of 1.68 m, say 1.7 m, which is entered in column 4 to give a total length of 2.7 m in column 5. For pipe run 3 there is one tee, 40, which gives an equivalent length of 40 × 28 = 1.12 m, say 1.1 m, to give a total equivalent length of 2.1 m. In pipe run 4 there are two elbows, 30 + 30, and one tap, 300, a total of 360 which multiplied by the assumed pipe diameter of 18 equals 6.48 m, say 6.5 m, to give a total equivalent length of 10.5m. With these more accurate totals of equivalent lengths in column 5 it is now possible to make a more accurate calculation of rate of loss of head. The head is 2 m and the total of equivalent lengths for pipe runs 1 to 4 in column 5 is 22.6. Thus the permissible rate of loss of head is 2 divided by 22.6 = 0.09, which is entered in column 8. With this more accurate rate of loss of head value, final pipe sizes are selected from Table 11.2.

For pipe run 1 the head loss and flow lines intersect just below the 42 mm pipe size line, confirming the assumed pipe size. The 42 mm pipe size line intersects the 1.98 flow line above a head loss value of 0.085 so that the actual head loss value in selecting the 42 mm pipe will be less than the permissible head loss value. The actual head loss in pipe run 1 will, therefore, be 0.085, which multiplied by the total equivalent length of 7.3 is 0.62, and this figure is entered in column 10 to give a figure of 1.38 head remaining in column 11. This figure of 1.38 head remaining at the end of pipe 1 will therefore be the head available for pipe 2 and this figure is entered in column 6 for pipe 2. This head remaining at the end of pipe 1 will also be available for pipe 7. Similarly, for pipe runs 2, 3 and 4, taking the permissible head loss value of 0.09 from Table 11.2, the required pipe sizes are 28, 28 and 18 mm respectively and the actual rate of head losses 0.06, 0.04 and 0.06. From these figures the head used and head remaining are calculated and tabulated. To calculate pipe sizes for pipe runs 5 and 6, tabulate pipe numbers, design flow and actual length. Assume pipe diameters of, say, 18 and 15 mm respectively, as these are common pipe sizes for branches to baths and WCs; then calculate equivalent lengths for fittings as before, based on these assumed pipe sizes, and tabulate total equivalent lengths. The head available for pipe 5 is taken from Table 11.2 as the head remaining at the end of the pipe run 3, that is 1.14. From the head available and the total equivalent length a permissible head loss value of 0.16 is calculated and a pipe size of 22 mm is taken from Table 11.2. The head available for pipe run 6 is the head remaining at the end of pipe run 2, that is 1.22, less half a metre the height of the top of pipe run 6 above that of pipe run 4. The head available is therefore 0.72, the permissible head loss value is 0.12 and the pipe size is 15 mm.

To determine the pipe size for pipe run 7, tabulate in columns 1, 2 and 3 as before. Now assume a pipe size for pipe run 7. As the flow in 7 is less than in pipe 1 and the head will be greater, assume that pipe run 7 will be the next size smaller than 1, i.e. 35 mm. On this assumption calculate equivalent length for fittings of 1.4 and total equivalent length of 4.4. The head available at the function of pipe runs 1 and 7 is the head at that point: 2.5 less the head used in pipe 1, which from Table 11.2, column 10, is 0.62. Therefore the head available at the junction of pipes 1 and 7 is 2.5 − 0.62, which is 1.88. The head along the length of pipe 7 is 3.0 so the head available in pipe run 7 is 1.88 + 3.0, or 4.88. The permissible head loss value is the head available divided by the total equivalent length, that is 4.88 ÷ 4.4 = 1.1. From Figure 11.23, for a rate of flow of 1.42 and a permissible head loss value of 1.1, a 28 mm pipe is selected.

Table 11.2 Table of calculation of pipe size

1	2	3	4	5	6	7	8	9	10	11	12
Pipe No.	Design flow	Length	Equiv. length	Total eq.length	Head available	Assumed diam.	Permissible H/L value	Actual H/L value	Head used	Head remaining	Final diam.
1	1.98	3.5	3.8	7.3	2.0	42	0.09	0.085	0.62	1.38	42
2	0.56	1.0	1.7	2.7	1.38	28	0.09	0.06	0.16	1.22	28
3	0.45	1.0	1.1	2.1	1.22	28	0.09	0.04	0.08	1.14	28
4	0.15	4.0	6.5	10.5	1.14	18	0.09	0.06	0.63	0.51	18
5	0.30	1.0	6.0	7.0	1.14	18	0.16	0.08			22
6	0.11	1.0	5.0	6.0	0.72	15	0.12	0.09			15
7	1.42	3.0	1.4	4.4	4.88	35	1.1	0.35	1.54	3.34	28
8	0.56	1.0			3.34						
9	0.45										
10	0.15										
11											

The actual head loss is then as before, and the head used and head remaining are calculated and tabulated. The head remaining will then be tabulated as available for pipe run 8.

Summary
The procedure outlined above is used in selecting pipe sizes for the rest of the pipe installation. The table is a record which can be used to check the calculations leading to the selection of pipe sizes, to confirm that actual head losses do not exceed the permissible head losses, and therefore that pressure is available to provide the required rates of flow at taps, and as a basis for subsequent calculations required by any change of plans. The calculations shown illustrate the method used to determine pipe sizes required to provide a reasonable rate of flow of water from taps. To reduce the labour of manual calculation there are various computer programs that will undertake the necessary calculations once the basic information and assumptions have been supplied.

In the example of selection of pipe sizes it was assumed that all the taps might be open at the same time and pipe sizes were selected for this. For small pipe installations, such as those for houses and other small buildings, and for branches from main pipe runs in large installations, it is usual to assume pipe sizes sufficient for simultaneous use of all taps. In these situations only small pipe diameters will be required and there would be no appreciable economic advantage to reduction of pipe sizes by making another assumption.

Table 11.3   Loading units

Appliances	Loading units
WC flushing cistern (9L)	2
Wash basin	$1\frac{1}{2}$ to 3
Bath tap of nom. size $\frac{3}{4}$	10
Bath tap of nom. size 1	22
Shower	3
Sink tap of nom. size $\frac{1}{2}$	3
Sink tap of nom. size $\frac{3}{4}$	5

In extensive pipe installations it is usual to assume a frequency of use for the taps to sanitary appliances so that smaller pipe sizes may be used than would be were it assumed that all taps were open simultaneously. Frequency-of-use values for individual sanitary appliances are expressed as loading or demand units, as set out in Table 11.3. The total of these units for sanitary appliances is used to determine notional rates of flow in pipes. The loading units of all the sanitary appliances shown in Figure 11.22 are:

❑ 3 WCs, $3 \times 2 = 6$
❑ 7 basins, $7 \times 11/2 = 10/1/2$

❏ 2 baths, 2 × 10 = 20
❏ Total of 36 $^1/_2$ loading units

This total of loading units would require a rate of flow of 0.68 litres per second in pipe run 1. Applying this figure to Figure 11.23, with a permissible head loss figure of 0.09 in pipe run 1, a pipe size of 28 mm would be selected, compared with the 42 mm pipe based on an assumption of simultaneous use of all taps.

If the same loading units are then applied to the pipe runs 2, 3 and 4, the sum total of the units is so small as to make no significant difference in the selection of pipe sizes, and the sizes previously selected will be used. From this example it will be seen that the use of loading units to determine rates of flow in pipes makes no significant economy in the selection of pipe sizes in pipe runs that serve a few sanitary appliances.

11.6 Recycled water systems

Up to 40% of household water consumption is, quite literally, flushed down the toilet. This is an unnecessary waste of treated water intended for drinking purposes. We also receive a great amount of 'free' water that falls as rain on our roofs throughout the year, and this too we throw away to soakaways or the main drainage system. The philosophy and practice of conserving water and recycling rainwater is well established in some countries, but has been slow to establish itself in the UK. Hot summers and water shortages are rare, and treated water is relatively cheap. There is, however, growing pressure on the mains water supply as the population of the UK grows, lifestyles change, and demand increases. Add to this a growing awareness of environmental issues and the practicalities of recycling water start to become a practical alternative.

Recycled water systems are designed to use recycled water, primarily rainwater, for tasks such as flushing the toilet and irrigating the garden. The term 'grey water' is also used to describe the reuse of water from domestic appliances such as washing machines, and these systems are usually combined with recycled rainwater. In a recycled water system the rainwater is collected from the roof in copper gutters and discharged to polypropylene containers, where it is stored. This water is then used to flush toilets and for other non-consumptive tasks such as bathing and irrigating the garden. Some systems are designed to also filter and purify the rainwater to provide drinking water (although this tends to be a relatively small amount of the total recycled water usage, approximately 5–10%).

To avoid contamination with domestic mains supply, the recycled water system is designed as a separate system to the main cold water distribution pipework. Pipework should be identified accordingly. This means that there may be some duplication of pipework and, depending on how the recycled water is stored, some need for additional pumps. However, the initial capital cost of the system may be recovered quite quickly given the savings in mains water consumption.

Ideally the rainwater should be stored at roof level. In this approach the toilet flushing system and other appropriate outlets are fed by gravity. Where the rainwater is stored at a lower level, for example in a basement, then the recycled water supply will need to be pumped to the outlets. In a pitched roof design there may be enough space for water storage vessels to be placed within unused roof space. Care should be taken to ensure that the additional loading from the water can be accommodated, i.e. the structure should be designed accordingly. The storage vessels should also be protected from frost and inspected on a regular basis. Filters are required to stop leaves and associated debris from getting into the water storage vessels. These filters require regular maintenance and cleaning to avoid blockages.

The storage system must be designed with an overflow system, to prevent flooding once the storage vessels become full, for example in a particularly wet period. Normally the overflow will run to a water storage system for irrigating the garden. Alternatively the overflow may run to the mains drainage system or a soakaway. Similarly, the recycled water system will need to be connected to a mains water feed to provide back-up water in particularly dry periods. The usual precautions are required to prevent contamination of the mains supply (see *Barry's Introduction to Construction of Buildings*).

Personnel with experience of these systems must carry out the design of an installation of recycled water systems. Building Regulation approval will be required for installation in new and existing properties.

11.7 Electronic communication systems: broadband

The majority of building users require cable-based electronic communication systems, such as the high speed Internet access 'broadband'. Installation to existing buildings is not always easy due to the unavailability of ready access for routing cables. The provisions in Approved Document Q of the Building Regulations aim to ensure that electronic communication services could be installed without inconvenience to the building owners and users, and without disruption to the building fabric and surrounding ground.

In practical terms this means that a duct should be laid from the site boundary to the building, into which appropriate cables may be easily installed and, if necessary, removed. A terminal chamber is required at the site boundary, and terminal boxes will be required externally and internally to the building (see Figure 11.24). Care is required to ensure that the external terminal box is positioned so as not to be too unsightly. Similarly, internal terminal boxes should be as small as practicably possible and positioned to suit anticipated room usage.

Installation

Commercial suppliers carry out installation of cable television and broadband service supply. The main supply will run under the pavement, with connections

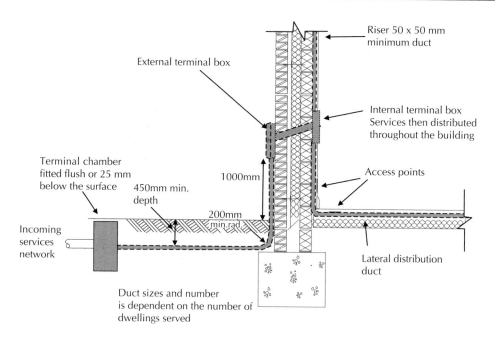

Figure 11.24 Means of supply to the building.

made to terminal chambers at the boundary of individual buildings. Connection will be made on payment of a subscription fee to the supplier. For new building work, the terminal chamber, ducts and terminal boxes should be installed as part of the general contract. This enables the service provider to add the cable to the duct and make the necessary connections quickly and easily.

Terminal chambers

The terminal chamber should allow for the connection of at least three electronic communication services' networks. The chamber should also allow for the connection of one or more customer ducts – see Figure 11.25. It is common practice to install the terminal chamber into a hole in the ground so that the top is flush with the finished ground level, or located 25 mm below a grassed or soiled area (but not covered over). Thus the terminal should be designed and constructed to be able to withstand a reasonable superimposed load. The chamber lid must be marked to clearly identify the purpose and contents of the terminal, i.e. electronic communications services. Chambers tend to be manufactured from plastics and are often identified by words such as 'Cable TV' or 'Broadband' on the lid.

Ducts

The size and number of ducts will be dependent on the number of dwellings served. One dwelling would typically be served by one duct with an internal diameter of 50 mm. A 90 mm diameter duct is used for provision to more

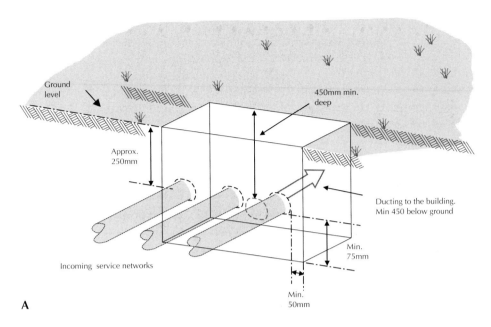

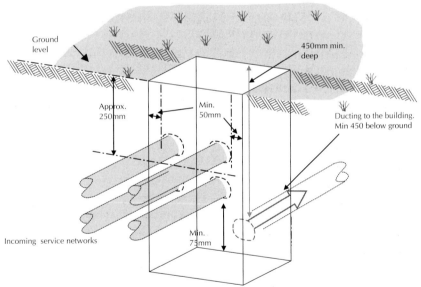

Figure 11.25   Terminal chamber configurations.

than one dwelling, with the number of ducts increasing to suit the number of
dwellings. Guidance is provided in Approved Document Q. The duct should
be installed a minimum of 450 mm below finished ground level and laid on a
bed of at least 25 mm well-compacted fine fill material. The duct should be surrounded with well-compacted fine fill material, to a minimum depth of 50 mm

above the duct – see Figure 11.26. The duct should be swept up to the external wall, or alternatively the internal face of the wall to the terminal box. For refurbishment work it is most likely that the terminal box will be placed externally (to avoid disturbing the foundations and floor construction). For new building work it is preferable to install the terminal box on the inside face of the external wall. Where the duct passes under foundations, it should be protected.

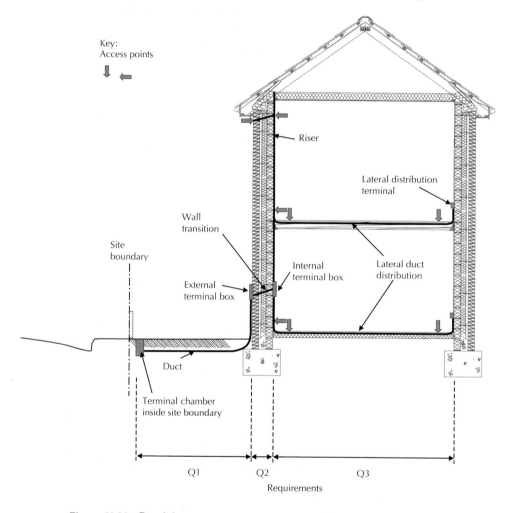

Figure 11.26   Provisions common to more than one of Part Q.

External and internal terminal boxes

An external terminal box and the duct leading to it may look unsightly, so consideration should be given to positioning on the wall. The external box will need to be connected to an internal terminal box (as illustrated in Figure 11.27A) via a conduit or duct passing through the external wall construction. Installing

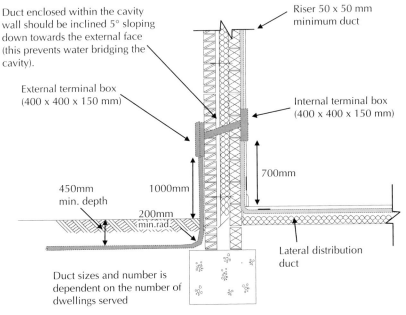

Duct enclosed within the cavity wall should be inclined 5° sloping down towards the external face (this prevents water bridging the cavity).

Riser 50 x 50 mm minimum duct

External terminal box (400 x 400 x 150 mm)

Internal terminal box (400 x 400 x 150 mm)

700mm

450mm min. depth

1000mm

200mm min.rad.

Lateral distribution duct

Duct sizes and number is dependent on the number of dwellings served

A External terminal box

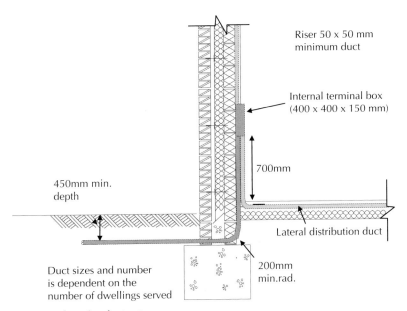

Riser 50 x 50 mm minimum duct

Internal terminal box (400 x 400 x 150 mm)

700mm

450mm min. depth

Lateral distribution duct

200mm min.rad.

Duct sizes and number is dependent on the number of dwellings served

B Internal service duct entry

Figure 11.27   Alternative arrangements for duct entry and distribution.

an internal box (Figure 11.27B) is a neater alternative. The sizes recommended for the wall-mounted terminal boxes in Approved Document Q are quite large (e.g. $400 \times 400 \times 150$ mm deep) and so some thought needs to be given to the position of the internal terminal box.

Distribution within the building

From the internal box will run the lateral distribution duct (20×40 mm). The duct should have access at a maximum spacing of 20 m. Access to vertical

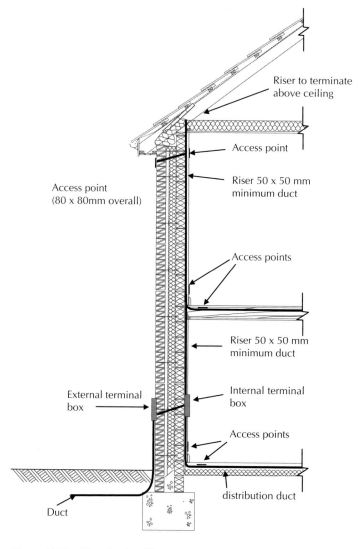

Riser to terminate above ceiling

Access point

Riser 50 x 50 mm minimum duct

Access point (80 x 80mm overall)

Access points

Riser 50 x 50 mm minimum duct

External terminal box

Internal terminal box

Access points

Duct

distribution duct

Figure 11.28   Riser in dwelling.

risers should be provided at each floor level. For dwelling houses Approved Document Q recommends that the distribution duct should be positioned to allow access at each floor level and at least one habitable room at each floor level in dwellings. Ducting should be spatially separated from mains electricity cables. The riser, lateral distribution ducts and access points are illustrated in Figure 11.28. Lateral distribution ducts can be run within timber floor cavities, within raised floors and/or around the perimeter of the room (e.g. fixed to or behind a suitable skirting board). Lateral ducts should terminate with a flush mounted box, the position of which should be so as to satisfy the provisions in Approved Document M. Additional guidance for flats and maisonettes can be found in Approved Document Q.

12 Alternative Approaches to Construction

Conventional construction methods rely on a plentiful supply of resources, many of which have started to become less plentiful and hence more expensive. In this chapter the focus is on the construction of resource-efficient, low-energy buildings, and a number of alternative, or non-conventional, approaches to construction are described. We have left these construction methods to this part of the book for two reasons. First, the techniques are primarily developed in response to a more sustainable approach to our built environment and therefore the discussion about energy usage in the preceding chapter provides useful background material to the methods described here. Second, the construction techniques are currently being developed (or rediscovered) and from a practical viewpoint new developments will be easier to incorporate in future editions of this book. For those readers with an interest in environmental design, the techniques described here should provide a sound basis from which to explore areas in more detail. For those readers who may need some convincing as to the merits of some of the less common approaches to the construction of buildings, our aim is to provide objective information on some of the more familiar approaches.

12.1 Functional requirements

The techniques described in this chapter do not differ fundamentally in their functional requirements to those set out in *Barry's Introduction to Construction of Buildings* and in the other chapters to this book (where appropriate). What does differ is the alternative approach (and attitude) to construction, in philosophy and use of materials, which has a lot to offer our built environment. At the risk of lecturing, we all should be making a greater effort to reduce the negative impact of construction activities on our natural environment. The mantra is:

❑ Reduce
❑ Reuse
❑ Recycle

An alternative approach

The energy expended in the extraction, working and transportation of materials to the site and the total resources used during construction should be included in the calculation of the structure's efficiency. Thus the integration of resource-friendly concepts into the design and construction processes can significantly reduce the environmental impact of the constructed works. Similarly, the occupants' habits and environmental ideals will affect the operating efficiency of the building. Adopting a less mechanised (and hence less conventional) approach to construction may be seen as a step in the wrong direction by some but, for many, natural materials and labour intensive methods provide an alternative approach to the construction of buildings. Primary drivers behind a non-conventional approach to construction may be one, or more likely, a combination of the following factors:

- ❑ Lower initial construction costs – affordability
- ❑ Energy efficiency – low heating (and cooling) costs throughout the life of the building
- ❑ Use of local materials
- ❑ Use of local (semi-skilled) labour, community involvement or self-build
- ❑ Cultural compatibility with the local environment
- ❑ Simplicity of design
- ❑ Easy to adapt as needs change
- ❑ Comfort
- ❑ Implementation of environmental ideals and principles
- ❑ Ease of disassembly and materials recovery at a future date

Most of these factors are covered in the descriptions that follow; however, the following underlying issues need a little more explanation.

Cost of labour
Labour costs continue to rise and make up a substantial part of the initial cost of most building projects. One way of mitigating labour costs is to employ a method of construction that is quick and efficient, although these tend to carry a high cost that is associated with the technologies and machinery required to manufacture and erect the building. Another approach is to engage in some form of self-build or self-help scheme, assuming that the self-builder has the time to invest in the project and has, or can readily acquire, the necessary skills to implement a quality product. For example, straw bale construction and rammed earth structures are attractive to owner-builders (self-builders) because of the cheap cost of the raw materials and the large savings in labour costs to be made by providing their own labour. Also new innovations, such as hollow polystyrene interlocking blocks (fitting together much like Lego bricks) that are filled with concrete to make a structural wall with low thermal conductivity, are attractive to the do-it-yourself builder. Experienced labour may, however, still be required for the foundation work, roof framing, electrical wiring and plumbing. Where possible the labour should be sourced locally, thus helping to stimulate the local economy.

Cost of materials

Compared with manufactured materials, the initial cost of materials for some of the non-conventional approaches may be considerably cheaper, although the increased use of manual labour may well offset this saving if some or all of the labour has to be paid for at the market rate. In the majority of cases there will be considerable life cycle cost benefits for the entire structure. Similarly, by using simple construction techniques the ease and hence cost of maintenance, repair and replacement should also be better than more conventional approaches. Adopting a passive design philosophy may help to reduce some of the services provision and need for integration; for example, passive ventilation instead of mechanical. Materials should be sourced locally, preferably from renewable resources.

Genius loci – the importance of site

The importance of the site and the manner in which the building is positioned on, or within, the ground becomes even more critical with some of the alternative approaches. Many of these materials are more sensitive to damage from moisture than conventional building products and they may be considerably less durable unless competently detailed and constructed. Site sensitivity is a crucial factor in ensuring a durable and trouble free building. The proposed site of the building must be carefully analysed in terms of the microclimate, soil type, position of water table etc. Then (and only then) should a decision be taken as to the most appropriate materials and construction techniques to employ. For example, some sites may be better suited to earth sheltered construction than straw bale construction and vice versa. In some cases a more traditional approach may be a better option once the data gleaned from a thorough site analysis has been collected and analysed. Readers with a strong desire to build using a particular material, e.g. straw bales, must first find an appropriate site.

12.2 Straw

Straw is a renewable and viable building material, being plentiful and inexpensive. The annual harvest of grain from barley, flax, oats, rye and wheat results in the generation of a considerable quantity of stalks (straw), which is mostly under utilised or even wasted. From an environmental perspective, straw contains carbon, which is trapped in the construction, rather than being released through burning/disposal. The embodied energy in the straw bale construction is also low compared with conventional walls, requiring little processing and little transportation (as use is usually local or regional). Straw is an organic tube made of cellulose that is structurally strong in compression and thus well suited to a number of applications in construction. Straw was first used to reinforce mud and prevent it from cracking, and then as a building block with the invention of the mechanical baler. Straw has reasonably good thermal insulating properties and, because bales are approximately 600 mm thick, a straw-bale wall has a high thermal resistance. Straw is currently used as a building material in bale form, or as a pressed panel.

Straw bale construction

Straw bale construction is a relatively old construction method that has become fashionable in recent years, especially in the American Southwest. The drive for a more sustainable approach to the construction of buildings has also resulted in considerable interest and practical application in many countries, including the UK. Straw bale buildings offer a simple and practical method of creating a building with excellent performance characteristics. Super insulated walls (thermal and acoustic), simple construction, low build costs and the conversion of an agricultural byproduct into building material are attractive characteristics. With the limited skills required to build a straw bale house, the technique tends to appeal to self-builders and community build groups who can realise low-cost, energy efficient dwellings, creating a relatively organic addition to the community and at the same time helping to generate income for local farms. Well-detailed, properly constructed and maintained, straw bale houses have a long life, which is comparable with other types of construction. Straw bale construction has also been used in conjunction with other materials in more complex designs for houses and commercial buildings. Here the economies of scale are not available to contractors, thus straw bale walls tend to be more expensive than more traditional approaches.

The bales

Automatic straw balers create tight blocks of straw that provide a relatively easy-to-manoeuvre building block. Sizes of bale vary depending on the baler; however, typical dimensions and weight when dry are:

- ❑ Two wire bale: 450 × 350 × 900 mm, approx 25 kg
- ❑ Three wire bale: 600 × 400 × 1050 mm, approx 35 kg

Although the smaller size bale is easier to handle, the medium sized three-wire bale provides better structural, thermal and acoustic performance. Larger cubical and round bales are also available but these require lifting by mechanical means. There have been moves in the USA to establish a 'construction-grade' straw bale. Ideally bales should be twice as long as they are wide to simplify and maintain a running bond in the courses.

Advantages of straw bale construction

In addition to simplicity of design and ease of adaptability there are a number of performance advantages a straw bale house has over conventional loadbearing masonry or framed construction.

Straw bale structures are highly fire resistant. The compressed bales contain enough air to provide good insulation values but because they are compacted firmly they do not hold enough air to permit combustion. Combined with render and plaster surface finishes, a high degree of fire resistance is possible. The type of straw and the moisture content of the bales will mainly determine the thermal characteristics of the wall. With infill construction the frame will also have a determining factor on the overall thermal performance of the wall. Acoustic insulation is considerably better than a conventional wall structure.

Closely packed straw bales, covered with render on the outside face and plaster on the inner face, provide good air leakage control. Coupled with simple geometric design, well-built and maintained straw bale construction provides a building with very little air leakage. The render and plaster finishes can be left unpainted and are inexpensive to maintain. The construction cost of a straw bale wall should be significantly less than that for a comparative wall with the same thermal and acoustic performance.

Disadvantages of straw bale construction

The main disadvantage of straw bale construction relates to concerns over durability, particularly moisture related damage. Straw bale construction is most suited to dry climates (hot and cold). In the UK the construction of straw bale houses is a relatively recent innovation, thus it is difficult to state with any certainty what the durability of the structure will be over the longer term. As more straw bale buildings are built in the UK and more research is conducted into their durability, designers and builders should have more information from which to make an informed decision.

Fungal rot represents the greatest threat to the life of a straw bale building, thus careful detailing is required to prevent the straw bales from becoming wet. Foundations and roof construction are critical in preventing unwanted rain and moisture penetration, and the construction of the wall must be done in such a way as to avoid any possibility of interstitial condensation. Fungi and mites can live in wet straw, therefore the bales must be bought when they are dry, kept dry until needed and then sealed into the wall construction with plaster and render to eliminate any chance of access for pests. Paint for interior and exterior wall surfaces should be permeable to water vapour so that moisture does not get trapped inside the wall. Services, especially sealing around outlets, need special attention.

Straw bale walls are considerably thicker than more traditional construction methods (at least double) and so where land is at a premium a straw bale house will provide less internal space for the same external dimensions as a traditional construction. This may be of little concern in rural locations. The appearance of straw bale buildings may not always appeal to a more urban environment, although there are a few examples of straw bale construction being used in densely populated urban environments, usually in conjunction with other materials as part of a composite construction.

Selection of straw bales

Straw bales should be purchased immediately after the harvest, when they are abundant, fresh and dry. Dealing directly with a farmer is the cheapest and perhaps most reliable method of selecting the best quality bales since quality is largely judged on visual appearance and touch. Transportation to the building site will need to be addressed, perhaps independently of the arrangement with the farmer if the bales are to be transported out of the local area. Straw bale merchants provide an alternative source of supply and will have an established transport infrastructure and storage facilities. Once again the selection of the bales, especially if sourced from more than one farmer, is an important consideration.

Bales should be tied tightly with polypropylene string or bailing wire and should not twist or sag when lifted. All bales should be uniformly compacted and contain thick, long-stemmed straw; any bales comprising short, thin straw should be rejected. Old and/or damaged bales should be rejected. If practical the construction of a straw bale building should be undertaken immediately after the main grain harvest when the bales are fresh and dry, with bales transported to site, positioned and protected from the weather as quickly as possible. If this is not an option then the bales should be selected and stored under dry, ventilated, conditions. Bales should be tested for moisture content, which must be 14% or below when purchased and when used for building. All bales must be stored under dry conditions until required for building purposes. Protection of the bales from the weather is also necessary during construction.

Foundation construction and site drainage

The base of the straw bale construction must be kept dry at all times, thus the manner in which the straw bale interacts with the foundation will be a crucial factor in determining the durability of the wall. The position of the lowest straw bale in relation to the finished ground level and the detailing of the external wall finish at this junction are particularly important. Similarly, site drainage must be designed to get water away from the base of the walls as quickly as possible (Figure 12.1).

Guidance on the minimum distance between the finished ground level and the lowest bale position varies from around 225 mm upwards to 300 mm and 450 mm. Knowledge of local rainfall patterns, the extent of the roof overhang in relation to the height of the wall and the ground finish adjacent to the wall will be determining factors. As a general guide the bales should be positioned at least 450 mm off the finished ground level to avoid splash back of rain, bouncing off the ground and on to the face of the wall. The material under the straw bales should be good quality stone, engineering brick or dense blockwork. The damp-proof course should be positioned between the solid foundation and the timber base or sole plate. The manner in which the external render is finished at the junction of foundation and wall is also important. A drip should be formed

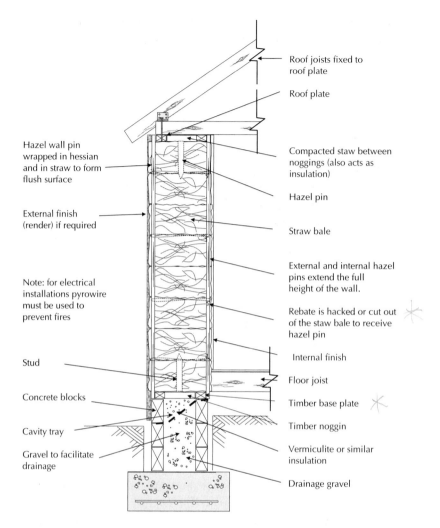

Roof joists fixed to roof plate

Roof plate

Hazel wall pin wrapped in hessian and in straw to form flush surface

Compacted staw between noggings (also acts as insulation)

Hazel pin

External finish (render) if required

Straw bale

External and internal hazel pins extend the full height of the wall.

Note: for electrical installations pyrowire must be used to prevent fires

Rebate is hacked or cut out of the staw bale to receive hazel pin

Internal finish

Stud

Floor joist

Concrete blocks

Timber base plate

Cavity tray

Timber noggin

Gravel to facilitate drainage

Vermiculite or similar insulation

Drainage gravel

Figure 12.1   Straw bale construction.

at the bottom of the render to help throw any water that runs down the wall off and away from the base of the wall.

Care should be taken so as not to compromise the site drainage through inappropriate surface finishes. The ground around the structure must be well drained and kept dry with adequate surface and below ground drainage. The position of the water table and history of the proposed site with regard to flooding should be investigated, and appropriate decisions taken to protect the construction from damp and/or wet conditions (see *Barry's Introduction to Construction of Buildings*). Self-draining foundations are one method of helping to keep the base of the wall free of water. Vermiculite is placed above the cavity tray. Below the cavity tray, the cavity should be filled with drainage gravel and weep holes provided at regular intervals (see Figure 12.2).

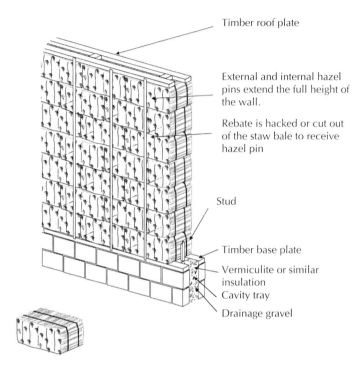

Timber roof plate

External and internal hazel pins extend the full height of the wall.

Rebate is hacked or cut out of the staw bale to receive hazel pin

Stud

Timber base plate

Vermiculite or similar insulation

Cavity tray

Drainage gravel

Figure 12.2   Straw bale construction.

Roof construction

The design and construction of the roof is another critical area. The roof should be constructed with a large overhang to provide protection to the wall below from rain and snow. Leaks in the roof and/or inappropriate detailing at the wall to roof junction will compromise the long-term durability of the structure. A large roof overhang on single-storey buildings will also provide good protection to the base of the wall. Roof pitch is another consideration, with steep pitches and a good quality roof covering to facilitate the rapid discharge of rainwater from the roof considered important to the durability of the roof. Timber roof construction is the usual method, with the roof loads transferred to the foundations via the loadbearing straw bale wall or via the timber frame construction (described in more detail below). Roof coverings may be of reed thatch, tile or lightweight finishes, with lightweight roof structures and coverings preferred for loadbearing straw bale walls.

Wall construction

The main principle of straw bale design is based on the use of full bales as a building module – simplicity is the key. Half bales will be required for bonding purposes, the same principle as that for brick and block wall construction. Bales smaller than a half size should not be used because of their lack of compressive strength. Careful planning is required before any drawings are finalised, simply because the actual size of bales varies between suppliers. Once a bale supplier and hence dimension has been determined, it is then possible to design and detail the building plan and also the opening sizes (doors and windows) and their position within the wall. The main principles relating to the strength and stability of the wall are similar to those for loadbearing or framed construction. Where possible, door and windows should be selected to suit the bale module size or be manufactured/built to suit the bale module. When using loadbearing straw bale construction, there will be some settlement of the wall after the roof has been completed. The straw bales must be sealed against the weather and pests. External surfaces are finished in lime render (stucco). Internal surfaces are plastered. The render and plaster provide an attractive surface finish, which also adds to the structural integrity of the wall.

Loadbearing straw bale construction

In loadbearing straw bale construction the straw bale walls directly support the loads from the roof. The quality of the bales is an important consideration. Hazel pins are used to tie the bales together, two per bale from the fourth course upwards, with staples used to tie the bales at changes of direction, such as corners. The bales are often tied together tightly with threaded metal rods, wire or plastic straps, which helps to compress them and hence minimise settlement of the bales when the roof structure is added. For small buildings, such as small single storey houses, with relatively short lengths of wall, there may be no need for additional support. For long walls it will be necessary to provide a system of vertical posts at 10 m (or less) intervals to help support the weight of the roof, in which case the infill method may be more suitable. The straw bales will provide lateral restraint at the corners, thus a minimum return of at least two bales is recommended before any openings are formed in the walls for doors or windows. This is a similar principle to that adopted for brick and block walls.

Settlement of the straw bales under load, which may be irregular, must be considered in the design. Settlement in a seven bale high wall may be up to 50 mm, and is difficult to predict because the extent of settlement will depend on the density of the bales (which varies between bales) and the amount of loading applied to them. Thus for practical reasons it is necessary to measure the opening sizes after the main structure is complete, and install doors and windows late in the construction process (after the initial settlement of the bales).

Mortar bale   *reinft*

A structural mortar made of Portland cement and sand is applied between the straw bales to create a lattice structure. When dry, the lattice forms a structural framework between the bales and thus provides a form of backup should a straw bale fail. Bales are finished with stucco on the external face and plastered on the interior to protect the bales and also to provide an attractive finish. The stucco and plaster also add to the structural integrity of the wall. Some designers and builders may feel that resorting to cement will compromise their environmental ideals.

Infill straw bale construction

Non-loadbearing, post and beam, framed and infill are terms used to describe a framed wall construction with the bales bearing their own weight. The wall-framing members carry the weight of the roof. This method is better suited to larger structures. The bales are attached to each other by piercing them with bamboo, hazel or metal reinforcing bars and attached to the timber framework. With the infill technique the settlement of the straw bales may be less (less load on them) and hence the settlement of the bales is less of a concern than with a loadbearing construction. Restricting the straw bales to an infill function only may be a less risky approach to straw bale construction. If there is a problem with any of the straw bales, the structural integrity of the structure will not be compromised. Framed construction provides an opportunity to construct the frame before the grain harvest, thus fresh bales can be taken directly from the field and installed. This helps to ensure dry bales are used and removes the need for storage and unnecessary weather protection.

Straw and clay building

A traditional and durable construction method is to use a 'batter' of clay and water stirred into loose straw to produce a straw-reinforced clay mud. The mixture is packed tightly into a lightweight timber ladder framework to create partition walls or as infill panels for framed wall construction.

Pressed straw panels

Pressed straw panels are made by compressing straw under temperature to produce a panel made of 100% straw. The combination of compressed straw, recycled paper lining and adhesive will form a board to comply generally with BS 4046. The result is a low cost, versatile product with environmental benefits. The panels can be used as a self-supporting, non-loadbearing partition system and also for roof decking. The technique was first introduced to the UK and developed into a sophisticated straw board product by Stramit, in Suffolk. Pressed straw panels may also be used in roofs and floors. Research and development is currently being conducted by a number of companies and research organisations into the use of pressed straw panels for use in a structural capacity.

12.3 Earth 27/1/16

Rocks and soils are some of the most economic building materials, available locally and with a long record of use in vernacular buildings in the UK, for example cob construction. There has been renewed interest in the use of earth as a building material in the UK, mainly associated with the drive for a more environment-friendly approach to construction. There remain a number of concerns about durability but, when detailed to suit the climate and topography of the site, there is no reason why earth-based construction cannot provide an alternative to more traditional materials and methods. The most common method used in the UK is earth-sheltered structures, with growing interest in adobe and rammed earth. To reduce transportation costs, and the associated pollution, earth should be used from the building site, and additional materials sourced locally. It may be an obvious point but earth construction will be specific to the locale, soil condition, techniques and materials available. There will be a strong reliance on the extent of local knowledge among the workers.

Earth sheltered construction

Earth-sheltered buildings are an energy efficient and environmentally conscious approach to construction that may suit certain sites and some clients. The concept of burying the building under the ground, or at least covering a significant proportion of the building with soil, is not new. Various designs exist, ranging from buildings completely buried beneath the ground, to buildings that have one or more walls protected by the earth.

When the entire structure is buried, the term 'underground' structure is used. When earth is banked up (bermed) against an external wall, this is known as a 'bermed' structure. A bermed structure may be built above, or partially above, natural ground level. In the majority of cases the roof will be covered with earth and vegetation to form a green roof (described below). The vegetation, grass or sedums, will help to reduce erosion and provide an attractive finish. Three generic house designs have been developed from the use of underground and/or bermed construction. They are:

❑ Courtyard (or atrium) design. An underground structure planned around a central atrium, from which rooms are accessed, and which forms the focal point of the house. Some designs are based on an open courtyard, which provides natural ventilation but poses some problems with removal of surface water and snow. Atrium designs are built with a glazed roof over the courtyard, which, depending on the design, can also help to provide some natural ventilation. Plan forms tend to be circular, ovoid or hexagonal in nature.

❑ Elevational design. A bermed structure, usually with a glazed south-facing wall. While the glazed facade is exposed, the other walls are covered with earth. The south-facing wall allows penetration of daylight and thermal gain, while also providing an opportunity for natural ventilation. Skylights can be used to alleviate the problems of poor light and poor air circulation further into the plan. To reduce the depth of the house, the south (and north) faces are considerably longer than the east and west walls, creating a long, thin house. To improve the space it is common to construct a curved elevation to the south.

❑ Penetrational design. Again a bermed structure that covers the entire house, with the exception of doors and windows. The house is usually built on a flat site with the earth bermed around the entire structure. This allows the opportunity for cross ventilation and for natural light to all rooms.

In the UK, earth sheltered construction has been employed on a small number of commercial buildings but tends to be favoured by clients wanting a bespoke house design. The primary concern is with protection from moisture. Careful choice of site (e.g. topography, ground water position, soil type) along with the provision of adequate drainage at and below ground and waterproofing to walls are crucial factors. Earth sheltered structures need to be well designed and constructed, and designers and contractors with experience of this type of work should be used.

A naturally sloping site is the best location for an earth sheltered building. The extent of the dig required to excavate a suitable area will be determined by the angle of the slope, thus a moderately steep site is usually preferable to a gently sloping site. Soil type is another determining factor. Granular soils compact well and are permeable, which allows ground water to drain away. Sites with cohesive soils (such as clays) should be avoided because the clay will expand when wet and has poor permeability. Site investigation is also necessary to establish the height of the water table (the lowest level of the building must be above the water table). Areas in which radon gas naturally occurs should be avoided and special measures may be required where the site is heavily contaminated.

Advantages of earth sheltered structures

Well-detailed and well-constructed, an earth sheltered house provides a number of benefits over more traditional structures. The biggest advantage is that the building is protected from the weather and hence is not subject to deterioration by, for example, solar radiation or wind damage. In very exposed locations, protection from adverse weather may be a considerable benefit. Protecting the building with earth will reduce or even eliminate the need for regular painting, cleaning of gutters and repair. Thus maintenance is considerably less onerous for the majority of earth sheltered designs.

The earth provides natural insulating properties to the structure, providing a relatively stable indoor temperature, which is less affected by external air temperature fluctuations. Thus less energy is required to maintain a stable internal temperature. The earth also provides considerable sound dampening in addition to that provided by the structure of the building.

Earth sheltered structures will blend into the landscape, and may be an ideal solution in areas where the visual appearance of a structure would cause a problem. This does not mean, however, that approval from the local town planning department will be forthcoming since other factors, such as land use and access, will also need to be considered.

Disadvantages of earth sheltered structures

There are a number of potential disadvantages associated with earth sheltered structures. Exclusion of moisture is perhaps the most obvious. Waterproofing is a primary concern (see also tanking, Chapter 3) and tends to carry a relatively high initial cost. Surface water drainage and underground drainage is key. Seasonal or regular surface water flows must be channelled away from the structure. Underground drainage should be designed to remove water (and water pressure on the walls) quickly. Regular maintenance of drainage channels to avoid blockages is also essential if problems are to be avoided.

The initial cost of an earth sheltered structure can be as much as 20% higher than a comparable structure built above the ground. Exact figures will depend on the nature of the ground and topography of the site (e.g. gently or steeply sloping). However, there may be considerable cost savings made during the life of the building, which must be considered in a whole life costing approach.

Earth sheltered houses tend to be characterised by high levels of humidity; therefore air exchange is important. Some designs may also result in pockets of stale air unless some form of air flow is introduced. Ensuring adequate air flow can be achieved through the use of passive or mechanical ventilation systems. Passive ventilation can be provided through vented skylights, which are required to get natural light into the internal spaces. Linking the air exchange system to an earth to air heat exchanger may help to keep energy costs to a minimum.

Repairs and remedial work to the structure of the building may be expensive, since it is difficult to get to walls without removing a great deal of earth. This places additional emphasis on the quality of the detailing and the quality of the work undertaken on site. Careful and systematic inspection of the waterproofing work as it proceeds is essential.

Materials

The most commonly used material for the retaining walls is concrete. Reinforced insitu concrete and/or prefabricated concrete can be used. Alternatively masonry (brick, block, stone) can be used, reinforced with steel bars. Attention must be given to joints (i.e. between precast concrete units) to ensure the walls remain watertight. In addition to their strength, walls must provide a good surface for waterproofing and for thermal insulation (where required). Waterproofing to the walls can be achieved with a number of systems. The main materials are:

❑ Rubberised asphalt. Sheets are applied directly to walls and roofs and have a long life expectancy.
❑ Plastic sheets. The integrity of these sheet systems relies on the seams between the sheet materials. The seam must be formed and sealed properly, otherwise the membrane will leak. A variety of sheets with different material properties are available.
❑ Liquid polyurethanes. Usually used in places where it is difficult to apply a membrane. Sometimes applied over the external side of insulation.
❑ Bentonite. A natural clay which expands on contact with water and forms a barrier to moisture penetration. Formed into panels attached to walls or spray applied in liquid form with a binding agent.

Thermal insulation
Insulation is usually placed on the exterior face of the walls, after the waterproofing system has been completed. In this way the heat generated, captured and absorbed within the earth sheltered envelope is retained by the building's interior. Rigid insulation sheets help to protect the waterproofing from punctures from sharp stones in the soil. Thin protective boards are often used as a barrier between the insulation and the earth.

Adobe

Adobe is compressed earth, rammed into moulds or pressed into blocks while damp and sun-dried to form a building block or brick. Adobe is a common construction method in many parts of the world (mainly places with a dry climate). The best adobes are made from soil that is high in clay content. Mechanical presses can be used to form adobe blocks directly from the building site's soil. These blocks are then used to build loadbearing (or infill) walls, laid without mortar. Blocks are laid in a walling bond, with the connecting surfaces wetted with water to help provide a bond between the individual units when dry. The walls are then rendered (stuccoed). Compaction of the earth is critical and over-compaction can lead to fracture of the material. Scientific analysis of the soil is therefore crucial.

Adobe and rammed earth walls absorb solar heat during the day and radiate heat back into the air at night, thus providing a comfortable and relatively constant internal temperature. Thus adobe tends to be used in dry and hot locations. Both exterior and internal adobe walls provide good thermal mass and can form part of a passive design. The principles of weather protection and durability of earth are similar to that of straw, namely protection from moisture.

Rammed earth

Rammed earth is a term used to describe the mixing of cement and earth, which is then packed into wall forms with a pneumatic tamper. Hence the term 'rammed' earth. The result is a material that resembles a sedimentary rock, with compressive strengths about half that of concrete. The strength of the material will vary, depending on the ratio of cement to soil (and soil type), thus testing of batches is necessary to determine the loadbearing capacity of the resulting wall. By using soil from the building site, rammed earth can provide a relatively cheap material with reasonable environmental credentials. The use of the material is similar to that described for concrete walls. By increasing the thickness of the walls, the loadbearing capacity may be increased.

12.4 Green roofs

The concept of the 'green' or living roof is not new but the technologies associated with the concept have evolved and matured. Largely associated with vernacular architecture until relatively recently, plant and soil layers (sod) have been used in cold climates to retain heat and in warm climates to keep buildings cool. The combined layers of soil and plants (usually grass or sedums) provide excellent insulation to roofs, both pitched and flat. By combining new technologies (hi tech) with old principles (low tech) it is possible to create a highly durable roof structure with good environmental credentials. Green roofs provide a habitat for insects, birds and other small animals. This is particularly useful in brownfield and urban environments to help improve biodiversity. The terms turf roof and green roof are used, sometimes interchangeably. A turf roof usually refers to a simple roof, usually pitched, with grass and wild flowers growing on it. A green roof usually refers to a roof, pitched or flat, with sedums and other larger plants growing on it. There are two basic types of green roof systems, the extensive system and the intensive system.

Extensive green roofs

Extensive green roofs are characterised by low weight and capital cost. They have limited plant diversity and usually require minimal maintenance once established. The growing medium is usually from 50 to 150 mm deep, comprising a mixture of sand, gravel, peat, organic matter and some soil. Typically these roofs are pitched (up to 30° from the horizontal) and are covered with turf or sedums, which grow well on a thin medium. This cheap and natural looking construction relies on rainfall for irrigation and can become dry and unattractive in long, dry summers. Similarly, the grass and sedums can look unattractive in the winter. These roofs may be constructed on timber, steel or concrete roof structures. Textured waterproofing membranes help to stop the turf or plant layer from slipping down the slope for gentle pitches. On slopes greater than 20° to the horizontal the sod or plant layer must be strapped horizontally to the roof structure to prevent the material slipping under its own weight when saturated. Some manufacturers provide support grid systems that have been specially designed for this purpose.

Intensive green roofs

Intensive green roofs tend to be characterised by deeper soil depth and greater weight. These roofs are built off a flat roof deck and are accessible to building users, thus increasing the amenity value of the building. Plant diversity tends to be greater than with extensive roofs, costs are higher and maintenance requirements more extensive. The growing medium tends to be from 200 to 600 mm deep, and is often soil (loam) based. The increased depth of the growing medium allows bushes and trees to be grown, creating a more extensive ecosystem. Regular watering is usually provided through automated irrigation systems. Some systems are designed to include storm water retention facilities. Typically, these roofs are constructed off reinforced concrete decks.

Typical construction

Green roofs are constructed in a series of layers (Figure 12.3), which typically include, from the top down:

❑ Vegetation layer, e.g. grass, sedums and specially selected plants
❑ Growing medium (usually 'engineered' lightweight material)
❑ Filter cloth (fleece), allows water in but helps to contain roots and the growing medium
❑ Drainage/water retention layer (may contain built-in reservoirs)
❑ Waterproof roofing membrane (with integral root repellent and/or metal foil between membrane layers and joints to prevent root damage)
❑ Rigid insulation
❑ Vapour control layer
❑ Structural deck
❑ Internal finish to underside of deck

The structural integrity of a green roof is dependent upon the interaction between the different layers and the quality of the waterproof membrane. Single-ply (heat-seamed reinforced plastic) roofing systems are regarded as the most efficient and cost effective. The material also provides additional protection against root penetration, which tends to offset concerns about the material's environmental credentials. Other suitable materials include rubber membranes (EPDM) and thermoplastic polyolifins (TPOs). Bitumen based roofing systems

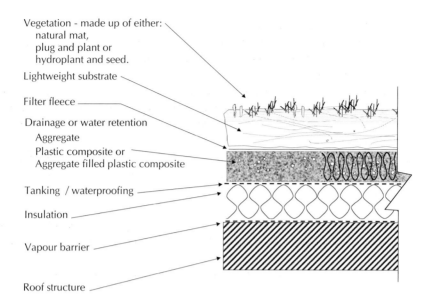

Vegetation - made up of either:
  natural mat,
  plug and plant or
  hydroplant and seed.

Lightweight substrate

Filter fleece

Drainage or water retention
  Aggregate
  Plastic composite or
  Aggregate filled plastic composite

Tanking / waterproofing

Insulation

Vapour barrier

Roof structure

Figure 12.3   Green roof construction (adapted from www.greenroof.co.uk).

must be isolated from the roots because the organic product is a potential food source for organisms.

Rainfall is largely maintained within the roof; however, proper drainage of the roof area is still important. Penetrations through the roof deck, for skylights, vents and chimneys must be well protected with flashings and a gravel skirt around the penetration. Similarly parapets need careful detailing.

Advantages

Green roofs provide a number of environmental benefits, which include the following:

❑ Increased energy efficiency. In the summer the vegetation layer shades the building from solar radiation, helping to keep the interior cool. In the winter the additional insulation provided by the growing medium helps to limit the amount of energy required to heat the building interior.

❑ Improved sound insulation. Primarily from the growing medium (low frequencies) but some reduction of higher frequency sound waves from vegetation.

❑ Longer life of the waterproof roof membrane. Green roof design helps to protect the waterproof roof membrane from temperature fluctuations, ultraviolet radiation and accidental damage from pedestrian traffic. Well constructed, green roofs will last at least twice as long as a conventional roof, thus reducing the need for replacement and reducing the associated waste of materials. Manufacturers provide guidance on life expectancy and replacement cycle of the waterproof membrane.

❏ Storm water runoff. Green roofs will retain some storm water and help to reduce the amount of storm water runoff from roofs, thus helping to reduce the risk of flooding. Intensive systems can be constructed to include storm water reservoirs.

❏ Environmental benefits. The vegetation layer will filter particulates from the air, helping to improve air quality. The vegetation will also help to reduce the 'urban heat island effect', the effect of solar radiation on hard surfaces in urban areas.

❏ Habitat creation and opportunities for biodiversity. The loss of natural habitat through building activity can be addressed through the use of a green roof. The vegetation layer provides habitat for wildlife and helps to encourage biodiversity.

❏ Amenity space for building users. Intensive systems provide an environment for building users to interact with nature, usually in an urban environment such as a city centre.

❏ Aesthetic benefits. Well designed and maintained green roofs are visually attractive.

Disadvantages

There are a few disadvantages associated with green roofs. They are:

❏ Fire resistance. While a saturated green roof may well help to limit the surface spread of flame, a dry roof can present a fire hazard, especially in a built up area. On large roof structures a series of fire breaks 600 mm wide, at 40 m intervals, made from non-combustible material, e.g. concrete pavers, should be used. The use of sedums can help (which have a high water content), although a sprinkler irrigation system, linked to a fire alarm, is a sensible precaution. Advice should be sought from the appropriate fire authority and building control office.

❏ Initial cost. Although extensive roof systems are a relatively cheap method of construction, the increased load of the roof requires a more substantial roof structure, for which there is an initial cost premium. Intensive roof systems are more expensive than a traditional roof construction; due to their weight and extensive planting. However, life cycle costing shows that green roofs are as cost effective as a traditional roof over the life of a building.

❏ Maintenance. More maintenance is required than that for a traditional construction.

❏ Access to the roof construction under the vegetation for repair and replacement of membranes can be difficult to do without removing the upper layers of the roof, and hence tends to be expensive. Membranes will need to be completely replaced after approximately 30–50 years, depending on the quality of the membrane and the design, construction and maintenance of the roof.

Structural considerations

The additional loading placed on the structure by the green roof construction is a primary consideration in assessing the viability of the roof. Wet soil weighs around 1,600 kg per cubic metre, which is quite a considerable loading to place on a structure. This has led to the development of many lightweight growing mediums, some of which weigh approximately 300 kg per square metre when saturated. For new build projects the loading can be considered at the design stage and the structure designed to accommodate any additional loading. When installing a green roof on an existing building, the structure and foundations must be checked to see what the maximum design load is, and how this affects the design of the green roof. Some structural upgrading should be anticipated.

Different systems use slightly different methods, but it is not uncommon for the green roof system to be built up from the finished waterproofing membrane by a specialist sub-contractor. There are a number of well-known proprietary systems on the market, so description here is limited to the basic principles. Manufacturers of proprietary systems provide an extensive technical design service and warranties for waterproof membranes and/or complete roofing systems.

12.5 Recycled materials

Architectural salvage, taking materials such as roof slates, bricks and internal fittings from redundant buildings for use on new projects, such as repair and conservation work, is a well-established business. The cost of the material tends to be higher than that for an equivalent new product because of the cost of recovery, cleaning and storage associated with the salvage operations. There is also a price premium for buying a scarce resource that will have a weathered quality that is difficult, if not impossible, to replicate with new products.

Quality of materials is difficult to assess without visiting the salvage yard and making a thorough visual inspection of the materials for sale, and even then there is likely to be some waste of material on the site. For example, the reuse of roof slates will be dependent upon the integrity of the nail hole, and many slates will need additional work before they are suitable for reuse. In some cases the slates will be unsuitable for reuse because of their poor quality. For work on refurbishment and conservation projects the use of reclaimed materials is a desirable option. However, the increased cost premium for using weathered materials with a reduced service life may not be a realistic option for the majority of projects. Another option is to use building products that have been made entirely from, or mostly from, recycled materials.

New products from recycled materials

Materials recovery from redundant buildings has occurred throughout history, with materials being reclaimed and reused in a new structure. Stone and timber were reused in vernacular architecture, while more recently steel and concrete have been recovered and reused. A more recent trend is for manufacturers to use materials recovered from household and industrial waste. Over recent years there has been a steady increase in the number of manufacturers offering new materials and building products that are manufactured partly or wholly from recycled materials. These innovative products offer greater choice to designers and builders keen to explore a more environmentally friendly approach to construction. Many of these products are also capable of being recycled at a future date, thus further helping to reduce waste. An area of growing interest is the manufacture of building products from household and industrial waste. A few examples are listed here:

- ❑ Glass. Recycled and used in the manufacture of some mineral thermal insulation products and as expanded glass granules in fibre-free thermal insulation.
- ❑ Salvaged paper. Used in the manufacture of plasterboards and thermal insulation.
- ❑ Plastics. Used for cable channels and sorted plastics recycled and used for foil materials and boards.

The issue of material choice and specification was discussed in Chapter 1, where the perception of risk associated with new products and techniques was discussed. The perception of risk associated with the use of new products is likely to be higher than that for the established and familiar products, which have a track record. The majority of products being manufactured from recycled materials are produced by relatively new manufacturers, offering products that may have little in the way of a track record in use. Thus the perception of risk is likely to be high until the products have been used (by others) and are known to perform as expected. This should not, however, stop designers, specifiers and builders from doing their own research and making informed decisions. Off-site production would appear to offer considerable scope for using recycled materials in the production of new materials and building elements.

12.6 New materials and products

There are a lot of new products and construction systems constantly entering the market, and a book such as this is unable to address all of these developments. We have, however, included a few examples here.

Phase Changing Materials: Wax microcapsules embedded in plaster

BASF (chemical company), Oxford University's Environmental Change Institute and the Tyndall Centre for Climate Change Research have developed building products that use Phase Changing Materials (PCM). PCMs are materials that change their form under certain known conditions. In this case, as wax capsules, which are embedded in materials, change from one form to another, they either take in or give out energy, in the form of latent heat.

The microcapsules, which measure less than a hundredth of a millimetre, make use of the latent heat required to change them from solid to liquid and liquid to solid. When the plaster gets hot, the wax within the plaster melts and as it melts it uses the excess heat within the building to transform the capsules from solid to liquid. This transformation helps to keep the building cool, and captures and stores the latent heat (Figure 12.4). When the weather becomes cold and the internal climate changes, the capsules will solidify releasing the latent heat. Phase changing materials can be integrated in both solid and liquid materials, thus they can be used in paints, plasters and concrete.

The plaster, which incorporates one-third phase changing material, has the same storage capacity as a 230 mm brick wall. The wax capsules (Phase Changing Material) help to equalise (stabilise) the temperature, making it cooler during hot periods and warmer during cold periods (Figure 12.4). The PCM material helps to optimise the room temperature. During warm periods, excess heat is taken from the internal environment and used as latent heat energy to transform the wax microcapsules from their solid form to liquid. The latent heat is stored and is released as the temperature drops and the material changes from liquid to solid, this time giving out heat energy as the material freezes.

BASF's real estate company has constructed 46 dwellings in the Ludwigshafen's Brunck district. It is estimated that the average fuel consumption of the BASF dwelling is one-twentieth of that of a conventional house; the CO_2 emissions in the prototypes were cut by 80%. The millions of tiny wax capsules

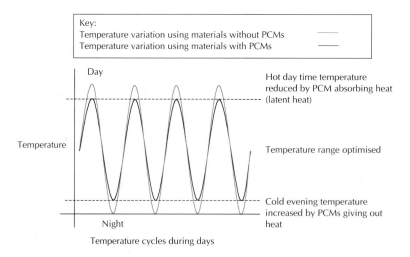

Figure 12.4   Graph showing room temperature adjustment using PCM – Phase Changing Materials (adapted from www.functionalpolymers.basf.com).

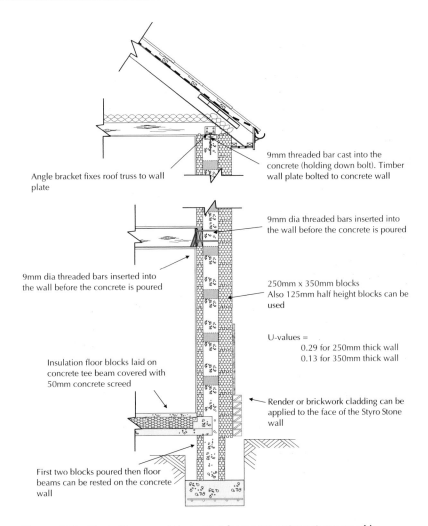

9mm threaded bar cast into the concrete (holding down bolt). Timber wall plate bolted to concrete wall

Angle bracket fixes roof truss to wall plate

9mm dia threaded bars inserted into the wall before the concrete is poured

9mm dia threaded bars inserted into the wall before the concrete is poured

250mm x 350mm blocks
Also 125mm half height blocks can be used

U-values =
0.29 for 250mm thick wall
0.13 for 350mm thick wall

Insulation floor blocks laid on concrete tee beam covered with 50mm concrete screed

Render or brickwork cladding can be applied to the face of the Styro Stone wall

First two blocks poured then floor beams can be rested on the concrete wall

Figure 12.5   Styro Stone wall (adapted from www.styrostone.co.uk).

embedded within the plaster is one of the key innovations in this project. The houses also incorporate triple glazed argon filled windows, 600 mm of external insulation and a special heat recovery ventilation unit.

Styro Stone: Interlocking polystyrene building blocks

The Styro Stone system uses hollow polystyrene blocks that are easily interlocked together. Styro Stone fits together much like the children's toy construction bricks, Lego, with interlocking nodules on the top of the bricks and rebates along the bottom to receive the nodules. The two skins of polystyrene that form the brick are tied together using plastic reinforcing spacers. The spacers ensure that the polystyrene forms retain their shape while the concrete fill is poured into the void. Ten polystyrene blocks can be laid on top of each other to form

Interlocking
styrostone block
with spacers

Reinforcement
inserted in void
prior to concrete
being poured

Photograph 12.1   Polystyrene and infill concrete walls: Styro Stone blocks (www.styrostone.co.uk).

a 2.5 m high wall, after which the polystyrene chamber needs to be filled with concrete (Photograph 12.1). Then, if necessary, the next lift of ten blocks can be placed ready to receive the concrete (Photograph 12.2). Special polystyrene blocks have also been made for use at corners and above window and doors (acting as lintels).

The polystyrene acts as a type of permanent or lost formwork, with the advantage of providing good thermal insulation. Reinforcement can be used to provide additional tensile strength where required (Photograph 12.3). Due to the lightness of the blocks and the ease with which they are assembled, the structural wall is quickly assembled. The walls of a single storey building can be fully constructed by three workers in just one day. The system has been tested and is capable of meeting current structural, thermal and acoustic requirements

Reinforced concrete floor using permanent polystyrene formwork

Props provide temporary support until the concrete has set

Photograph 12.2   Polystyrene and infill concrete floors: Styro Stone blocks (www.styrostone.co.uk).

The blocks can be used with reinforcement to form water resisting basement structures

Photograph 12.3   Polystyrene and infill concrete basement walls: Styro Stone blocks (www.styrostone.co.uk).

Photograph 12.4   Polystyrene and infill concrete house: Styro Stone blocks (www.styrostone.co.uk).

with considerable savings on many traditional forms of construction. U values range from 0.29 for a 250 mm wall to 0.12 for a 350 mm wall. In Germany the method has been used to construct five-storey houses. To ensure that the dwellings fit in with the local vernacular, the Styro Stone (polystyrene and concrete) structural membrane can be clad with render, stucco, stone or brick (Photograph 12.4). Specially designed brick slips have already been developed so that the Styro Stone wall can be clad to give the appearance of brickwork, without the cost of a traditional masonry wall. The Styro Stone blocks are of particular interest to the self-build market.

Appendix A: Web Sites

Chapter 2

Construction Industry Council	www.safetyindesign.org
Doka Scaffolding	www.doka.com
Heath and Safety Executive	www.hse.gov.uk
National Federation of Demolition Contractors	www.demolition-nfdc.com
RMD Formwork and support systems	www.rmdkwikform.net
SGB Scaffolding	www.sgb.co.uk
The Institute of Demolition Engineers	www.ide.org.uk

Chapter 3

Coastal erosion in East Anglia;	
Digital worlds, historic maps and images, current photographs to demonstrate the loss of land to the sea	www.digitalworlds.co.uk
Environment agency (database on landfill sites, floodplains, subsidence and contaminated land)	www.environment-agency.gov.uk
Floodplains in the UK	www.antiflood.com and www.homecheck.co.uk
Roger Bullivant	www.rogerbullivant.co.uk
The Hadley Centre provides information on global warming and environmental changes	www.met-office.gov.uk/research/hadleycentre

Chapter 5

Metal housing and steel frame solutions Redrow and Corus	www.framing-solutions.co.uk

Chapter 6

Doka formwork	www.doka.com
Peri formwork	www.peri.ltd.uk
RMD formwork and scaffolding	www.rmdkwikform.com

Chapter 8

Corus steel   www.corusconstruction.com
Roger Buillivant pile foundations and
   soil stabilisation   www.roger-bullivant.co.uk
Sustainable steel
   construction   www.sustainablesteelconstruction.com/s/sustainability
Yorkon   www.yorkon.com

Chapter 9

Doka lift shaft construction   www.doka.com
Kone lifts   www.kone.com
Otis lifts   www.otis.com
Peri lift shaft construction   www.peri.ltd.uk
The Lift and Escalator Industry Association   www.leia.co.uk

Chapter 12

BASF Micronal PCM   www.functionalpolymers.basf.com
Carbon trust   www.carbontrust.co.uk
Styrostone: Polystyrene building blocks   www.styrostone.co.uk
World changing   www.worldchanging.com/archives/002404.html

Appendix B: Further References

Chapter 1

Bett, G., Hoehnke, F. and Robinson, J. (2003) *The Scottish Building Regulations Explained and Illustrated (Third edition)*, Blackwell Publishing, Oxford.

Billington, M.J., Simons, M.W. and Waters, J.R. (2003) *The Building Regulations Explained and Illustrated (Twelfth edition)*, Blackwell Publishing, Oxford.

Emmitt, S. and Yeomans, D. (2001) *Specifying Buildings: a design management approach*, Butterworth-Heinemann, Oxford.

Chapter 2

BSI (2000) BS 6187:2000, *Code of Practice for Demolition*, British Standards Institution, London.

BSI (1993) BS 5973:1993, *Access and working scaffolds*, British Standards Institution, London.

BSI (1991) BS 1139-2.1:1991, EN 74:1988, *Metal Scaffolding – Part 2: Couplers, Section 2.1 Specification of steel couplers, loose spigots and base-plates for use in working scaffolds and falsework made of steel tubes*, British Standards Institution, London.

BSI (1991) BS 1139-2.2: 1991, *Metal Scaffolding – Part 2 Couplers, Section 2.2 Specification for steel and aluminium couplers, fittings and accessories for use in tubular scaffolding*, British Standards Institution, London.

Bussell, M., Lazarus, D. and Ross, P. (2003) *Retention of Masonry Facades – Best Practice Guide*, CIRIA, London.

CIRIA (1995) *Temporary Access to the Workface*, Special Publication 121, CIRIA, London.

Highfield, D. (1987) *The Rehabilitation and Re-Use of Old Buildings*, E & FN Spon, London.

Highfield, D. (1991) *The Construction of New Buildings Behind Historic Facades*, E & FN Spon, London.

Highfield, D. (2000) *Refurbishment and Upgrading of Buildings*, E & FN Spon, London.

Riley, M. and Cotgrave A. (2005) *The Technology of Refurbishment and Maintenance*, Palgrave Macmillan, Basingstoke.

Chapter 3

Gaba, A.R., Simpson, B., Powrie, W. and Beadman, D.R. (2003) *Embedded Retaining Walls – Guidance for Economic Design*, CIRIA, London.

Harrison, H.W. and Trotman, P.M. (2002) *BRE Elements, Foundations, Basements and External Works*, BRE, Garston.

NHBC (2000) *Standards, Chapter 4.2 Building near trees*, National House-building Council, Buckinghamshire.

NHBC (2000) *Standards, Chapter 4.4 Strip and trench fill foundations*, National House-building Council, Buckinghamshire.

NHBC (2000) *Standards, Chapter 4.2 Raft, pile, pier and beam foundations*, National House-building Council, Buckinghamshire.

Warren, D.R. (1996) *Civil Engineering Construction*, Macmillan, London.

Chapter 4

Brookes, H. (1997) *The Tilt-Up Design and Construction Manual*, HBA Publications, Ohio.

Chilton, J. (2003) *Space Grid Structures*, Architectural Press, Oxford.

Curtin, W.G., Shaw, G., Beck, K. and Bray, W.A. (1982) *Design of Brick Diaphragm Walls, Design Guide 11*, BDA, London.

Gilbertson, A. (2004) *CDM Regulations – Work Sector Guidance for Designers*, CIRIA, London.

HSE (1998) *Health and Safety in Roof Work*, HSE, Norwich.

ODPM (2002) *The Building Regulations 2000, Fire Safety – Approved Document B*, ODPM, London.

Southcott, M.F. (1998) *Tilt-Up Concrete Buildings, Design and Construction Guide*, British Cement Association, Berkshire.

SPRA (2003) *Design Guide for Single Ply Roofing*, SPRA, London.

Chapter 6

BCA (2002) *Concrete Practice Third edition*, British Cement Association, Berkshire.

COFI (1996) *Concrete Formwork*, COFI, Canada.

Chapter 7

Harrison, P., Masat, J. and Peric-Matthews, A. (2000) *Cladding Fixing – Good Practice Guidance*, CIRIA, London.

Chapter 8

Stirling, C. (2003) *Off-Site Construction: An Introduction. Good Building Guide 56*, Building Research Establishment, Watford.

Wilson, D.G. Smith, M.H. and Deal, J. (1999) *Prefabrication and Preassembly: Applying the Techniques to Building Engineering Services*, BISRIA, London.

Chapter 9

Demetri, G. (1999) 'Lifts, Stairs and Escalators', *Architects' Journal – Focus*, June, 33–46.

Howkins, R. (2004) 'Lifts and Escalators', *Architects' Journal – Focus*, April, 18–28.

Ogden, R.G. (1994) *Electrical Lift Installations in Steel Frame Buildings*, National Association of Lift Makers, Berkshire, The Steel Construction Institute, London.

Chapter 10

NHBC (2000) *NHBC Standards, Internal Walls*, Chapter 6.3, January 2000, NHBC.

Index